A Companion to

ASTRONOMY AND ASTROPHYSICS

Chronology and Glossary with Data Tables

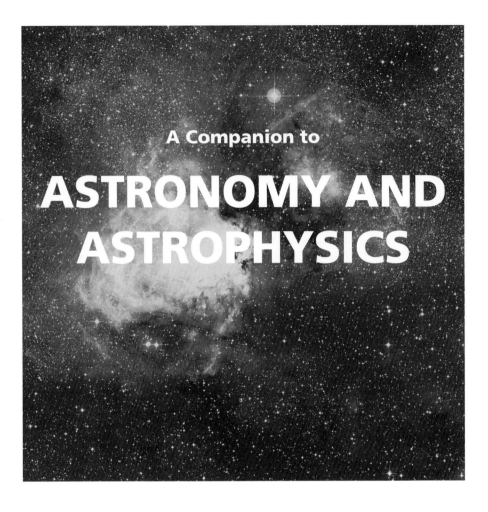

A Companion to

ASTRONOMY AND
ASTROPHYSICS

Kenneth R. Lang

Chronology and Glossary with Data Tables

 Springer

Kenneth R. Lang
Department of Physics and Astronomy
Robinson Hall
Tufts University
Medford, MA, 02155
USA
ken.lang@tufts.edu

Cover illustration: The Crab Nebula supernova remnant(M1 or NGC 1952) is an expanding cloud of interstellar gas that was ejected by the supernova explosion of a massive star, about 9 solar masses, in 1054 A. D. The filamentary gases are expanding at a velocity of 1.5 million meters per second. The red-yellow wisps of gas shine primarily in the light of hydrogen, while the blue-white light is the non-thermal radiation of high-speed electrons spiraling in magnetic fields. The south-westernmost (bottom-right) of the two central stars is the remnant neutron star of the supernova explosion, and a radio pulsar with a period of 0.33 seconds. This famous supernova remnant is also a source of intense emission at radio and X-ray wavelengths. (Courtesy of Rudolph Schild, Harvard-Smithsonian Center for Astrophysics.)

Library of Congress Control Number: 2005937183

ISBN-10: 0-387-30734-6   e-ISBN: 0-387-33367-3
ISBN-13: 978-0387-30734-3

Printed on acid-free paper.

Printed in Singapore.   (Apex/KYO)

9 8 7 6 5 4 3 2 1

springer.com

DEDICA A PROVVIDENZA

PREFACE

A Companion to Astronomy and Astrophysics is a comprehensive, fundamental, up-to-date reference book. It is filled with vital information and basic facts for amateur astronomers and professional astrophysicists, and for anyone interested in the Universe, from the Earth and other planets to the stars, galaxies and beyond. Although serious and thorough, the language and ideas will attract the general reader, as well as students and professionals.

A Companion to Astronomy and Astrophysics consists of two main parts, a *Chronology* and a *Glossary*. The *Chronology* is a concise history, arranged chronologically, which provides the complete story of cosmic discovery from early Chinese and Greek astronomy to the latest findings of modern astrophysics and robotic spacecraft. It provides a sense of destination and flow in our growing awareness of the Universe. Each entry in the unfolding narrative is written in a concise, light and friendly style that will be appreciated by all, without being weighted down with incomprehensible specialized terms. And the reader can either trace out our historical journey of realization, or sample the entries at random, dipping in and out of the fascinating ideas and findings.

The story line of science is one of individuals, as well as discoveries and ideas. So our *Chronology* includes the people who have made major scientific contributions that are related to astronomy and astrophysics. Nobel Prize winners for all the relevant discoveries are also included, in the year of the award.

The *Chronology* also contains line drawings selected to demonstrate a key discovery, including the abundance and origin of the elements in the Sun, annual parallax, the blackbody radiation spectrum, the Bohr atom, dark matter around a spiral galaxy, the electromagnetic spectrum, globular clusters and the extent of the Milky Way, Great Walls of galaxies, the Hertzsprung-Russell diagram, the Hubble diagram – original and recent, the mass-luminosity relation of stars, precession of Mercury's perihelion, the proton-proton chain, retrograde planetary loops, the Roche limit, space curvature, and Thomas Wright's model of the Milky Way.

The *Glossary* defines all terms and acronyms that deal with our celestial science, from important concepts to individual objects or observatories and space missions. It is arranged alphabetically and complemented by numerous tables of fundamental data. Line drawings are also used to forcefully compact a scientific insight, with clear labels and captions.

Individual names are not given in the *Glossary*, but the reader can use the *Author Index* to find their significant contribution in the *Chronology*. The exceptionally thorough *Subject Index* cross-references concepts, discoveries, ideas, people, history and time in both the

Chronology and *Glossary*. The combined result is a unique, comprehensive, stand-alone reference volume in which the reader can quickly locate information, while also discovering new and unexpected knowledge. *A Companion to Astronomy and Astrophysics* is therefore an indispensable book for any library and all persons interested in astronomy and astrophysics, as well as the related fields of cosmology, geology and physics.

Astronomy is the oldest of all sciences, with origins dating back at least several thousand years. The *Chronology* therefore begins with early astronomy from China to India, and continues with the ancient Greeks. The word astronomy comes from two Greek words, *astron*, meaning "star", and *nomos*, for "law." So our concise historical *Chronology* includes the discoveries of the laws of physics that help describe the energy, ingredients, magnetism, motion, origin and destiny of cosmic objects. These developments include the quantum theory of the very small, with its exclusion and uncertainty principles, atomic structure and sub-atomic particles, electromagnetic radiation and the speed of light, and the special and general theories of relativity.

Major scientific breakthroughs are highlighted in the *Chronology*, beginning with the invention of the telescope, which enabled astronomers to glimpse distant worlds that cannot otherwise be seen or to resolve details on nearby ones. Key discoveries include the concept of universal gravitation, new planets in our Solar System and around other stars, determination of the chemical ingredients of the stars, a realization that stars shine by nuclear processes deep within their cores, observations of stars that are being born, evolving and perishing before our very eyes, and the knowledge that the Universe is expanding.

New ideas concerning our home planet Earth are woven into the fabric of our account, including its age, size and mass, continental drift, global warming and the ice ages.

Radio waves, which lie beyond the range of visual perception, have revealed an explosive Universe, from exploding stars to the relic radiation of the Big Bang that set the expanding Universe in motion. Radio astronomers have also discovered the cold hydrogen of interstellar space, rapidly spinning neutron stars called pulsars, a pair of neutron stars headed for collision and emitting gravitational radiation in the process, and the compact, distant and energetic radio galaxies and quasars, apparently powered by supermassive black holes.

Spacecraft have played an important role in astronomical discovery in recent decades. Men were sent to the Moon, helping us understand its origin and evolution. Spacecraft have traveled throughout the Solar System, obtaining close-up views that have changed the planets from moving points of light to fascinating worlds and transformed their satellites into unique objects with complex and disparate surfaces. Robot spacecraft have landed on Mars and Venus, and probes have been dropped into the atmospheres of Jupiter and Saturn's largest satellite, Titan. Spacecraft have also obtained images of the icy nucleus of three comets and scrutinized the battered, broken surfaces of several asteroids.

The **Hubble Space Telescope** has been lofted above the Earth's obscuring atmosphere to view cosmic objects in visible light with a clarity never achieved before.

X-ray telescopes have similarly been propelled into space, first in rockets and then aboard long-lived satellites. They have opened a new window to the previously unknown worlds of million-degree gases, black holes and hot matter hidden in the space between galaxies. Cosmic gamma rays have also been detected using space-borne instruments, including gamma ray bursts that briefly emit more energy than the rest of the Universe combined.

More recently, substantial amounts of dark matter have been inferred from rapid motions of stars or galaxies, outweighing all of these shining visible objects. Mysterious dark energy has also been found, filling apparently empty space. Scientists have even speculated that everything in the observable Universe could have originated from the nothingness of space, during a rapid growth spurt, or inflation, soon after the Big Bang, or by a host of microscopic, oscillating strings at earlier times closer to the Big Bang.

The *Glossary* provides an alphabetical listing of the terms, names and acronyms relevant to *A Companion to Astronomy and Astrophysics,* along with a discussion of their meanings and applications. Here the reader can find that long-forgotten definition together with an introductory background to the concept. Informative tables accompany the *Glossary,* some of them created for this book and others extracted from the author's widely used reference books, *Astrophysical Formulae I, II* and *Astrophysical Data,* published by Springer.

These tables include the atmospheres, escape velocities, magnetic fields and orbital parameters of the planets, as well as physical data for each planet, such as their angular diameter, mass, mass density, radius, rotation and temperature. Comprehensive physical information is also given in tables of the largest asteroids, aurora spectral features, comets, the corona, conversion factors, the cosmic microwave background radiation, cosmic rays, the Crab Nebula, Fraunhofer lines, the Galilean satellites, greenhouse gases, Halley's comet, our Milky Way Galaxy, meteor showers, radioactive isotopes, satellites of the planets, SN 1987A, solar flares, stellar classification, and the Sun. Tables of bright celestial objects are also included in the *Glossary,* providing physical parameters and the celestial coordinates required to observe them. They include the constellations, the brightest galaxies, emission nebulae, the Local Group of galaxies, the Messier objects, planetary nebulae, the brightest stars, the nearest stars, supernovae, and supernova remnants. Other tables include space missions by NASA, ESA and ISAS, Nobel Prize winners related to astronomy and astrophysics, and the major optical, or visible-light, telescopes.

The *Glossary* additionally contains line drawings, carefully selected to convey an idea, including the asteroid belt, bremsstrahlung, comets – anatomy and trajectory with tails, coronagraph, the Earth's layered atmosphere, an ellipse, Io's interaction with Jupiter, an interferometer, Jupiter's winds, Kepler's laws, the Kirkwood gaps, a lunar eclipse, the magnetosphere, the Moon's phases, the Oort cloud, pulsars – radio and X-ray, the seasons, a solar eclipse, solar flare, spectroheliograph, SS 433, sunspot cycle, synchrotron radiation, telescopes and the tides.

Thanks to Gayle Grant at Tufts University for preparing some crucial tables, to Librarian Regina F. Raboin at Tufts University for help in obtaining some of the dates that were so hard to locate, and to Sue Lee for producing the wonderful line drawings. The preparation of *A Companion to Astronomy and Astrophysics: Chronology and Glossary with Data Tables* was made possible through funding by NASA Grant NNG05GB00G with NASA's Applied Information Systems Research Program.

Kenneth R. Lang
Tufts University *and* Anguilla, B.W. I.
January 1st 2006

TABLE OF CONTENTS

LIST OF FIGURES

CHRONOLOGY

GLOSSARY

LIST OF TABLES

CHRONOLOGY:
A CONCISE HISTORY OF COSMIC DISCOVERY

2800 BC to 1600 AD

c. 2800 BC

Stonehenge is built in England. It was probably an astronomical observatory with religious functions; the motions of the Sun and Moon can be followed with the aid of carefully aligned rocks.

c. 2700 BC

A lunar calendar is developed in Mesopotamia in which new months begin at each new Moon and the year is 354 days long.

2296 BC

The Chinese make the earliest recorded sighting of a comet.

c. 1500 BC

The *Rig-Veda,* the oldest of the Hindu sacred texts, includes the idea that the Earth is a globe. In the same millennium, the subsequent *Yajur-Veda* mentions that the Earth circles the Sun.

1361 BC

Chinese astronomers make the first recording of an eclipse of the Moon.

c. 1300 BC

The Shang dynasty in China establishes the solar year at 365.25 days, introducing a calendar with 12 months of 30 days each.

1217 BC

Chinese astronomers make the first recording of an eclipse of the Sun.

763 BC

Assyrian archivists record an eclipse of the Sun on 15 June 763 BC; the same event is recorded in the *Bible* (Amos 8:9).

747 BC

A continuous record of solar and lunar eclipses is begun in Mesopotamia.

585 BC

Greek philosopher **Thales of Miletus (c. 624–c. 547 BC)** correctly predicts the eclipse of the Sun on 28 May 585 BC. He also proposes that water is the material constituent of all things.

c. 530 BC

Greek mathematician and philosopher **Pythagoras (c. 572–c. 479 BC),** born on the island of Samos, proposes the notion of a spherical Earth. He also supposed that concentric, transparent, crystalline spheres carry the Sun, Moon, planets and stars in perfect circles, rotating at constant speed around the centrally placed, unmoving Earth. As quoted by Aristotle in his *Metaphysics,* Pythagoras said, "there is music in the spacing of the spheres."

c. 500 BC

The Greek philosopher **Heraclitus of Ephesus (c. 540–c. 475 BC)** maintains that permanence is an illusion, and that all things are flowing, in a state of becoming. He also thought that most people sleepwalk through life, not understanding what is going on about them.

c. 450 BC

The Greek philosopher **Anaxagoras (c. 500–c. 428 BC)** born in Clazomenae, Lydia (now Turkey), proposes that the Moon shines by reflected sunlight, and is thus able to explain total eclipses of the Sun and Moon. He was imprisoned for claiming that the Sun was a "red-hot stone", and not a god, and that the Moon reflects the Sun's light.

Greek philosopher **Empedocles (c. 490–c. 430 BC),** of Acragas (Agrigentum) in Sicily is one of the earliest to propose that terrestrial objects are made up of four elements or basic principles – fire, air, water, and earth. He viewed these as united or divided by attraction and repulsion, or more poetically, love and strife.

c. 435 BC

Greek philosopher **Leucippus of Miletus (c. 480–c. 420 BC)** is the first to propose the atomic theory, in which matter is composed of very small, invisible particles now called atoms. His pupil **Democritus (c. 460–c. 370 BC)** subsequently develops the theory.

c. 420 BC

Greek philosopher **Democritus (c. 460–c. 370 BC),** born in Abdera, Thrace, states that all matter consists of an infinite number of eternal, invisible and indivisible particles, which he called *atomon,* or atoms, and that the space between the atoms is a vacuum, or void, which gives the atoms a place to move into.

c. 366 BC

The Greek mathematician and astronomer **Eudoxus of Cnidus (408–435 BC)** builds an observatory and constructs a model of nested, geocentric, rotating spheres to explain the motions of the Sun, Moon, and planets as viewed from Earth, which was at the center of the system. By using a total of twenty-seven concentric spheres he was able to calculate the Sun's annual motion through the zodiac, the Moon's motion including its wobble, and the planets' motions, including the retrograde movements of some of them.

352 BC

Chinese astronomers make the earliest known record of a guest star, subsequently known as an exploding star or supernova.

c. 350 BC

Greek philosopher **Aristotle (384–322 BC)** rejects the notion of a vacuum, since an object moving in a vacuum would have no sense of direction and might move impossibly fast.

Greek philosopher **Aristotle (384–322 BC)** proves the Earth is spherical, since it always casts a circular shadow on the Moon during lunar eclipses. **Parmenides (c. 515–c. 450 BC),** of Elea in southern Italy, is said to have been the first Greek philosopher known to have asserted that the Earth is spherical, but an alternative tradition states that it was **Pythagoras (c. 572–c. 479 BC).** Parmenides also proposed that movement is illusory, belonging to the domain of the senses, and that the ultimate reality was invariant, immobile and unchanging. The Indians of Asia also proposed that the Earth is a globe around 1500 BC.

Greek philosopher **Aristotle (384–322 BC)** presents a geocentric, or Earth-centered cosmology of rotating, transparent spheres in *De caelo,* or *Concerning the Heavens.* The outermost sphere contains the fixed stars, then follows the spheres of Saturn, Jupiter, Mars, the Sun, Venus, Mercury, and closest to the Earth, the Moon. Each of these has several other spheres to account for all their movements. To explain the circular motion of the heavenly spheres, Aristotle proposes a fifth element, the aether or quintessence, in addition to the four natural elements, earth, air, water and fire, which either fall or rise.

Babylonian astronomers have learned enough about the Moon's motion that they can predict the occurrence of lunar eclipses.

c. 330 BC

Greek philosopher and astronomer **Heraklides of Pontus (388–315 BC)** proposes that the Earth turns on its axis, from west to east, once every 24 hours, while the stars are at rest. He also thought that the observed motions of Mercury and Venus suggest that they orbit the Sun rather than the Earth, and considered each planet to be a world with an Earth-like body and with an atmosphere.

c. 300 BC

Greek philosopher **Epicurus of Samos (341–270 BC)** argues that the innumerable atoms in an infinite Universe would come together to create other worlds, either resembling our Earth or different from it.

The Greek mathematician **Euclid of Alexandria (lived c. 300 BC)** writes *The Elements,* a book of plane and solid geometry, which formed the basis for mathematical study for the next 2,000 years. It was not until the 19th century that a different form of non-Euclidean geometry was considered.

c. 270 BC

Greek mathematician and astronomer **Aristarchos [Aristarchus] of Samos (c. 320–c. 250 BC)** moves the center of the Universe from the Earth to the Sun, and supposes that the Earth and other planets travel in circular orbits around the stationary Sun. He further stated that the fixed stars do not move, and that their apparent daily motion is due to the Earth's rotation on its axis. Aristarchos also used observations and trigonometry to estimate the size and distances of the Sun and Moon, showing that the Sun is much farther away than the Moon.

c. 240 BC

Greek scholar **Eratosthenes (c. 276–c. 194 BC),** born in Cyrene (now known as Shahhat, part of Libya) calculated the diameter of the Earth by measuring noontime shadows of the Sun at Alexandria and Syene, using the known distance between these places to infer the circumference of the Earth at just over 250,000 stades, and close to the modern value of roughly 40 thousand kilometers. The two sites are separated by 800 kilometers, and the measured angle between them was seven degrees. Assuming that the Earth is a sphere, its circumference is 50 times 800, or about 40,000 kilometers.

240 BC

Chinese astronomers make the first recorded sighting of the comet that would later become known as Halley's Comet.

c. 200 BC

The Greeks invent the astrolabe, used to measure the angular distance between any two objects, including the elevation in the sky of planets and stars.

c. 165 BC

Chinese astronomers first observe and record sunspots. Imperial astronomers keep continuous records of sunspots from 28 BC to AD 1638.

129 BC

The Greek astronomer **Hipparchus (c. 190–c. 120 BC),** born in Nicaea, in Bithynia (now in Turkey) creates the first known star catalog. It gave the celestial latitude and longitude and brightness of nearly 850 stars and was later reproduced by the Egyptian astronomer **Ptolemy (c. 87–c. 165 AD)** in his writings, and eventually used by **Edmond Halley (1656–1742)** to show that at least three stars move. Hipparchus also discovered the precession of the equinoxes, in which the locations of the spring and fall equinoxes move slowly, or precess, around the stellar background; the equinoxes are the two points where the Sun crosses the celestial equator. The Earth's rotation axis traces out a circle in the sky, and a cone in space, once every 26,000 years, causing a slow, steady precessional shift in the positions of objects on the celestial sphere at the rate of about 50 seconds of arc per year.

55 BC

Roman poet **Lucretius (99–55 BC),** author of the philosophical epic *De Rerum Natura,* or *On the Nature of Things,* proposes a plurality of Earth-like worlds in an infinite Universe.

c. 100

The Greek writer **Plutarch (c. 45–c. 125)** compares the Moon's face to the Earth's surface, and advocates the plurality of worlds.

150

Egyptian astronomer **Ptolemy (c. 87–c. 165),** or **Claudius Ptolemaeus,** writes his *Mathematical Syntaxis;* awed Arabic translators of the ninth century called it "The Greatest Composition" – *Almageste,* or, as it became known, the *Almagest.* Ptolemy supposed that each planet moves with constant speed on a small circle, while the center of that circle rotates uniformly on a larger one, with the Earth located slightly to one side of its center. By selecting suitable radii and speeds of motion, Ptolemy could use this system of uniform motion around two circles to reproduce the apparent motions of the planets with remarkable accuracy. He succeeded so well that his model was still being used to predict the locations of the planets in the sky more than a thousand years after his death.

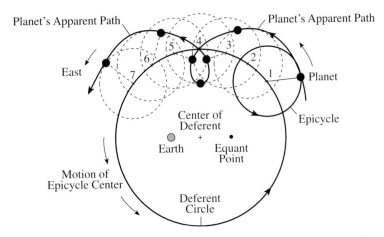

FIG. 1 Retrograde loops in Ptolemy's Universe An Earth-centered model of the Universe described by the Egyptian astronomer **Claudius Ptolemaeus (lived 2nd century)** or **Ptolemy** in the year 150. To explain the occasional retrograde loops in the apparent motions of Mars, Ptolemy imagined that the planet travels with uniform speed around a small circle, known as the epicycle. The epicycle's center moves uniformly on a larger circle, the deferent. In the Ptolemaic system, the Earth was displaced from the center of the large circle, and each planet traveled with uniform motion with respect to another imaginary point, the equant, appearing to move with variable speed when viewed from the Earth.

c. 450

Greek philosopher **Proclus [Proklus] Diadochus (410–485)** states that astronomers do not arrive at conclusions by starting from hypotheses, as is done in the other sciences, but instead use observations to construct hypotheses, which are tested against other observations to save the appearance of the phenomena.

800

The Peruvian city of Machu Picchu is built, containing an astronomical altar used to accurately measure solar and lunar movements.

c. 950

The Persian astronomer **al-Sufi, Abd al-Rahman (903–986)** revises Ptolemy's star catalogue, providing approximate positions, magnitudes and colors for these stars, and includes the first recorded reference to the Andromeda nebula.

987

Toltec conquerors of the Central American Mayan city of Chichén Itzá construct monuments with ritual astronomical alignments to the rising and setting of the Sun and the sacred planet Venus.

c. 1000

Arab philosopher, physician and scientist **Avicenna (980–1037),** in Arabic Abu 'Ali Al-Husain Ibn Abdallah Ibn Sina, proposes that terrestrial mountains might be caused by upheavals of the Earth's crust, such as those accompanying earthquakes, and that mountains could be noticeably cut and worn away by water flowing over long periods of time.

1054

Chinese astronomers record a bright "guest" star that appeared in the constellation Taurus on 4 July 1054, becoming as bright as Venus and remaining visible to the unaided eye for 22 months. Its expanding remnant, the Crab Nebula, is still detected using telescopes today.

1079

The Persian poet, mathematician and astronomer **Omar Khayyam (1048–1131)** contributes to calendar reform, measuring the length of the year as 365.24219858156 days, which surpassed the accuracy of the Julian value and agrees with modern determinations to the sixth decimal place, currently at 365.242190 days.

1267

English philosopher and scientist **Roger Bacon (c. 1220–c. 1292)** asserts that the only basis for scientific certainty is through the use of observations and experiment to prove an argument.

1328

English theologian **William of Ockham (c. 1280–c. 1349)** proposes a rule in science and philosophy, now known as Ockham's razor, which states that the simplest of two or more competing theories is preferable.

1444

German philosopher and theologian **Nikolas of Cusa (1401–1464)** argues that the Earth cannot be at the center of an unbounded Universe, suggests that the Earth moves around the Sun, and proposes that the stars are other Suns with inhabited planets.

1492

The Genoese navigator **Christopher Columbus (1451–1506)** lands in Española, an island now consisting of Haiti and the Dominican Republic, forging a path to the New World.

1521

Ferdinand Magellan (1480–1521) observes the bright nebulae subsequently named the large and small Magellanic Clouds; they are two irregular galaxies that are satellites of our Milky Way Galaxy, visible only from the equatorial regions and the southern hemisphere. His ship Victoria completes the first circumnavigation of the world the following year, after the death of Magellan.

1543

The Polish astronomer and cleric **Mikolaj Kopernik**, better known as **Nicolaus Copernicus (1473–1543),** publishes his influential book *De revolutionibus orbium coelestium*, or *Concerning the Revolutions of the Celestial Bodies*, reviving the Sun-centered model of the Universe introduced by the Greek philosopher **Aristarchos (c. 320–c. 250 BC)** eighteen centuries earlier, and by the Indians of Asia about a millennium before that. Copernicus proposed that the Earth is one of several planets moving at constant velocities in perfect circles around the Sun, in the same direction but at different distances and various speeds. In order of increasing distance from the Sun, they are Mercury, Venus, Earth, Mars, Jupiter and Saturn. As Copernicus noticed, the farther a planet is from the Sun, the slower it moves and the longer it takes to complete one circuit.

1572

The Danish astronomer **Tycho Brahe (1546–1601)** observes a brilliant "new star" or "nova" in the constellation Cassiopeia, which was brighter than Venus and had not been seen before. The debris from this stellar explosion, or supernova as it is now known, is still observed today; it is called Tycho's supernova remnant. Tycho subsequently builds an observatory, named *Uraniborg*, or "celestial castle", on the island of Hven near Copenhagen. He designed his own ingenious measuring instruments that resembled large gun sights with graduated circles, and used them at Uraniborg for two decades, compiling exceptionally accurate observations of the positions and motions of the planets.

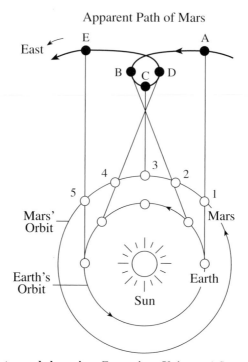

Apparent Path of Mars

FIG. 2   Retrograde loops in a Copernican Universe A Sun-centered model of the Universe proposed in 270 BC by the Greek philosopher **Aristarchos (c. 320–c. 250 BC)** and revived in 1543 by the Polish astronomer and cleric **Nicolaus Copernicus (1473–1543).** It explains the looping path of Mars in terms of the relative speeds of the Earth and Mars. The Earth travels around the Sun more rapidly than Mars does. As Earth overtakes and passes the slower moving planet (*points 2 to 4*), Mars appears to move backward (*points B to D*) for a few months.

1582

The Julian calendar is switched over to the Gregorian calendar in Catholic countries, by decree of **Pope Gregory XIII (1502–1585);** Protestant German countries adopted the Gregorian reform in 1700 and England in 1752.

1584

Italian philosopher and priest **Giordano Bruno (1548–1600)** proposes a decentralized, infinite and infinitely populated Universe that is filled with countless habitable planets revolving around stars other than the Sun, publishing these ideas in *De l'infinito universo e mondi*, or *The infinite universe and worlds*.

1600 to 1700 AD

1600

William Gilbert (1544–1603), physician to Queen Elizabeth I of England, writes *De magnete, magneticisque corporibus, et de magno magnete tellure,* or *Concerning Magnetism, Magnetic Bodies, and the Great Magnet, the Earth,* in which he shows that the Earth is itself a great magnet, explaining the orientation of compass needles. It is as if there is a colossal dipolar magnet at the center of the Earth.

1604

Italian astronomer **Galileo Galilei (1564–1642)** discovers his law of falling bodies, proving that gravity acts with the same strength on all objects, independent of their mass.

German astronomer **Johannes Kepler (1571–1630)** observes a bright "new star" where no star had been seen previously. In the same year he wrote *Ad Vitellionem Paralipomena, Quibus Astronomiae Pars Optica Traditure,* or *Supplement to Witelo* (his treatise on optics), *In Which is Expounded the Optical part of Astronomy.* In this work Kepler states that the intensity of light varies inversely with the square of the distance from the source. He also states that vision is caused by the formation of an image on the retina by the eye's lens.

1608

In October 1608 the national government of the Netherlands discussed the patent applications of first the Dutch spectacle maker **Hans Lippershey (c. 1570–1619)** of Middleburg, and then **Jacob Metius (1571–1635)** of Alkmaar, on a device for "seeing faraway things as though nearby." It consisted of a concave and a convex lens in a tube, and the combination magnified three or four times. Such spyglasses soon became very common, and no patent was awarded.

1609

Italian astronomer **Galileo Galilei (1564–1642)** builds his own telescopes, which had been invented in Holland the year before, and is the first to use telescopes for astronomical observations.

German astronomer **Johannes Kepler (1571–1630)** publishes *Astronomica nova,* or *New Astronomy,* which contains his first two laws of planetary motion and accurately describes the orbit of Mars. His first law states that planets travel in elliptical orbits around the Sun, with the Sun at one of the two foci of the ellipses. The second law states that the line joining the Sun and a planet sweeps our equal areas of space in equal periods of time, so that planets move more quickly when they are nearer the Sun.

1610

Italian astronomer **Galileo Galilei (1564–1642)** publishes *Sidereus nuncios*, or *The Starry Messenger*, reporting his pioneering observations with the newly invented spyglass, or telescope. They included his discoveries of previously unseen mountains, valleys and craters on the Moon, four previously unknown moons that circle Jupiter, and numerous stars in the Milky Way that cannot be seen with the unaided eye. In the same year he began systematic studies of sunspots through a telescope.

1611

German astronomer **Johannes Kepler (1571–1630)** discusses the optics of the telescope in his *Doptrice* or *Dioptrics*, including a description of a new type of telescope using two convex lenses.

German astronomer **Simon Marius (1573–1624)** is the first to observe the Andromeda Nebula with a telescope. He also uses a telescope to observe the four moons of Jupiter independently of Galileo, and names them Io, Europa, Ganymede, and Callisto, which we use today.

1619

German astronomer **Johannes Kepler (1571–1630)** publishes *De harmonices mundes*, or *The Harmonies of the Worlds*, which includes his third law of planetary motion, in which the square of a planet's orbital period about the Sun is proportional to the cube of its average distance from the Sun.

1620

Great Britain establishes the first permanent European settlement in New England at Plymouth.

1621

Italian astronomer **Galileo Galilei (1564–1642)** discovers that the acceleration of a falling body is proportional to the time and independent of weight and density.

1630

Italian astronomer **Galileo Galilei (1564–1642)** publishes his book *Dialogo sopra i due massimi sistemi del mondo, tolemaico e copernicano*, or *Dialogue Concerning The Two Chief World Systems, Ptolemaic and Copernican*, presenting persuasive arguments for the heliocentric theory of planetary motion. The book was banned and Galileo was taken to Rome to face trial on a charge of heresy in 1633.

1631

The French philosopher and physicist **Pierre Gassendi (1592–1655)** observes the 1631 transit of Venus across the Sun, which had been predicted by Kepler, and establishes that the orbit of Venus lies closer to the Sun than does the Earth's orbit.

1639

English astronomer and clergyman **Jeremiah Horrocks (1617–1641)** observes the 1639 transit of Venus, which he had predicted from the 1631 transit noting that the transits occur in pairs eight years apart, and uses Kepler's laws to prove that the Moon orbits the Earth in an elliptical orbit as if it was "continually falling toward Earth".

1655

The Dutch astronomer **Christiaan Huygens (1629–1695)** discovers Titan, Saturn's largest satellite.

1656

Dutch astronomer **Christiaan Huygens (1629–1695)** realizes that Saturn is surrounded by a ring, publishing in 1659 a Latin anagram that, when interpreted, explains Saturn's handle-shaped appearance by: "It [Saturn] is surrounded by a thin, flat ring, nowhere touching it and inclined to the ecliptic."

1665

The Italian-born French astronomer **Gian (Giovanni) Domenico (Jean Dominique) Cassini (1625–1712)** discovers the Great Red Spot on Jupiter.

1666

English physicist and mathematician **Isaac Newton (1643–1727)** uses a prism to disperse white sunlight into its component colors, and then uses a second prism to combine the colors in the spectrum and form white light again. In this way, Newton proved that the colors are a property of light and not of the prism.

1667

Construction of the Observatoire de Paris (Paris Observatory) begins. It was founded by **Louis XIV (1638–1715), The Sun King,** and is the oldest operating astronomical observatory in existence. **Gian (Giovanni) Dominico (Jean Dominique) Cassini (1625–1712)** was the first of four generations of his family to hold the post of director of the observatory. While director, Cassini discovered the dark gap in Saturn's rings in 1675, now named Cassini's gap, and also discovered four satellites of the ringed planet – Iapetus, Rhea, Tethys and Dione, in 1671, 1672, 1684 and 1684, respectively. The Observatoire de Paris is now located in both Paris and nearby Meudon.

1668

English physicist and mathematician **Isaac Newton (1643–1727)** constructs the first reflecting telescope. He used a primary concave spherical mirror of speculum metal, an alloy of copper and tin, which reflected light up the tube of the telescope to a flat secondary mirror of speculum, angled at 45 degrees to direct the light sideways. Nowadays the primary mirror is made of glass and of parabolic shape.

1672

Italian-born French astronomer **Gian (Giovanni) Domenico (Jean Dominique) Cassini (1625–1712)** and the French astronomer **Jean Richer (1630–1696)** observe the opposition of Mars, from Paris and Cayenne, South America, respectively, triangulating the distance between the Earth and Mars and hence determining the mean distance between the Earth and the Sun to be 138 million kilometers, or about 11.6 kilometers short of the modern value.

1675

The Italian-born French astronomer **Gian (Giovanni) Domenico (Jean Dominique) Cassini (1625–1712)** discovers a dark separation in Saturn's rings, now called the Cassini gap.

Charles II (1630–1683) founds the Royal Observatory of England at Greenwich, England, and the English astronomer **John Flamsteed (1646–1719)** is appointed as the first Astronomer Royal. The Royal Greenwich Observatory was built and operated to improve navigation at sea and to find the east-west position, or longitude, while at sea and out of sight of land, by astronomical means.

1679

Danish astronomer **Ole Römer (1644–1710)** announces that light travels at a finite speed, which he proved from observations of the inconstant interval between Jupiter's eclipses of its satellite Io. When the distance between Earth and Jupiter was least, the interval between eclipses was smallest, and *vice versa*. Römer estimates that the speed of light is 225,000 kilometers per second and that the Sun's rays reach the Earth in 11 minutes, when traveling at this velocity, compared with the currently accepted values of 299,792 kilometers per second and of 8.3 minutes or 499 seconds.

1680

English physicist **Isaac Newton (1643–1727)** calculates that an inverse-square law of gravitational attraction between the Sun and planets would explain the elliptical planetary orbits discovered by **Johannes Kepler (1571–1630).**

1686

French philosopher and writer **Bernard de Fontenelle (1657–1757)** publishes his book *Entretiens sur la Pluralité des Mondes*, which popularizes the idea that the Universe contains many inhabited worlds.

The English astronomer **Edmond Halley (1656–1742)** publishes the first meteorological chart of the world, showing the directions of prevailing winds in different regions.

1687

English physicist **Isaac Newton (1643–1727)** publishes his *Philosophiae naturalis principia mathematica*, or *Mathematical Principles of Natural Philosophy*, known as the *Principia* for short, with the financial support of English astronomer **Edmond Halley (1656–1742)**, in which Newton showed that the same unchanging physical laws apply to everything in the Cosmos from the terrestrial to the celestial, the apple to the Moon. The *Principia* includes a mathematical presentation of his laws of motion and his concept of universal gravitation, explaining the motions of the Moon around the Earth and the planets around the Sun. In his words, written in the *Principia*, "The [Sun-centered] Copernican system of the planets stands revealed as a vast machine working under mechanical laws here understood and explained for the first time." Nevertheless, Newton did not derive Kepler's third law in the *Principia*, but instead used the law to infer that the gravitational force must fall off with the inverse square of the distance. In the 1713 edition of the Principia, Newton introduced the idea that material objects are attracted by the force and action of the aether, a "subtle, all-pervasive spirit."

1693

English astronomer **Edmond Halley (1656–1742)** discovers the formula for the focal distance of a lens; its reciprocal is the sum of the reciprocals of the distances of the object and image from the lens.

1700 to 1800 AD

1704

English physicist **Isaac Newton (1643–1727)** publishes his *Optics*, which describes refraction, reflection, and other properties of lenses and mirrors. He also elaborates his theory that light is composed of a stream of particles.

1705

The English astronomer **Edmond Halley (1656–1742)** finds that the retrograde orbit of the comet of 1682 is similar to those of the comets observed in 1607 and 1531, and notices that the Great Comet of 1456 also traveled in the retrograde direction. He concludes that all four comets were returns of the same comet in a closed elliptical orbit around the Sun with a period of about 76 years, predicting its return in December 1758 and noting that he would not live to see it. Following the predicted reappearance, it became known as Comet Halley.

1718

English astronomer **Edmond Halley (1656–1742)** compares the observed positions of bright stars on the celestial sphere with those found in Ptolemy's *Almagest*, borrowed from the list compiled by **Hipparchus (c. 190–c. 120 BC)**, concluding that at least three stars had changed position and were not "fixed" in the firmament. The exceptionally bright stars, Aldebaran (called Palificium or the Bulls Eye), Sirius and Arcturus, had changed position, by between 15 and 20 minutes of arc since Hipparchus' time and were apparently moving. The stellar motion that Halley detected is across the sky, transverse or perpendicular to the line of sight, and it is known as proper motion.

1724

The Englishman **George Graham (1674–1751)** discovers large, irregular fluctuations in compass needles; the Swedish astronomer **Anders Celsius (1701–1744)** saw some of them at about the same time. Three years later, in 1727, they both showed that a disturbance on the Sun produced a magnetic field fluctuation on Earth. A century later the German geophysicist and writer **Alexander von Humboldt (1769–1859)** named them magnetic storms, and instituted a worldwide program for compiling magnetic and weather observations.

1728

The English astronomer **James Bradley (1693–1762)** discovers stellar aberration, which is a tilting of the apparent position of a star in the direction of the motion of the observer. The size of the effect depends on the velocity of the Earth, the velocity of light, and the angle between the direction of observation and the direction of motion, with a maximum value of 20.47 seconds of arc. Bradley's discovery proved that the Earth moves through space, and enabled him to determine a value of 308,300 kilometers per second for the velocity of light. In 1679, the Danish astronomer **Ole Römer (1644–1710)** had measured the velocity of light using observations of Jupiter's moons, obtaining a value of 225,000 kilometers per second. The modern value is 299,792 kilometers per second.

1735

The English horologist and instrument maker **John Harrison (1693–1776)** submits his marine chronometer to the British government's Board of Longitude, which had offered an award of 20,000 pounds to anyone who could invent an instrument that could determine longitude at sea to an accuracy of half a degree. Harrison's chronometers were subsequently used by the British Navy, obtaining an accuracy of about one minute of longitude, but he did not receive the full award until 1773.

1750

English astronomer and teacher **Thomas Wright (1711–1786)**, of Durham, publishes *An Original Theory or New Hypothesis of the Universe*, in which the Sun is assumed to be similar to the rest of the stars, that there is a multitude of worlds, and that all the stars are located in a thin spherical shell. The entire Milky Way is just an edgewise view of a small segment of the shell. Wright also suggested that the nebulae might be other collections of stars, resembling the Milky Way but lying outside it, and that life might be present on other worlds. His alternative, less preferred model portrayed the Milky Way as a flat ring, and it was taken up in 1755 by

Sphere of Stars   Milky Way

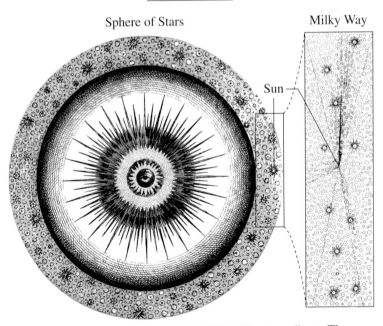

Sun

FIG. 3   Thomas Wright's model of the Milky Way According to **Thomas Wright of Durham (1711–1786)**, the Milky Way is composed of a large number of stars arranged in a slab-like layer, with the Sun placed at the center of the slab. A cross section of this stellar layer is shown here. It is a small segment of a much larger spherical shell of stars, with the Divine Presence signified by an eye located at the center.

the German philosopher **Immanuel Kant (1724–1804),** converting the ring into a disk-shaped system.

1755

The German philosopher **Immanuel Kant (1724–1804)** publishes his book *Allgemeine Naturgeschichte und Theorie des Himmels* or *Universal Natural History and Theory of the Heavens*, in which he proposes that the Milky Way is a flattened, spinning disk of stars, and argues that the nebulae are disk-shaped island universes, or galaxies, similar to the Milky Way and located far outside it, but seen at various angles with different shapes. Kant also proposed that the Sun and planets formed together out of a single collapsing, rotating cloud of interstellar gas, now called the solar nebula, that new planets would be discovered beyond Saturn, and that other planets in the Solar System are either inhabited now or will be in the future. The French astronomer and writer **Pierre Simon Marquis de Laplace (1749–1827)** popularized this nebular hypothesis for the origin of the Solar System in 1796, extending it to the formation of rings and moons around the planets.

1761

Astronomers travel to remote locations to observe the transit of Venus across the Sun, repeating the observations during the 1769 transit. Both sets of observations were used to place limits to the mean distance between the Earth and the Sun, the astronomical unit, at between 148 and 154 million kilometers.

1766

English natural philosopher **Henry Cavendish (1731–1810)** demonstrates the existence of "inflammable air", a substance that was named hydrogen by the French chemist **Antoine Laurent Lavoisier (1743–1794)** in 1783; both scientists showed that hydrogen combustion in oxygen produces water.

The German mathematician **Johann Daniel Titius (1729–1796)** first notices the formula that approximately describes the distances of all the then known planets from the Sun. It was brought to prominence by the German astronomer **Johann Elert Bode (1747–1826)** in his popular book on astronomy entitled *Anleitung zur Kenntnis des gestirnten Himmels*, or *Instruction for the Knowledge of the Starry Heavens*, and is now known as the Titius-Bode law. The law states that the relative distances of the planets from the Sun can be approximated by taking the sequence 0, 3, 6, 12, 24, . . . adding 4, and dividing by 10. The Titius-Bode law predicts the distances of Uranus and the first asteroids, which were discovered after the law was stated; it also predicts the approximate distance of Neptune.

1771

French astronomer and comet hunter **Charles Messier (1730–1817)** creates his first *Catalogue des nebuleuses et des amas d'etoiles* or *Catalog of Nebulae and Star Clusters*, consisting of 45 objects. By 1784 the catalogue included just over 100 objects. Astronomers still designate these objects by the capital letter M followed by the number in Messier's list. The Crab Nebula is, for example, known as M1, and the Andromeda Nebula as M31.

1781

German-born English astronomer **William Herschel (1738–1822)** discovers the planet Uranus using a homemade reflecting telescope on the night of 13 March 1781.

1784

In unpublished work, the reclusive English scientist **Henry Cavendish (1731–1810)** calculates the gravitational deflection of light rays by the Sun using Newtonian gravity and the assumption that a light ray is a material corpuscle.

The English astronomer **John Goodricke (1764–1786)** discovers a periodic variation in the intensity of light from the star Delta Cephei. It is the prototype of the Cepheid variable stars that are used to infer cosmic distances through the period-luminosity relation discovered by the American astronomer **Henrietta Swan Leavitt (1868–1921)** in 1912.

English clergyman and astronomer **John Michell (1724–1793)** proposes that an invisible, black star might be so large and massive, and its gravitational pull so intense, that light could not escape from it. Such dark and invisible stars are now known as stellar black holes. Michell also speculated that the unseen star might betray its presence by its gravitational effect on a nearby luminous star in orbit around it. French astronomer **Pierre Simon Marquis de Laplace (1749–1827)** subsequently developed the idea that the most massive stars might be unable to send light outside because their gravity is too strong.

1787

German-born English astronomer **William Herschel (1738–1822)** discovers the two largest satellites of Uranus, and, two years later, two satellites of Saturn.

1789

German-born English astronomer **William Herschel (1738–1822)** completes construction of his 40-foot reflector, with a 1.2-meter (48-inch) mirror, which remained the largest telescope of its time for half a century. It soon gained a reputation as the "eighth wonder of the world", and was visited by many people, including **Franz Joseph Haydn (1732–1809)** whose visit was the direct inspiration for "The Heavens Are Telling" passage in his great oratorio *The Creation*. Although the 40-foot reflector brought many faint stars into view and pushed the edge of the known Universe further into space, Herschel eventually concluded that his new telescope could not fathom the depth of the Milky Way.

1795

Scottish natural philosopher **James Hutton (1726–1797)** publishes the definitive version of his *Theory of the Earth*, in which geological change is exceptionally gradual, and hence the Earth must be incalculably old.

1796

French astronomer and writer **Pierre Simon Marquis de Laplace (1749–1827)** popularizes the nebular hypothesis for the origin of the Solar System in his *Exposition du système du monde*, arguing that the planets and Sun formed together during the gravitational collapse of a spinning nebula.

1797

English natural philosopher **Henry Cavendish (1731–1810)** uses an apparatus devised by the English clergyman and astronomer **John Michell (1724–1793)** to measure the force between two objects in the laboratory, and to thereby determine Newton's gravitational constant and the mass of the Earth. He obtains a mass density of about 5.5

times that of water and a mass of about 6 billion trillion tons (6×10^{21} tons or 6×10^{24} kilograms).

1800 to 1900 AD

1800–1849

1800

German-born English astronomer **William Herschel (1738–1822)** discovers invisible caloric rays or radiant heat, later named infrared radiation, by noticing a rise in temperature when a thermometer is placed beyond the red end of the visible spectrum of sunlight.

1801

Sicilian astronomer and monk **Giuseppe Piazzi (1746–1826)** discovers the first asteroid on the night of 1 January 1801, naming it Ceres after the patron goddess of Sicily. He followed it for six weeks, until it moved too close to the Sun to be observed and was lost. The German mathematician, physicist and astronomer **Karl Friedrich Gauss (1777–1855)** developed a least-squares approximation method of calculating an orbit from just a few observations, and using Piazzi's data predicted the location of Ceres. The German astronomer **Heinrich Wilhelm Olbers (1758–1840)** rediscovered the asteroid where Gauss had predicted it would be found. While following Ceres, Olbers discovered a second asteroid, Pallas, in 1802; the German astronomer **Karl L. Harding (1765–1834)** discovered a third, Juno, in 1804.

The German physicist **Johann Wilhelm Ritter (1776–1810)** discovers ultraviolet light in the spectrum of sunlight, at wavelengths just a bit shorter than the wavelength of violet light.

German astronomer **Johann Georg von Soldner (1766–1833)** uses Newton's theory of gravity to calculate the bending of light by the Sun under the assumption that light is a stream of material particles, obtaining a deflection of 0.84 seconds of arc for a grazing light ray. The German physicist **Albert Einstein (1879–1955)** confirmed this value in 1911, but he obtained twice this amount in 1915 using his *General Theory of Relativity* in which the Sun curves nearby space. The greater light bending was confirmed during solar eclipse observations in 1919.

English physician and physicist **Thomas Young (1773–1829)** discovers the interference of light when he observes that light passing through two closely spaced pinholes produces alternating bands of light and dark in the area of overlap. He thereby establishes the wave theory of light, proposing that light waves are transmitted as undulatory motions in an all-pervasive aether.

1802

German-born English astronomer **William Herschel (1738–1822)** discovers binary stars, consisting of two stars that revolve around each other, and eventually catalogues 848 of them.

English astronomer, chemist and physicist **William Hyde Wollaston (1766–1828)** discovers dark gaps, or absorption lines, in the spectrum of sunlight.

1803

English chemist **John Dalton (1766–1844)** proposes that gases are made of atoms, and that atoms of different elements have different weights.

1804

The Sicilian astronomer and monk **Giuseppe Piazzi (1746–1826)** notices that one relatively undistinguished star, known as 61 Cygni, has an exceptionally large proper motion of 5.2 seconds of arc per year. Also called the "Flying Star" due to its rapid motion, 61 Cygni became the first star other than the Sun whose distance was reliably determined, by the German astronomer **Friedrich Wilhelm Bessel (1784–1846)** in 1838.

1808

German mathematician, physicist and astronomer **Karl Friedrich Gauss (1777–1855)** discovers a method of determining an orbit from just three observations, describing it in his *Theoria motus corporum coelestium in sectionibus conicis solem ambientum*, or *Theory of motion of the celestial bodies moving in conic sections around the Sun*, which also includes his presentation of the least-squares method.

1814

German astronomer **Joseph von Fraunhofer (1787–1826)** allows light to pass through a narrow slit and then a prism, detecting and cataloguing more than 300 absorption lines in the spectrum of visible sunlight, assigning Roman letters to the most prominent; these absorption lines are now sometimes called Fraunhofer lines.

1823

The Hungarian mathematician **János Bolyai (1802–1860)** describes a non-Euclidean geometry in a paper entitled *The Absolute True Science of Space*. Bolyai invented a geometry that violates Euclid's fifth postulate, that there is only one line through a point outside another line that is parallel to it, showing that there might be many such lines drawn through the point.

German astronomer **Heinrich Wilhelm Olbers (1758–1840)** discusses the paradox that now bears his name, in a paper entitled *Ueber die Durchsichtigkeit des Weltraumes*, or *On the Transparency of Space*, presented in 1823 and published in English translation in 1826. According to his paradox, the sky in an infinite, uniform Universe should be covered by stars shining as brightly as our Sun, so the night sky should be as bright as the day. Olbers explained the darkness of the night sky by assuming that space is not absolutely transparent, and that absorbing matter distributed throughout space obscures our view of distant stars. Olbers' paradox is now resolved by the expansion of the Universe; the Doppler effect of the distant moving galaxies redshifts their light out of the visible part of the spectrum.

1826

Russian mathematician **Nikolai Lobachevsky (1792–1856)** discovers a hyperbolic, non-Euclidean geometry in which more than one line can be drawn through a point outside another line parallel to it, and the sum of the angles in a triangle is always less than 180 degrees. This was similar to the non-Euclidean geometry introduced by the Hungarian mathematician **János Bolyai (1802–1860)** in 1823, but the work of both of them was largely ignored at the time.

1827

In a paper entitled *Microscopical Observations of Active Molecules*, the Scottish botanist **Robert Brown (1773–1858)** states that very fine pollen grains move about in a continuously random manner when suspended in water and observed under the microscope. This phenomenon, now known as Brownian motion, is true for any small solid particles suspended in a liquid or a gas. In 1905 the German physicist **Albert Einstein (1879–1955)** explained the effect in terms of the random motion of molecules of water, which will knock a sufficiently small particle into Brownian motion. In 1909 the French physicist **Jean Baptiste Perrin (1870–1942)** demonstrated the existence of atoms by careful measurements of Brownian motion.

The French mathematical physicist **Joseph (Jean Baptiste) Fourier (1768–1830)** develops new mathematical tools to describe the transfer of heat, known today as Fourier analysis and the Fourier series, and in a *Mémoire sur les Températures du Globe Terrestre et des Espaces Planétaires* describes how the Sun's heat is retained by the Earth's atmosphere. Visible sunlight passes through our transparent atmosphere to warm the Earth's land and oceans, and some of this heat is reradiated in infrared form. Our atmosphere absorbs some of the infrared heat radiation, emitting part of it downward to warm the planet's surface and the air immediately above it. This ability of the atmosphere to hold heat near a planet's surface, and warm it to higher temperatures than possible without an atmosphere, is popularly known as the greenhouse effect.

1830

Scottish geologist **Charles Lyell (1797–1875)** publishes *The Principles of Geology*, in three volumes from 1830 to 1833, with regular revisions until 1875. He proposes that the Earth's geological structure evolved gradually over eons of time through the slow continuous action of forces still at work today.

1836

The English astronomer **Francis Baily (1774–1884)** describes the total eclipse of the Sun on 15 May 1836, noting that immediately before the Sun disappeared behind the Moon the sunlight appeared as a discontinuous line of brilliant spots of light, now known as Baily's beads. They are caused by sunlight shining through the valleys between mountains on the Moon's edge or limb, while the mountains block the sunlight in other places along the limb.

1838

German astronomer **Friedrich Wilhelm Bessel (1784–1846)** measures the annual parallax of the star 61 Cygni, also known as the "Flying Star", thereby obtaining the first reliable measurement of the distance to a star other than the Sun. He used a special telescope, known as a heliometer because of its previous use to establish the Sun's apparent diameter. Bessel's heliometer consisted of a lens, just 0.15 meters (6 inches) in diameter, cut exactly in half. When the two halves were moved apart, their separation could be used to precisely measure the angle between 61 Cygni and a nearby, fainter star, which Bessel did with the utmost care every clear night for more than a year, usually repeating his observations at least sixteen times a night. He found that the star 61 Cygni has a parallax of 0.31 seconds of arc, very close to the modern value of 0.287 seconds of arc, corresponding to a distance of 11.4 light-years. The following year the Scottish astronomer **Thomas Henderson (1798–1844)** announced his measurement of the annual parallax of Alpha Centauri, which is just 4.29 light-years away, and in 1840 the German-born Russian astronomer **Friedrich Georg Wilhelm Struve (1793–1864)** reported his measurement of the parallax of the bright star Vega.

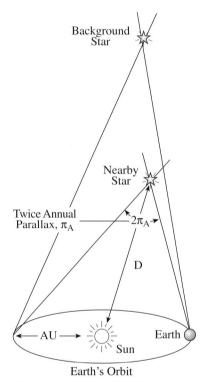

German mathematician and astronomer **Karl Friedrich Gauss (1777–1855)** publishes a mathematical description of the Earth's dipolar magnetic field, using his calculations with observations to show that the magnetism must originate within the Earth's core. A unit of magnetic strength is subsequently named Gauss in his honor.

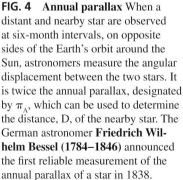

Earth's Orbit

FIG. 4   Annual parallax When a distant and nearby star are observed at six-month intervals, on opposite sides of the Earth's orbit around the Sun, astronomers measure the angular displacement between the two stars. It is twice the annual parallax, designated by π_A, which can be used to determine the distance, D, of the nearby star. The German astronomer **Friedrich Wilhelm Bessel (1784–1846)** announced the first reliable measurement of the annual parallax of a star in 1838.

1842

German astronomer **Friedrich Wilhelm Bessel (1784–1846)** attributes periodic variations in the proper motions of the bright stars Sirius and Procyon to the gravitational attraction of unseen companion stars that revolve about the visible ones. The invisible objects are now known to be white dwarf stars, about the size of the Earth.

1842

Austrian physicist **Christian Johann Doppler (1803–1853)** publishes *Über das farbige Licht der Doppelsterne*, or *On the Colored Light of Double Stars*, in which he describes how the frequency, or wavelength, of sound and light waves change with the motion of their source relative to the observer, a phenomenon now known as the Doppler effect. The observed wavelength depends on the relative velocity of the source and observer, with shorter waves detected when the source is moving toward the observer and longer wavelengths when moving away. The size of the change in wavelength caused by the motion can be used to infer the radial velocity, along the line of sight.

1843

German pharmacist and amateur astronomer **Samuel Heinrich Schwabe (1789–1875)** announces that the number of sunspots varies from a maximum to a minimum and back to a maximum again in a period of about ten years. This periodicity is now known as the sunspot cycle, and more generally called the cycle of solar magnetic activity. His discovery only became widely known in 1851 when the German explorer and naturalist **Alexander von Humboldt (1769–1859)** published it in his book *Kosmos*.

1845

Irish astronomer **William Parsons (1800–1867),** the third **Earl of Rosse,** completes his reflecting telescope, known as the Leviathan of Parsonstown, with a metallic mirror that is 1.8 meters (72 inches) in diameter, and uses it to resolve the nebula M51 into a spiral shape that he attributed to its rotation. He went on to find 15 spiral nebulae, and resolved other nebulae into clusters of stars, while also naming the famous Crab Nebula. His telescope remained the largest in the world until the 2.5-meter (100-inch) Hooker telescope was completed at the Mount Wilson Observatory in 1917.

1846

French astronomer **Urbain Jean Joseph Leverrier (1811–1877)** uses the differences between the observed and predicted location of the planet Uranus to predict the position of a large unknown planet, located far beyond Uranus, whose gravitational tug on Uranus causes it to deviate from its expected position. German astronomer **Johann Gottfried Galle (1812–1910)** and his student **Heinrich Louis d'Arrest (1822–1875)** discover the planet Neptune on the night of 23 September 1846, near the position predicted by Leverrier. The English astronomer **John Couch Adams (1819–1892)** performed similar calculations, also predicting the existence of the unknown planet at about the same time as Leverrier.

1847

The German physicist **Hermann von Helmholtz (1821–1894)** formulates the law of conservation of energy in the general form that we use today.

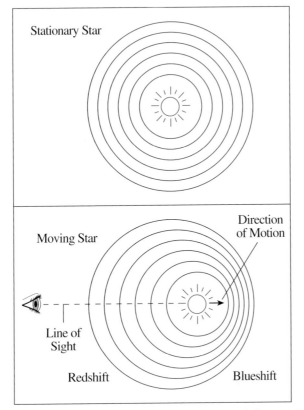

FIG. 5   Doppler effect A stationary star (*top*) emits regularly spaced light waves that get stretched out or scrunched up if the star moves (*bottom*). Here we show a star moving away (*bottom right*) from the observer (*bottom left*). The stretching of light waves that occurs when the star moves away from an observer along the line of sight is called a redshift, because red light waves are relatively long visible light waves; the compression of light waves that occurs when the star moves along the line of sight toward an observer is called a blueshift, because blue light waves are relatively short. The wavelength change, from the stationary to moving condition, is called the Doppler shift, and its size provides a measurement of radial velocity, or the velocity of the component of the star's motion along the line of sight. The Doppler effect is named after the Austrian physicist **Christian Doppler (1803–1853)**, who first considered it in 1842.

The French mathematician **Edouard Roche (1820–1883)** derives the distance at which a large satellite is torn to pieces by the differential gravitational forces of a planet. Now known as the Roche limit, it has a radius of about 2.5 times the radius of the planet. Most planetary rings lie within the Roche limit. The planet's gravitational forces also prevent small particles or satellites from coalescing to form a larger moon within the Roche limit. Roche additionally considered the gravitational interaction of binary star systems. The gravitation of each star has its own zone of influence now called a Roche lobe. And when the stars are close together, the two Roche lobes touch at a point where their gravitational forces exactly cancel. When the largest component star expands and fills its Roche lobe, matter is transferred to the other star.

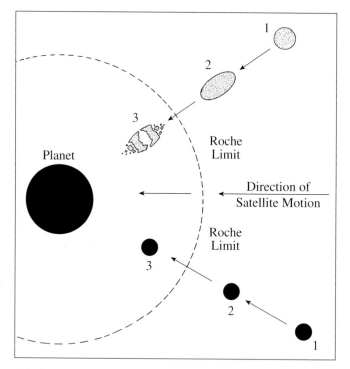

FIG. 6   *Roche limit* A large satellite (*top*) that moves well within a planet's Roche limit (*dashed curve*) will be torn apart by the tidal force of the planet's gravity, in an effect first investigated in 1847 by the French mathematician **Edouard Roche (1820–1883).** The side of the satellite closer to the planet feels a stronger gravitational pull than the side farther away, and this difference works against the self-gravitation that holds the body together. A small solid satellite (*bottom*) can resist tidal disruption because it has significant internal cohesion in addition to self-gravitation.

1848

Irish-born Scottish physicist **William Thomson (1824–1907),** later **Lord Kelvin,** the first Baron Kelvin of Largs, devises the absolute temperature scale, now known as the Kelvin scale, and defines absolute zero as −273 degrees Celsius, where the energy of molecules is zero and motion stops.

1850–1899

1850

The German physicist **Rudolf Clausius (1822–1888)** introduces the concept of entropy, derived from the Greek word *entrop-* for "transformation," showing that entropy arises from the conversion of heat into work or the transfer of heat from low to high temperature. He concluded that entropy must inevitably increase in the Universe, a formulation of the second law of thermodynamics. It implies that heat can never pass of its own accord from a colder to a hotter body.

1851

French physicist **Jean Bernard Léon Foucault (1819–1868)** proves that the Earth rotates by showing that a pendulum, some 67 meters (220 feet) long, always swings in the same plane as the Earth rotates beneath it. In the previous year he used a rotating mirror to measure the velocity of light in air, obtaining a value of 298,000 kilometers per second.

1852

English astronomer and geodesist **Edward Sabine (1788–1833)** demonstrates that global magnetic disturbances of the Earth, now called geomagnetic storms, vary in tandem with the 11-year sunspot cycle.

1854

English astronomer **George Biddell Airy (1801–1892)** measures the Newtonian constant of gravitation and calculates the mass of the Earth by swinging a pendulum at the top and bottom of a deep coal mine and measuring the increase in gravity with descent below the Earth's surface.

The German physicist **Hermann von Helmholtz (1821–1894)** proposes that slow gravitational collapse could supply the Sun's luminous output at the presently observed rate for about 25 million years. If the Sun, or any other star, was gradually shrinking, then its gravitational potential energy could be slowly converted into kinetic energy, heating the stellar gas to incandescence with a total lifetime now known as the Kelvin-Helmholtz time, as the result of the Helmholtz calculation in 1854 and similar conclusions by **William Thomson (1824–1907),** subsequently **Lord Kelvin,** about a decade later. In 1854 Helmholtz also predicted the heat death of the Universe on the basis of thermodynamic theory.

The German mathematician **Bernhard Riemann (1826–1866)** summarizes non-Euclidean geometry, showing that parallel lines can meet in spherical space and developing the mathematical description of curved space that later enabled the German physicist **Albert Einstein (1879–1955)** to develop his *General Theory of Relativity*.

1857

Scottish physicist **James Clerk Maxwell (1831–1879)** demonstrates mathematically that Saturn's rings are made up of a vast number of small particles in orbit around the planet, each one moving like a tiny moon with a speed that decreases with the orbit's distance from Saturn. In 1895, the American astronomer **James Edward Keeler (1857–1900)** proved that the inner parts of Saturn's rings move faster than the outer parts, using observations of the Doppler effect of spectral features in sunlight reflected from the rings.

1858

English amateur astronomer **Richard C. Carrington (1826–1875)** shows that sunspots move from high solar latitudes to the Sun's equator during the 11-year sunspot cycle. He also demonstrates that sunspots at high latitudes are carried around the Sun more slowly than sunspots at low latitudes, so they rotate differentially with slower speeds at higher latitudes.

1859

English amateur astronomers **Richard C. Carrington (1826–1875)** and **Richard Hodgson (1804–1872)** independently observe a solar flare in the white light of the photosphere. Seventeen hours after the flare, a large magnetic storm begins on the Earth.

English naturalist **Charles Darwin (1809–1882)** publishes his book *The Origin of Species by Means of Natural Selection or the Preservation of Favored Races in the Struggle for Life*, which supposes that all life had a common ancestor and that the origin and evolution of species was the result of natural selection, and reproduction, of the most adapted members of a population.

German physicist **Gustav Kirchhoff (1824–1887)** and German chemist **Robert Bunsen (1811–1899)** show that the dark lines in the solar spectrum can be produced in the terrestrial laboratory in both emission and absorption, and that the wavelengths of the lines can be used to identify the chemical element that produced them, including sodium, calcium and iron.

French mathematician **Urbain Jean Joseph Leverrier (1811–1877)** reworks the theory of the orbit of Mercury, considered by him since 1837, and uses recent accurate observations to show that the point in Mercury's orbit closest to the Sun, the planet's perihelion, is moving ahead of its predicted location at the rate of 38 seconds of arc per century. Leverrier proposes that this anomalous motion is due to the gravitational pull of a hitherto unrecognized planet, which he named Vulcan, revolving at somewhat less than half of Mercury's mean distance from the Sun. The American astronomer **Simon Newcomb (1835–1909)** made more refined calculations in 1895, showing that Mercury's perihelion is progressively moving ahead of the place predicted using

Newton's theory of gravitation by 43 seconds of arc per century. The hypothetical planet Vulcan was never found, and in 1915 the German physicist **Albert Einstein (1879–1955)** explained the advance in Mercury's perihelion as a consequence of the Sun's curvature of nearby space.

1861

Irish physicist **John Tyndall (1820–1893)** publishes a paper *On the Absorption and Radiation of Heat by Gases and Vapours . . .*, concluding that significant heat is absorbed by water vapor and carbon dioxide, which are minor ingredients of our atmosphere. The main constituents of the atmosphere, nitrogen and oxygen play no part in the greenhouse effect.

1862

Swedish astronomer and physicist **Anders Jonas Ångström (1814–1874)** announces his discovery of hydrogen in the spectrum of the Sun.

American astronomer **Alvan Clark (1804–1887)** observes the faint companion star of the bright star Sirius.

German physicist **Gustav Kirchhoff (1824–1887)** introduces the concept of a perfect blackbody that would absorb and emit radiation at all wavelengths, reaching an equilibrium that depends on temperature and not on the nature of the surface of the body.

The Irish physicist **William Thomson (1824–1907)**, later **Lord Kelvin,** estimates that the Earth cooled from its primordial molten state in 20 million to 400 million years, with 100 million years the most likely age of the Earth. This was before it was learned that radioactivity warms the inside of the Earth, which can therefore be much older than Thomson supposed. William Thomson took the title Baron Kelvin of Largs in 1892.

1864

English astronomer **William Huggins (1824–1910)** obtains the spectra of nebulae. He discovers three emission lines radiated from the planetary nebula now known as NGC 6543, thus showing that this nebula is made of a low-density gas. One of the lines was attributed to hydrogen, but the other two suggested the existence of an undiscovered element that became known as "nebulium". They were not explained until 1928, by the American astronomer **Ira S. Bowen (1893–1973)** who showed that the nebulium lines arise from ionized atoms of oxygen and nitrogen undergoing "forbidden" transitions in the low-density gas. Huggins also showed that the Orion Nebula consists of gases that radiate emission lines, and that the spectrum of the Andromeda Nebula contains stellar absorption lines, indicating that it is composed of stars. Huggins additionally obtained stellar spectra that showed that other stars have the same ingredients as the Sun.

Scottish physicist **James Clerk Maxwell (1831–1879)** presents the fundamental equations that describe electromagnetic radiation in *A Dynamical Theory of the Electromagnetic Field*, showing that radiation consists of electric and magnetic waves that oscillate at right angles to each other and to the direction of travel. He also notices that a magnetic field can be generated by a changing electric field.

1866

The American astronomer **Daniel Kirkwood (1814–1895)** announces his discovery of regions in the asteroid belt, between the orbits of Mars and Jupiter, which contain very few asteroids. These Kirkwood gaps, as they are now called, are located at places where an asteroid would have an orbital period that is a rational fraction, such as 1/4, 1/3 and 1/2, of Jupiter's orbital period, and repeated gravitational interactions with Jupiter seems to have tossed the asteroids out of these places. The American scientist **Jack Wisdom (1953–)** provided a complete explanation in the 1980s, by showing that Jupiter induces a chaotic zone in the vicinity of the Kirkwood gaps, eventually removing asteroids from them.

Italian astronomer **Giovanni Schiaparelli (1835–1910)** proposes that all meteor showers, or shooting stars, are the result of the disintegration of comets, using observational evidence to suggest that the meteors that give rise to the luminous showers travel in orbits around the Sun that are similar to those of comets.

1867

The French astronomers **George Rayet (1839–1906)** and **Charles Wolf (1827–1918)** discover a relatively rare and new class of stars whose visible-light spectra contain broad and intense emission lines of hydrogen and helium. These kinds of stars are now known as Wolf-Rayet stars.

1868

Swedish astronomer **Anders Jonas Ångström (1814–1874)** provides measurements of the wavelengths of more than one hundred dark lines in the Sun's visible spectrum, in units that become known as the Ångström, abbreviated Å, where $1\text{Å} = 10^{-10}$ meters.

French astronomer **Pierre Jules Janssen (1824–1907)** and British astronomer **Joseph Norman Lockyer (1836–1920)** discover a previously unknown line in the yellow region of the Sun's spectrum during the solar eclipse on 18 August 1868. Lockyer named the element helium after the Greek Sun god *Helios*. Helium was not found on Earth until 1895, when the Scottish chemist **William Ramsay (1852–1916)** discovered it as a gaseous emission from a heated uranium mineral called cleveite.

The Italian astronomer **Angelo Pietro Secchi (1818–1878)** classifies 4,000 stars by their color and spectra, dividing them into four main groups with the colors white, yellow, orange and red, and with the respective spectral features of strong absorption lines of hydrogen, strong calcium lines, strong metallic lines and broad absorption lines in the violet.

1869

The Irish physicist **John Tyndall (1820–1893)** discovers that light is scattered by very small particles suspended in a medium, and explains the blue color of the sky by the preferential scattering of the blue component of sunlight by air molecules.

The American astronomers **Charles A. Young (1834–1908)** and **William Harkness (1849–1900)** independently discover a bright green emission line in the spectrum of the solar corona during the solar eclipse of 7 August 1869. The conspicuous spectral feature remained unidentified with any known terrestrial element for more than half a century. In 1941 the Swedish astronomer **Bengt Edlén (1906–1993)** and the German astronomer **Walter Grotrian (1890–1954)** identified the green line with iron atoms that had become highly ionized due to an unexpectedly high million-degree temperature.

1872

American amateur astronomer **Henry Draper (1837–1882)** takes the first photograph of the spectrum of a star, that of Vega. His widow established a fund to support further spectral studies of stars at the Harvard College Observatory, resulting in the comprehensive *Henry Draper Catalogue* of the positions, brightness and spectra of stars.

1873

American geophysicist **Elias Loomis (1811–1889)** demonstrates a correlation between intense auroras and the number of sunspots, both recurring with an 11-year periodicity and with similar times of maximum.

1876

The Lick Observatory is constructed between 1876 and 1887, from a bequest by the California businessman **James Lick (1796–1876).**

1877

An unemployed Scottish astronomer **David Gill (1843–1914)** travels to the island of Ascension near the Earth's equator to measure the parallax of Mars during its close approach at the opposition of 1877, deriving a value of 149.8 million kilometers for the astronomical unit, the mean distance between the Earth and the Sun, with an uncertainty of only 200 thousand kilometers.

During a favorable opposition of Mars, when the planet was even closer to the Earth than during most other oppositions, the American astronomer **Asaph Hall (1829–1907)** discovers the two small moons of Mars, named Phobos and Deimos.

During the favorable opposition of Mars in 1877, the Italian astronomer **Giovanni Schiaparelli (1835–1910)** reports that a maze of dark, narrow straight lines traverses the planet's surface. He called them *canali*, the Italian word for "channels" or "canals", assuming that they were natural features and giving the broadest ones the names of large terrestrial rivers, such as the Ganges and Indus.

1879

The Austrian physicist **Josef Stefan (1835–1893)** shows that the radiant energy of a thermal source, such as a star, increases with the fourth power of the temperature. The absolute luminosity of a star is proportional to the square of its radius and the fourth power of the temperature, a relation known as the Stefan-Boltzmann law. The Austrian physicist **Ludwig Boltzmann (1844–1906)** gave a theoretical explanation of the law in 1884, based on the theory of a blackbody.

1882

American physicist **Henry Rowland (1848–1901)** invents the concave diffraction grating, in which up to 20,000 lines to the inch are engraved on a spherical concave mirrored surface. The grating revolutionized spectrometry by widely dispersing light and permitting numerous spectral lines to be investigated with great accuracy and efficiency. Rowland used his grating to investigate the visible-light spectrum of the Sun, publishing the wavelengths for 14,000 lines.

1884

The meridian at the Royal Greenwich Observatory, England, is chosen to be the Prime Meridian of the World, with exactly zero degrees longitude, dividing the eastern and western hemispheres of the Earth. It is defined by the position of the Observatory's large Transit Circle telescope built in 1850 by the seventh Astronomer Royal **George Biddell Airy (1801–1892)**.

1885

Swiss mathematics teacher **Johann Balmer (1825–1898)** publishes an equation that describes the regular spacing of the wavelengths of the four lines of hydrogen detected in the spectrum of visible sunlight, and predicts the wavelengths of other solar lines at invisible ultraviolet and infrared wavelengths. The American physicist **Theodore Lyman (1874–1954)** observed the ultraviolet lines of hydrogen in 1906, announcing in 1914 that he had observed three of the lines specified by the Balmer formula. The German physicist **Friedrich Paschen (1865–1947)** confirmed the infrared lines in 1908.

The German astronomer **Ernst Hartwig (1851–1923)** observes a new star, or nova, in the Andromeda Nebula, M31, whose peak intensity was as great as all the rest of the nebula. The outburst was a supernova, also named S Andromedae.

The Austrian geologist **Eduard Suess (1831–1914)** publishes *Das Antlitz der Erde* or *The Face of the Earth*, describing the Earth's geographical evolution and suggesting that the southern continents were once joined together in a great supercontinent that he named Gondwanaland. His results and speculations were amplified and extended by the German geologist **Alfred Wegener (1880–1930)** in his theory of continental drift, proposed in 1929.

1887

German physicist **Heinrich Hertz (1857–1894)** discovers radio waves, generating them with an oscillating electric current, and demonstrates that the invisible long-wavelength rays can be sent through the air, traveling at the velocity of light. He also demonstrates that the radio waves can be reflected, refracted and polarized like light.

German-born American physicist **Albert Michelson (1852–1931)** and American chemist **Edward Morley (1838–1923)** use an interferometer to detect any difference in

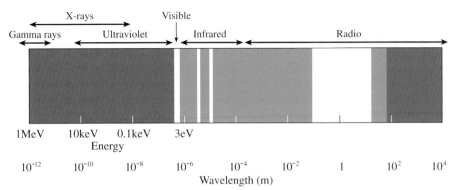

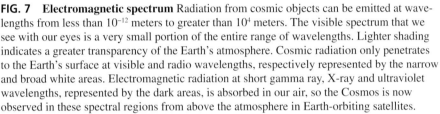

FIG. 7 **Electromagnetic spectrum** Radiation from cosmic objects can be emitted at wavelengths from less than 10^{-12} meters to greater than 10^{4} meters. The visible spectrum that we see with our eyes is a very small portion of the entire range of wavelengths. Lighter shading indicates a greater transparency of the Earth's atmosphere. Cosmic radiation only penetrates to the Earth's surface at visible and radio wavelengths, respectively represented by the narrow and broad white areas. Electromagnetic radiation at short gamma ray, X-ray and ultraviolet wavelengths, represented by the dark areas, is absorbed in our air, so the Cosmos is now observed in these spectral regions from above the atmosphere in Earth-orbiting satellites.

the velocity of light in two directions at right angles to each other, but find no difference in the values. The measured velocity was therefore unaffected by the motion of the Earth through the hypothetical aether. So this discredited the idea of the aether and led to the conclusion that the speed of light is a universal constant. This experiment was resolved by the *Special Theory of Relativity*, anticipated by the French mathematician **Jules Henri Poincaré (1854–1912)** in 1904 and developed by the German physicist **Albert Einstein (1879–1955)** in 1905.

1888

The Danish-born Irish astronomer **John Louis Emil "J.L.E." Dreyer (1852–1926)** publishes his *New General Catalogue*, abbreviated NGC, of the apparent brightness and celestial positions of thousands of nebulae and star clusters. The catalogue was new when first compiled in 1888. Dreyer extended the data in two *Index Catalogues*, or IC for short, published in 1895 and 1908. Many galaxies, as well as emission nebulae and star clusters, are still designated by their NGC or IC numbers.

1889

Irish physicist **George FitzGerald (1851–1901)** shows that the Michelson-Morley experiment, of 1887, could be explained if a fast-moving object diminishes in length. The Dutch physicist **Hendrik Lorentz (1853–1928)** independently developed the concept in 1895, and it became known as the Lorentz-FitzGerald contraction.

1890

American physicist **Albert Michelson (1852–1931)** describes the use of interference methods in measuring the angular size and the one-dimensional brightness distribution of cosmic sources that are too small to be resolved by a single telescope. A telescope that examines the interference of radiation detected by two or more smaller telescopes is called an interferometer, or interference meter. Michelson was awarded the 1907 Nobel Prize in Physics for his optical precision instruments and the spectroscopic and metrological investigations carried out with their aid.

The first version of the *Henry Draper Catalogue* is published. Produced by astronomers at Harvard College Observatory, it lists the position, magnitude and spectral type of over 10,000 stars, and begins the alphabetical system of classifying stars according to temperature. Subsequent editions increased the listing to 400,000 stars.

1892

French astronomer **Camille Flammarion (1842–1925)** releases the first volume of his book *La Planéte Mars et ses Conditions d'Habitabilité*, published by Gauthier-Villars et Fils, advocating the existence of intelligent Martian life that constructed canals to redistribute increasingly scarce water across the planet. The wealthy Bostonian **Percival Lowell (1855–1916)** advocated similar ideas in his book entitled *Mars*, published in 1895 by Houghton Mifflin, Boston. We now know that the channels, first observed by the Italian astronomer **Giovanni Schiaparelli (1835–1910)** in 1877, are an illusion created when the eye arranges minute, disconnected details into lines.

American geologist **Grove Karl Gilbert (1843–1918)** presents persuasive arguments for the origin of lunar craters, reasoning that they are the result of impacts that occurred before the process of erosion began on the Earth.

1895

American astronomer **James Edward Keeler (1857–1900)** makes spectral observations of sunlight reflected from Saturn's rings, showing that the inner parts of the rings move around Saturn faster than the outer parts.

The Swedish industrial chemist and philanthropist **Alfred Nobel (1833–1896),** who invented dynamite, leaves most of his fortune in his last will to the Nobel Foundation that would bestow annual awards, beginning in 1901, on "those who, during the preceding year, shall have conferred the greatest benefit on mankind."

Scottish chemist **William Ramsay (1852–1916)** discovers that a uranium mineral named cleveite produces the chemically inert gas helium when heated. Helium was known from spectrographic evidence to be present on the Sun but had not yet been found on Earth. Ramsay was awarded the 1904 Nobel Prize for Chemistry for his discovery of the inert gases argon, helium, neon and xenon; they are also known as rare or noble gases.

German physicist **Wilhelm Konrad Röntgen (1845–1923)** discovers X-rays on 8 November 1895, which he named X because of their unknown origin, using them to reveal the bones inside his wife's hand. The X-rays were produced by an electrical discharge in a glass vacuum tube, invented by the English physicist **William Crookes (1832–1919)** in 1871. The energetic X-rays are able to penetrate skin and muscle, detecting human bones and revolutionizing medicine. For this achievement Röntgen was awarded the first Nobel Prize for Physics in 1901.

1896

Swedish chemist **Svante August Arrhenius (1859–1927)** shows that a doubling of the Earth's atmospheric carbon dioxide will produce a global warming of 5 to 6 degrees Centigrade (9 to 11 degrees Fahrenheit), comparable to modern estimates. In 1905 Arrhenius expressed concern about global warming as a result of burning fossil fuels.

French scientist **Henri Becquerel (1852–1908)** discovers gamma rays when leaving uranium and a photographic plate in a drawer; the radioactive uranium emitted gamma rays that fogged the plate. Becquerel was awarded the 1903 Nobel Prize in Physics for his discovery of spontaneous radioactivity, sharing the prize with the French scientists **Pierre Curie (1859–1906)** and **Marie Curie (1867–1934)** for their investigations of the radiation phenomenon that he discovered.

1897

English physicist **Joseph John "J. J." Thomson (1856–1940)** demonstrates that an electrical current in a glass vacuum tube, the Crookes tube, is made up of negatively charged particles of sub-atomic size. The Dutch physicist **Hendrik Lorentz (1853–1928)** gave the name electron to this fundamental particle that conducts electricity; a term previously introduced by the Irish physicist **George Johnstone Stoney (1826–1911)**. Thomson received the 1906 Nobel Prize in Physics for his theoretical and experimental investigations of the conduction of electricity by gases.

Dutch physicist **Pieter Zeeman (1865–1943)** observes the splitting of spectral lines into three components by an intense magnetic field in the terrestrial laboratory. This Zeeman effect, as it is now called, permits the measurement of the strength and direction of cosmic magnetic fields. Zeeman and **Hendrik Lorentz (1853–1928),** also from the Netherlands, were awarded the 1902 Nobel Prize in Physics for their researches into the influence of magnetism upon radiation phenomena.

1898

American astronomer **James Edward Keeler (1857–1900)** begins systematic photography of nebulae using the 0.90-meter (36-inch) refracting telescope at Lick Observatory, demonstrating the ubiquitous nature of spiral nebulae. He estimated that over one hundred thousand spiral nebulae could be detected photographically with his telescope.

Irish scientist **George Johnstone Stoney (1826–1911)** shows that the stability of a planetary atmosphere of a given composition depends upon the temperature of the atmosphere, the mass of the atmospheric molecules, and the mass of the planet, or on the thermal motions of the individual molecules as well as the gravitational pull of the planet. Lighter, faster molecules are more likely to escape, but planets with larger mass and gravity are able to hold onto substantial atmospheres. In 1930, the American astronomer **Donald Menzel (1901–1976)** used these arguments to propose that the cold and massive giant planets are mainly composed of hydrogen.

1899

New Zealand-born English physicist **Ernest Rutherford (1871–1937)** discovers alpha and beta rays, produced by the radioactivity of uranium. This led to his explanation of radioactivity, with English chemist **Frederick Soddy (1877–1956)** in 1903, in terms of the transformation of elements. In 1907 Rutherford was able to show that alpha particles are helium nuclei. He was awarded the 1908 Nobel Prize for Chemistry for his investigations into the disintegration of the elements and the chemistry of radioactive substances, while Soddy received the 1921 Chemistry award for his contributions to our knowledge of the chemistry of radioactive substances, and his investigations into the origin and nature of isotopes.

The German astronomer **Hermann Carl Vogel (1841–1907)** discovers spectroscopic binary stars when studying the periodic displacements of the spectral lines of the eclipsing binary stars Algol and Spica, whose companions could not then be directly observed. From his spectra, Vogel derived the distances between the visible and invisible components, their diameters, orbital velocity, and combined mass.

1900 to 2010 AD

1900–1909

1900

The photographs and initial publications of the *Cape Photographic Durchmusterung Star Catalogue* are completed. It is the first major astronomical work to use photographic observations to catalogue stellar positions and brightness. It was based on Scottish astronomer **David Gill's (1843–1914)** photographs of 450,000 stars of 11th magnitude or brighter in the southern hemisphere from South Africa between 1885 and 1890.

German physicist **Max Planck (1858–1947)** derives the formula for the radiation spectrum of a perfect absorber, or blackbody, introducing the idea that it radiates energy in fundamental indivisible units, which he called quanta, whose energy is proportional to the frequency of the radiation. The constant of proportionality between the frequency and the energy of the radiation is now known as Planck's constant, designated by the letter h. Planck's discovery marked the foundation of the quantum theory, and led to the award of the 1918 Nobel Prize for Physics for his achievement.

1901

The Italian electrical engineer **Guglielmo Marconi (1874–1937)** sends a long-wavelength radio signal in Morse code across the Atlantic Ocean on 12 December 1901. The

FIG. 8 **Blackbody radiation** The spectral plot of blackbody radiation intensity as a function of wavelength depends on the temperature of the gas emitting the radiation. The German physicist **Max Planck (1858–1947)** derived the formula that describes the shape of this spectrum in 1900, proposing that the radiation energy was quantized and providing a foundation for quantum theory. At higher temperatures the wavelength of peak emission shifts to shorter wavelengths, and the thermal radiation intensity becomes greater at all wavelengths. At a temperature of 6,000 degrees Kelvin, or 6,000 K, the thermal radiation peaks in the visible, or V, band of wavelengths. A hot gas with a temperature of 100,000 K emits most of its thermal radiation at ultraviolet, or UV, wavelengths, while the emission peaks in X-rays when the temperature is 1 to 10 million K.

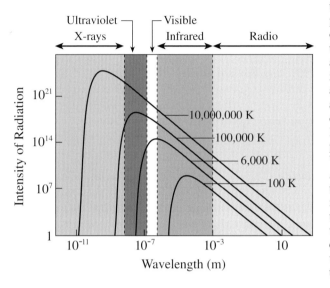

radio waves get around the curvature of the Earth by reflection from the ionosphere. Marconi was awarded the 1909 Nobel Prize in Physics, jointly with the German physicist **Karl Ferdinand Braun (1850–1918),** for their development of wireless telegraphy.

1902

British physicist and electrical engineer **Oliver Heaviside (1850–1925)** and American electrical engineer **Arthur Kennelly (1861–1939)** independently predict the existence of a conducting layer in the atmosphere that reflects radio waves. This layer, produced by the Sun's X-ray radiation, is now called the ionosphere, but it was known for a time as the Kennelly-Heaviside layer.

The English physicist **James Jeans (1877–1946)** derives the conditions for gravitational instability, showing that concentrations of matter larger than a limiting size, now called the Jeans wavelength, and more massive than a limiting mass, known now as the Jeans mass, will undergo gravitational collapse.

1903

American aeronautical engineeers **Orville Wright (1871–1948)** and **Wilbur Wright (1867–1912)** achieve the first sustained, controlled flight in a powered flying vehicle, the heavier-than-air machine named an aeroplane, on 17 December 1903 at Kill Devil Hills near Kitty Hawk, North Carolina.

1904

The German astronomer **Johannes Hartmann (1865–1936)** discovers interstellar calcium, which creates stationary absorption lines in the light of the binary star system Delta Orionis.

French mathematician **Jules Henri Poincaré (1854–1912)** proposes a theory of relativity in which the laws of physics remain unchanged with uniform motion and nothing moves faster than the velocity of light. This would explain why American scientists **Albert Michelson (1852–1931)** and **Edward Morley (1838–1923)** failed, in 1887, to measure the velocity of the Earth though the "aether" by measuring the speed of light in two directions. The German physicist **Albert Einstein (1879–1955)** independently developed the *Special Theory of Relativity* during the following year.

1905

German physicist **Albert Einstein (1879-1955)** proposes that light can be considered as particles of energy when it interacts with matter. He called the particles light quanta, and they later became known as photons. The light energy is given by the expression first used in 1901 by the German physicist **Max Planck (1858-1947)** to explain blackbody radiation. Einstein used his interpretation of light as quantum particles to explain the photoelectric effect, proposing that light particles striking the surface of certain metals cause electrons to be emitted. Einstein was awarded the 1921 Nobel Prize for Physics for this discovery.

German physicist **Albert Einstein (1879–1955)** develops his *Special Theory of Relativity*, proposing that space and time are one, and that in a system in motion relative to an observer length would decrease, time would slow and mass would increase. In the same year, Einstein used the postulates of the Special Theory to show that energy radiated is equivalent to mass lost, concluding that the mass of a body is a measure of its energy content.

German physicist **Albert Einstein (1879–1955)** explains Brownian motion in terms of the random motion of molecules of water. Any sufficiently small particle suspended in the water will be knocked into Brownian motion by molecular impacts of random strength coming from random directions.

Danish astronomer **Ejnar Hertzsprung (1873–1967)** shows that the absolute, or intrinsic, luminosity of many stars decreases systematically with spectral class and lower disk temperature, and that other stars with unusually narrow and sharp spectral lines have much higher luminosities. In this way Hertzsprung first recognized what have come to be called dwarf, or main sequence, and giant stars.

Dutch astronomer **Jacobus Kapteyn (1851–1922)** announces a pattern in the proper motions of nearby stars, indicating that they are apparently moving in two large, intermingled, and oppositely directed streams that pass through each other at a relative velocity of about 40 kilometers per second. One star stream carries roughly half the nearby stars toward the constellation Sagittarius at a speed of about 20 kilometers per second, while the other stream carries the remaining closest stars away from this direction at a comparable speed. As explained by the Swedish astronomer **Bertil Lindblad**

(1895–1965) in 1925, the two star streams are explained if all the stars in the Milky Way are whirling in differential rotation about a remote galactic center in Sagittarius.

1906

French geologist **Bernard Brunhes (1867–1910)** finds ancient lava flows in France whose magnetization appears to be reversed from that in more recent flows.

Irish geologist **Richard Oldham (1858–1936)** uses observations of seismic waves to prove that the Earth has a molten core.

1907

The Swedish physical chemist **Svante August Arrhenius (1859–1927)** proposes that tiny spores of living matter can be transported through space from one planet to another, and that life on Earth could have descended from interstellar microorganisms pulled in by the planet's gravity. This hypothesis is known as *panspermia*, meaning "life everywhere."

American chemist **Bertram Boltwood (1870–1927),** working with **Ernest Rutherford (1871–1937)** at the Cavendish Laboratory in England, estimates the decay rates of radioactive elements and uses the abundances of these elements and their decay products to show that some terrestrial rocks are about three billion years old.

German physicist **Albert Einstein (1879–1955)** derives the exact expression for the equivalence of mass and energy, namely his celebrated equation $E = mc^2$, where E is the energy, m is the mass, and c is the velocity of light.

1908

American astronomer **George Ellery Hale (1868–1938)** records the Zeeman splitting of the spectral lines of sunspots and compares them with those lines produced in the laboratory by intense magnetic fields, showing that the magnetic fields in sunspots are as strong as 2,900 Gauss and thousands of times stronger than the Earth's magnetism.

The German mathematician **Hermann Minkowski (1864–1909)** proposes that space and time be combined to define the separation of two events, in what is now known as the four-dimensional, space-time Minkowski metric.

The American astronomer **Frank W. Very (1852–1927)** suggests that the atmospheres of the planets will allow optically visible sunlight to pass through to the ground, which will heat up and reradiate at infrared wavelengths. Because the atmospheres are opaque to the infrared spectrum, this radiation will be trapped beneath the atmosphere where it heats up the planetary surface. Such a greenhouse effect had been proposed by the French mathematical physicist **Joseph (Jean Baptiste) Fourier (1768–1830)** in 1827, and the main molecules that absorb the infrared identified by the Irish physicist **John Tyndall (1820–1893)** as water and carbon dioxide in 1861. The American astronomers **Walter Adams (1876–1956)** and **Theodore Dunham Jr. (1897–1984)** identified the absorption lines of carbon dioxide in the atmosphere of Venus in 1932. Eight years later, the German-born American astronomer **Rupert Wildt (1905–1976)** proposed that the carbon dioxide in the thick atmosphere of Venus would make its surface much hotter than its outer atmosphere, and in 1958 the American astronomer **Cornell H. Mayer (1921–)** and his colleagues reported that observations with radio waves, which penetrate the atmosphere, indicated a very hot surface on Venus of at least 600 degrees kelvin. In 1971 the Soviet probe *Venera 7* descended to the surface of Venus and measured a temperature of 735 degrees kelvin.

1909

The German physicist **Albert Einstein (1879–1955)** introduces his idea that light exhibits both wave and particle characteristics.

The French physicist **Jean Baptiste Perrin (1870–1942)** demonstrates the existence of atoms by careful measurements of Brownian motion.

Croatian physicist **Andrija Mohorovicic (1857–1936)** uses seismic data to discover the Mohorovicic discontinuity that forms the boundary between the Earth's crust and mantle at a depth of about 35 kilometers below the Earth's surface.

The American astronomer **Vesto M. Slipher (1875–1969)** photographs the spectra of Jupiter, Saturn, Uranus and Neptune in the red and infrared, discovering absorption bands that were eventually interpreted, in 1931, by the German-born American astronomer **Rupert Wildt (1905–1976)** as absorptions by ammonia and methane.

1910–1919

1911

English astronomer **Jacob Halm (1866–1944)** shows that the masses of stars are correlated with spectral type and therefore with their absolute, or intrinsic, luminosities.

Danish astronomer **Ejnar Hertzsprung (1873–1967)** publishes a diagram of the absolute, or intrinsic, luminosity of stars in the Pleiades and Hyades star clusters, plotted as a function of their spectral type or effective temperature. In 1914 the American astronomer **Henry Norris Russell (1877–1957)** published a similar diagram for field stars not in clusters, and such plots are now known as Hertzsprung-Russell diagrams.

The New Zealand-born British physicist **Ernest Rutherford (1871–1937)** proposes the concept of the nuclear atom, in which the mass of the atom is concentrated in a nucleus which is 100,000 times smaller than the atom and has a positive charge balanced by surrounding electrons.

1912

Austrian physicist **Victor F. Hess (1883–1964)** discovers cosmic "rays" during balloon flights that show an increase of atmospheric ionization at high altitudes, attributing the increase to energetic "radiation" that comes from outer space. His balloon flight during the solar eclipse of 12 April 1912 additionally proved that the ionization was not caused by the Sun. Hess was awarded the 1936 Nobel Prize in Physics for his achievement.

American astronomer **Henrietta Swan Leavitt (1868–1921)** discovers a period-luminosity relation for twenty-five Cepheid variable stars in the Small Magellanic Cloud, in which the longer the period of light variation the greater the intrinsic brightness of the star. Leavitt was working at the Harvard College Observatory in a routine, menial task paid with low wages, analyzing photographs of stars obtained with Harvard's 0.6-meter (24-inch) refractor at Arequipa, Peru. The period-luminosity relation for Cepheid variable stars permitted a new method of determining a variable star's distance, from the star's observed apparent luminosity and the absolute, intrinsic luminosity inferred from the observed variation period. It was used by the American astronomer **Harlow Shapley (1885–1972)** in 1918 to establish the large size and distant center of our stellar system, the Milky Way, and in 1925 by the American astronomer **Edwin P. Hubble (1889–1953)** to discover that the spiral nebulae M31 and M33 are galaxies that lie outside our Milky Way Galaxy.

German meteorologist and geologist **Alfred Wegener (1880–1930)** develops a theory of continental drift to explain the close jigsaw-fit between the coastlines on either side of the Atlantic Ocean and the geological and fossil similarities between Brazil and Africa, proposing that all the continents were once united in a single landmass, named Pangaea, that fragmented into today's continents, which started drifting apart about 200 million years ago. The evidence for continental drift was further developed and published in 1915 in Wegener's book *Die Enstehung der Kontinente und Ozeane*, first translated into English in 1924 as *The Origin of Continents and Oceans*.

1913

The Norwegian physicist **Kristian Birkeland (1867–1917)** proposes that the Earth's dipolar magnetic field guides energetic electrons into the polar regions where they produce auroras.

FIG. 9 Bohr atom In this model, proposed in 1913 by the Danish physicist **Niels Bohr (1885–1962)**, a hydrogen atom's one electron revolves around the hydrogen nucleus, a

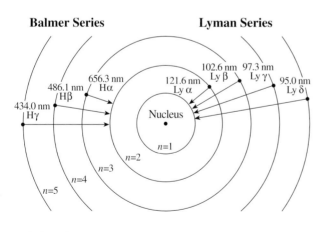

single proton, in well-defined orbits described by the integer n = 1, 2, 3, 4, 5, …. An electron absorbs or emits radiation when it makes a transition between these allowed orbits. The electron can jump upward, to orbits with larger n, by absorption of a photon of exactly the right energy, equal to the energy difference between the orbits; the electron can jump down to lower orbits, of smaller n, with the emission of radiation of that same energy and wavelength. Transitions that begin or end on the n = 2 orbit define the Balmer series that is observed at visible wavelengths. They are designated by Hα, Hβ, Hγ, …. The Lyman series, with transitions from the first orbit at n = 1, is detected at ultraviolet wavelengths. The orbits are not drawn to scale for the size of the radius increases with the square of the integer n.

Danish physicist **Niels Bohr (1885–1962)** applies quantum theory to the radiation emitted by atoms, developing an atomic model that explains the wavelengths of the spectral lines of hydrogen, described in an equation by the Swiss mathematician **Johann Balmer (1825–1898)** in 1884. According to this model, now known as the Bohr atom, the lone electron in a hydrogen atom revolves about the atomic nucleus, a proton, in specific orbits with definite, quantized values of energy, and an electron only emits or absorbs radiation when jumping between these allowed orbits. Bohr was awarded the 1922 Nobel Prize in Physics for his investigations of the structure of atoms and the radiation emanating from them.

American astronomer **Vesto Slipher (1875–1969)** shows that the visible-light spectra of some nebulae exhibit absorption lines identical to those found in their central stars, indicating that these "reflection" nebulae are reflecting the starlight.

English chemist **Frederick Soddy (1877–1956)** discovers that different forms of the same element have the same chemical characteristics, differing only in the atomic weight, or the number of neutrons and protons, but having the same atomic number, or the number of protons. He named the multiple forms *isotopes*, meaning "same place" because they occupied the same place in the periodic table. Soddy was awarded the 1921 Nobel Prize in Chemistry for his contributions to our knowledge of the chemistry of radioactive substances, and his investigations into the origin and nature of isotopes.

1914

American astronomer **Walter S. Adams (1876–1956)** shows that the faint companion to the star 40 Eridani is a hot white star of spectral type A0. The high temperature and low intrinsic luminosity meant that the faint companion star had to be very small, only about the size of the Earth. Such stars became known as white dwarf stars.

American astronomer **Walter S. Adams (1876–1956)** and German astronomer **Arnold Kohlschütter (1883–1969)** show that the relative intensities of certain neighboring spectral lines can be used to determine the absolute, or intrinsic, luminosities of both dwarf, or main sequence, and giant stars, making it possible to estimate the distances of millions of stars from their spectra and apparent brightness alone.

German-born American geologist **Beno Gutenberg (1889–1960)** discovers the boundary between the Earth's mantle and top of the Earth's liquid outer core, located at an average depth of 2,885 kilometers below the surface.

The American physicist **Theodore Lyman (1874–1954)** announces that he has observed three ultraviolet spectral lines of hydrogen whose wavelengths are described by the formula derived for the four visible-light hydrogen lines in 1885 by the Swiss mathematics teacher **Johann Balmer (1825–1898)**.

American astronomer **Henry Norris Russell (1877–1957)** publishes a diagram of the absolute, or intrinsic, luminosity of field stars, not in star clusters, plotted as a function of their spectral type, color index, or effective temperature, showing two well-defined classes of stars that he called dwarf and giant stars. Most stars, including the Sun, are dwarf stars that lie on the main sequence, which is a diagonal band that extends from the bright blue stars in the upper left of the diagram to the faint red ones in the lower right. The more luminous giant stars are found in the upper part of the diagram above the main sequence. Such a plot is known as a Hertzsprung-Russell, or H-R, diagram, after the pioneering investigations of Russell and **Ejnar Hertzsprung (1873–1967)** who first plotted such a diagram for the Pleiades and Hyades star clusters in 1911. The H-R diagram became an important tool for the understanding and interpretation of stellar evolution.

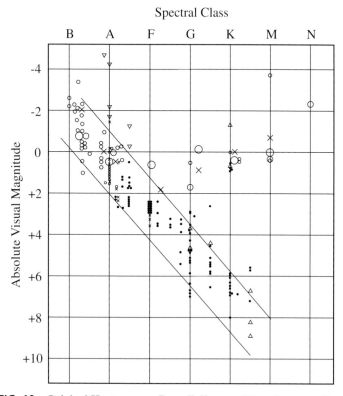

FIG. 10   Original Hertzsprung-Russell diagram When data on stellar parallax enabled the distance and absolute luminosity of stars to be determined, **Henry Norris Russell (1877–1957)** published, in 1914, this plot of the absolute visual magnitudes (*left axis*) of stars against their spectral class (*top axis*). The two diagonal lines mark the boundaries of **Ejnar Hertzsprung's (1873–1967)** sequence, inferred in 1911 for the Pleiades and Hyades open star clusters; this is now known as the main sequence along which most stars, including the Sun, are located. The giant stars are located at the upper right. A single faint anomaly is found in the lower left-hand corner; it is now known to be a white dwarf star. The larger dots denote stars whose absolute magnitudes had a probable error of 1 magnitude, and the small dots represent results with larger errors. The larger open circles denote mean results for about 120 bright stars of small proper motion; the smaller open and filled circles denote, respectively, stars whose parallaxes had been measured once or more than once.

American astronomer **Harlow Shapley (1885–1972)** suggests that the brightness variations of Cepheid variable stars are due to the pulsations of isolated individual stars, rather than eclipses of binary star systems.

American astronomer **Vesto Slipher (1875–1969)** discovers the large radial velocities of a few nearby spiral nebulae when attempting to measure their rotations under the assumption that the spiral nebulae were nascent planetary systems.

1915

American astronomer **Walter S. Adams (1876–1956)** shows that the faint companion of the bright star Sirius is a hot white star of spectral type A0. The high temperature and low intrinsic luminosity meant that the faint companion star had to be very small, only about the size of the Earth. Such stars became known as white dwarf stars.

The German physicist **Albert Einstein (1879–1955)** explains the anomalous motion of Mercury by means of his *General Theory of Relativity*, in which space is curved by a nearby mass, and predicts that the curvature will bend a light ray passing the Sun's edge by 1.75 seconds of arc. The predicted light bending was observed, within the uncertainties of measurement, by a change in the positions of stars seen near the Sun during the total solar eclipse of 29 May 1919. The French astronomer **Urbain Jean Joseph Leverrier (1811–1877)** presented accurate compilations of the unexplained anomaly in the motion of Mercury in 1859, and Einstein read about it in a 1915 article by the German astronomer **Erwin Freundlich (1885–1964).**

1916

English astronomer **Arthur S. Eddington (1882–1944)** shows that radiation pressure must stand with gravitation and gas pressure as the third major factor in maintaining the equilibrium of a star. In the same year, he shows that a star cluster will only remain in equilibrium if the kinetic energy of the moving stars balances their mutual gravitational pull.

The German physicist **Albert Einstein (1879–1955)** publishes his comprehensive *General Theory of Relativity*, in which he postulates that space is curved by mass.

German astronomer **Karl Schwarzschild (1873–1916)** solves the exact expressions that describe the curved space external to an isolated, spherical, non-rotating mass, using the field equations of Einstein's *General Theory of Relativity*, and shows that the equations have a singularity at the gravitational radius, also now called the Schwarzschild radius.

1917

English chemist and physicist **Francis W. Aston (1877–1945)** demonstrates that the helium atom is slightly less massive, by a mere 0.7 percent, than the sum of the masses of the four hydrogen atoms which enter into it, which led to the English astronomer **Arthur Eddington (1882–1944)** proposing, in 1920, that hydrogen is converted into helium in stellar cores, with the mass difference released as energy to make the stars shine.

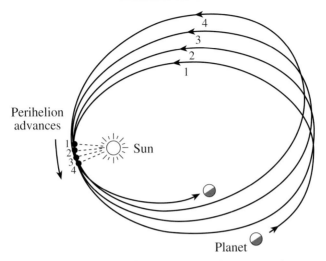

FIG. 11   Precession of Mercury's perihelion Instead of
always tracing out the same ellipse, the orbit of Mercury pivots
around the focus occupied by the Sun. The point of closest
approach to the Sun, the perihelion, is slowly rotating ahead of
the point predicted by Newton's theory of gravitation. This was
at first explained by the gravitational tug of an unknown planet
called Vulcan, but we now know that Vulcan does not exist.
The German physicist **Albert Einstein (1879–1955)** explained
Mercury's anomalous motion in 1915 by inventing a new theory
of gravity in which the Sun's curvature of space makes the
planet move in a slowly revolving ellipse.

Heber D. Curtis (1872–1942), at the Lick Observatory in California, notices a jet of
visible light emerging from the nuclear center of M87.

American astronomers **Heber D. Curtis (1872–1942)** and **George W. Ritchey (1864–
1945)** independently discover novae in spiral nebulae, including Andromeda or M31,
suggesting that the brightest novae are at least 100 times as far away as the novae previously observed in our Milky Way Galaxy. The American astronomer **Harlow Shapley
(1885–1972)** estimates a distance of a million light-years for the bright nova observed
in the Andromeda nebula in 1885.

Dutch astronomer **Willem de Sitter (1872–1934)** considers the astronomical consequences of **Albert Einstein's (1879–1955)** *General Theory of Relativity*, and proposes
a solution to his field equations that describes a non-static, expanding Universe devoid
of all matter. It subsequently became known as the "de Sitter Universe", which contains
motion without matter, as distinct from the static, non-moving "Einstein Universe",
which contains inert matter without motion. The Belgian astrophysicist and Catholic
priest **Georges Lemaître (1894–1966)** found a solution that permits both matter and
motion, using it to explain the receding velocities of extra-galactic nebulae in 1927.

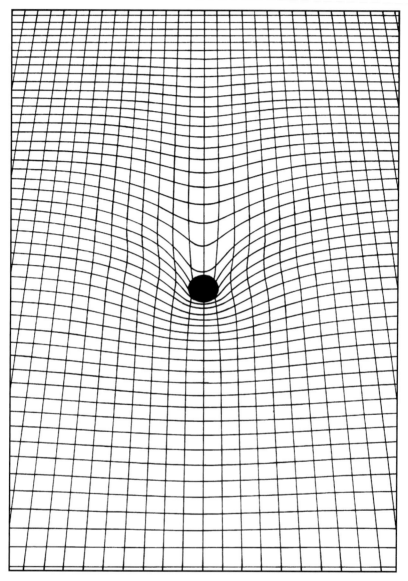

FIG. 12 **Space curvature** A massive object creates a curved indentation upon the "flat" space described by Euclidean geometry, which applies in a world without significant amounts of matter. Notice that the amount of space curvature is greatest in the regions near the object, while further away the effect is lessened.

The German physicist **Albert Einstein (1879–1955)** publishes a paper entitled "Cosmological Considerations of the General Theory of Relativity," explaining a static, non-moving spherical Universe. To counteract the combined inward gravitational pull of matter, and keep the Universe from collapsing on itself, Einstein modified his equations by adding an extra adjustable parameter, now known as the cosmological constant, which opposes the gravitational attraction of matter at large distances. Einstein abandoned the concept in 1931, after the discovery that the Universe is expanding.

American chemist **William D. Harkins (1881–1938)** notices that elements of low atomic weight are more abundant than those of high atomic weight, proposes that the relative abundances of the elements depend on nuclear rather than chemical properties, and suggests that the heavy elements must have been synthesized from the light ones.

The 2.5-meter (100-inch) **Hooker telescope** is completed on Mount Wilson near Pasadena, California, primarily funded by **John D. Hooker (1838–1911).** It remained the world's largest telescope from 1917 to 1948. The 100-inch Hooker telescope was used by **Edwin Hubble (1889–1953)** to discover the large distances and extragalactic nature of spiral nebulae in 1925, and to obtain the first evidence of the expanding Universe in 1929.

The American astronomer **Vesto Slipher (1875–1969)** publishes the large, outward velocities of 21 bright spiral nebulae, based on spectroscopic observations with Lowell Observatory's 0.61-meter (24-inch) refractor; an additional four seemed to be approaching the Sun with lower velocities. Five years later, in 1922, Slipher supplied the English astronomer **Arthur Eddington (1882–1944)** with the radial velocities of 40 of the 44 known spiral nebulae, which Eddington included in his book *The Mathematical Theory of Relativity*, published in 1923.

1918

American "computer" **Annie Jump Cannon (1863–1941)** classifies the spectra of 225,300 stars and publishes them in the first volume of the *Henry Draper Catalogue* of stellar spectra. In her lifetime Cannon classified over 500,000 stars, culminating in the ninth volume of the stellar catalogue published in 1924. This work was important in understanding the diversity of stars and their evolution.

American astronomer **Harlow Shapley (1885–1972)** determines the distances of globular star clusters using Cepheid variable stars in them, and shows that they are distributed in a roughly spherical system that is centered far from the Sun and envelops the flattened stellar disk of the Milky Way.

1919

English chemist **Francis W. Aston (1877–1945)** develops the mass spectrograph, which he used to study atomic masses and to establish the existence of isotopes. In 1919 he used his instrument to discover the neon isotope, and eventually discovered 200 of them. Aston also showed that the mass of the helium atom is less than that of four hydrogen atoms, leading his colleague **Arthur Eddington (1882–1944)** to declare

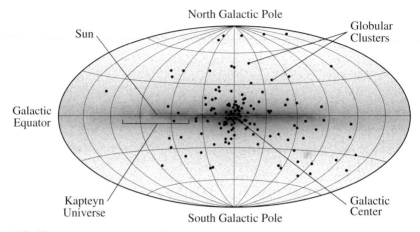

FIG. 13   Edge-on view of our Galaxy As shown in 1918 by the American astronomer **Harlow Shapley (1885–1972)**, the globular star clusters are distributed in a
roughly spherical system whose center coincides with the core of our Galaxy. The Sun
is located in the galactic disk, about 27,700 light-years away from the center of our
Galaxy. The galactic disk and central bulge are shown edge-on as a negative print of an
infrared image taken from the *InfraRed Astronomical Satellite*. The infrared observations can penetrate the obscuring veil of interstellar dust that hides the distant Milky
Way from observation at visible wavelengths. It is this dust that limited astronomers'
view of the surrounding stars to a much smaller Kapteyn Universe, centered on the
Sun.

in 1920 that the mass difference in synthesizing hydrogen into helium could release
the energy that makes stars shine. Aston received the 1922 Nobel Prize in Chemistry
for his discovery, by means of the mass spectrograph, of isotopes, in a large number of
non-radioactive elements.

American astronomer **Edward Barnard (1857–1923)** publishes his first photographic
atlas of dark markings, which were subsequently found to contain vast quantities of interstellar dust and molecules. The second atlas was published in 1927 after Barnard's death.

The Greenwich Observatory and England's Royal Society jointly send expeditions to
test Einstein's *General Theory of Relativity* by observing star positions during the total
solar eclipse on 29 May 1919, from Sobral, Brazil and Principe, a small island off the
coast of West Africa. The English astronomers **Frank W. Dyson (1868–1939), Arthur
Eddington (1882–1944)** and **Charles Davidson (1875–1970)** published this determination of the deflection of light by the Sun's gravitational field in 1920 in the *Philosophical Transactions of the Royal Society*. The observed change in the stellar positions,
produced by the Sun's gravitational bending of starlight passing near it, agreed with
Einstein's prediction, making him world famous nearly overnight.

American astronomer **George Ellery Hale (1868–1938)** and his colleagues show that
sunspots occur in bipolar pairs with an orientation that varies with a 22-year period.

Polish mathematician **Theodor Kaluza (1885–1954)** suggests that the Universe might
have more than three spatial dimensions, and proposes the existence of a fourth spatial

dimension that might connect the *General Theory of Relativity* with the theory of electromagnetism. Although his ideas did not succeed in their original purpose, they were refined by others and used decades later in string theory.

Irish physicist **Joseph Larmor (1857–1942)** proposes that the magnetic fields of the Earth and the Sun can be generated by the internal motions of conducting material through the action of a self-sustaining dynamo.

British physicist **Frederick A. Lindeman (1886–1957),** later **Lord Cherwell,** proposes that electrically neutral plasma ejections from the Sun are responsible for intense, non-recurrent geomagnetic storms.

New Zealand-born British physicist **Ernest Rutherford (1871–1937)** announces that he has bombarded nitrogen with alpha particles, or helium nuclei, causing the nitrogen nuclei to disintegrate and producing hydrogen and oxygen nuclei. This was the first artificially produced splitting of atomic nuclei and the first artificial transformation of one element into another. Rutherford found that other elements also eject hydrogen nuclei when similarly bombarded with alpha particles, and speculated that the nucleus of any atom contains them. In 1920 Rutherford gave the name proton to the hydrogen nucleus, speculating in the same year that uncharged particles, which were later called neutrons, must also exist in the nucleus to keep the protons from repelling each other.

The Scottish engineer **Robert Watson-Watt (1892–1973)** patents his first "radio locator", the predecessor of radar, which was fully developed in Britain and America during World War II (1939–1945).

1920–1929

1920

In an article entitled *The Internal Constitution of the Stars*, the English astronomer **Arthur Eddington (1882–1944)** suggests that nuclear processes in the hot cores of stars might provide the energy that makes them shine, and specifically describes the possibility that hydrogen could be transformed to helium, with the resultant mass difference released as energy to power a star. In the same year, Eddington published his book *Space, Time and Gravitation*, which outlines **Albert Einstein's (1879–1955)** *General Theory of Relativity*.

The New Zealand-born, British physicist **Ernest Rutherford (1871–1937)** announces that the massive nuclei of all atoms are composed of hydrogen nuclei, which he named protons. He also postulates the existence of an uncharged nuclear particle, later called the neutron, which was required to keep the positively charged protons from repelling each other. The English physicist **James Chadwick (1891–1974)** discovered the neutron in 1932.

Indian astrophysicist **Meghnad Saha (1893–1956)** publishes his ionization equation, which relates the degree of ionization of an atom to temperature and pressure. Within a year he had used this equation to specify the temperatures of stars of different spectral types, and paved the way for the interpretation of stellar spectra in terms of both excitation and chemical composition.

The American astronomers **Harlow Shapley (1885–1972)** and **Heber D. Curtis (1872–1942)** participate in the now-famous Shapley-Curtis debate over "The Scale of the Universe" at a meeting of the National Academy of Sciences on 26 April 1920. Shapley defended his interpretation of a much larger Milky Way system with a remote center, which proved essentially correct, but he wrongly supposed that the spiral nebulae are embedded in the Milky Way. Curtis incorrectly defended a smaller Sun-centered stellar system, but provided cogent arguments for the island Universe theory of spiral nebulae as remote stellar systems comparable to the Milky Way, which was subsequently proven correct. Shapley was misled by **Adriaan van Maanen's (1884–1946)** reports of systematic angular motions of spiral nebulae, which would mean they would have implausible velocities if they were not nearby. Shapley accepted these observations at face value, but Curtis discounted them as probably erroneous – which they were.

1921

American physicist **Albert Michelson (1852–1931)** and American astronomer **Francis G. Pease (1881–1938)** publish the first successful measurement of the angular diameter of a star other than the Sun. They used the interferometer method with two mirrors, including the 100-inch reflector of the Hooker telescope on Mount Wilson, obtaining an angular diameter of about 0.05 seconds of arc for the supergiant star α Orionis, also known as Betelgeuse. Modern measurements give an angular diameter of 0.045 seconds of arc, which at a distance of 430 light-years corresponds to a physical radius of about 435 million kilometers and 625 times the radius of the Sun.

1922

The Swedish astronomer **Carl Vilhelm Ludvig "C.V.L." Charlier (1862–1934)** maps the distribution of 12,000 nebulae outside the plane of the Milky Way, and shows that they are grouped together in concentrations now known as clusters of galaxies.

The American physicist **Arthur H. Compton (1892–1962)** shows that X-rays scattered from various elements exhibit an increase in wavelength. Compton explained this discovery in 1923 by supposing that X-rays behave like particles, which he called photons, and lose some of their energy in collisions with electrons. This phenomenon, which is now known as the Compton effect, confirmed the wave-particle duality of radiation, which was important for quantum theory.

The Russian mathematician **Aleksandr A. Friedman (1888–1925)** derives solutions to the field equations of Einstein's *General Theory of Relativity* that describe an isotropic, homogeneous non-stationary Universe whose mean density and radius vary with time, indicating either an expanding or contracting Universe. In 1922 he considered a non-stationary world with positive space curvature, and in 1924 he discovered an expanding solution with negative spatial curvature. Such possibilities have become known as "Friedman Universes," but astronomers of the time were generally unaware of his results, and Friedman did not realize their possible astronomical consequences.

The Dutch astronomer **Jacobus Kapteyn (1851–1922)** publishes his analysis of star counts, proposing that the stars are concentrated in a disk with the Sun near the center. The length of the stellar disk was estimated at about 40,000 light-years, and the width at 5,000 light-years.

English astronomer **E. Walter Maunder (1851–1928)** provides a full account of the 70-year dearth of sunspots, from 1645 to 1715, previously noticed by the German astronomer **Gustav F. W. Spörer (1822–1895)** in 1887. This interruption in the normal sunspot cycle is now referred to as the Maunder Minimum. A prolonged period of unusually cold weather in Europe, known as the Little Ice Age, overlaps the Maunder Minimum and another period of unusually low sunspot numbers, the Spörer Minimum, from 1420 to 1570.

Estonian astronomer **Ernst Öpik (1893–1985)** deduces the distance to the spiral galaxy Andromeda, or M31, from its rotation velocity under the assumption that its mass to luminosity ratio is comparable to that of the stars in the Milky Way, showing that M31 lies outside our Galaxy and is a galaxy in its own right. Öpik obtained a distance of 480,000 parsecs to M31, or about 1.565 million light-years.

1923

English astronomer **Arthur Eddington (1882–1944)** publishes his book *The Mathematical Theory of Relativity*, which includes the radial velocities of 40 spiral nebulae supplied by the American astronomer **Vesto Slipher (1875–1969)**.

German astronomer **Maximilian Wolf (1863–1932)** shows that dark nebulae in the Milky Way are obscuring clouds composed of solid dust particles, which absorb the light of distant stars.

1924

French physicist **Louis de Broglie (1892–1987)** develops the principle that an electron or any other sub-atomic particle can be considered to behave as a wave as well as a particle. He showed that the electrons in an atom do not occupy orbits but exist as standing waves around the nucleus. For this discovery of the wave-particle duality, de Broglie was awarded the 1929 Nobel Prize for Physics.

English astronomer **Arthur Eddington (1882–1944)** shows that the luminosity of a star is a function of its mass, a result achieved by treating stars as gaseous spheres in hydrostatic and radiative equilibrium, with giant stars supported by radiation pressure and dwarf, or main sequence, stars supported by gas pressure. This paved the way to the realization that the evolutionary history of a star is determined by its mass, age and initial composition.

American astronomer **George Ellery Hale (1868–1938)** develops the spectrohelioscope that enables the entire Sun to be scanned visually at selected wavelengths, most notably in the light of the hydrogen-alpha spectral line at 656.3 nanometers that is sensitive to solar activity.

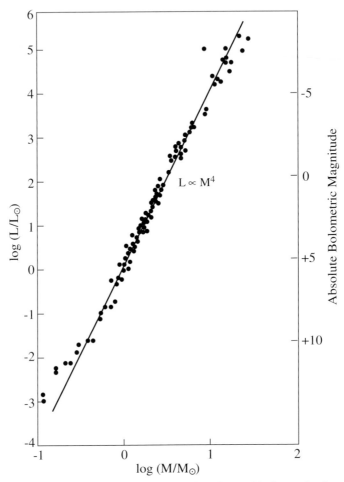

FIG. 14 **Stellar mass-luminosity relation** The empirical mass-luminosity relation for main-sequence stars of absolute luminosity, L, in units of the solar luminosity, $L_\odot$, and mass, M, in units of the Sun's mass, $M_\odot$. The straight line corresponds to a luminosity that is proportional to the fourth power of the mass. The English astronomer **Arthur Eddington (1882–1944)** proposed a theoretical explanation for this relation in 1924.

1925

The English physicist **Edward V. Appleton (1892–1965)** and his New Zealand graduate student **M. A. F. "Miles" Barnett (1901–1979)** use reflected radio transmissions to verify the existence of the electrically conducting ionosphere. The American physicists **Gregory Breit (1899–1981)** and **Merle A. Tuve (1901–1982)** confirmed the existence of the ionosphere the following year using pulsed radio waves. A height of about 250 kilometers was inferred for the radio-reflecting layer by measuring the time delay between the transmission of the radio signal and the reception of its echo. Appleton was awarded the 1947 Nobel Prize in Physics for his investigations of the physics of the upper atmosphere, especially for the discovery of the so-called Appleton layer.

The American astronomer **Edwin Hubble (1889–1953)** announces his discovery of Cepheid variable stars in two nearby spiral nebulae, the Andromeda Nebula, or M31, and the great spiral in Triangulum, or M33, and shows that they are remote galaxies separated from our Milky Way by wide gulfs of apparently empty space. Hubble used the period-luminosity relation for Cepheid variable stars to derive a distance of about 285,000 parsecs, or about 0.93 million light-years, for the two nebulae, demonstrating that they lie far beyond our Milky Way and are thus galaxies in their own right. This important result was presented in a paper read *in absentia* at a meeting of the American Astronomical Society in Washington, D.C. on New Year's Day 1925. Hubble was using the 100-inch (2.5-meter) **Hooker telescope** on Mount Wilson to make this discovery of the large distances and extragalactic nature of spiral nebulae.

Swedish astronomer **Bertil Lindblad (1895–1965)** proposes that the stars in the Milky Way are revolving in circular orbits about a remote, massive galactic center, in the direction of the constellation Sagittarius, with orbital speeds that decrease with increasing distance from the center. The nearby stars that are a little closer than the Sun to the center seem to move in one direction, while those a bit farther away appear to move in the opposite direction, thereby explaining the two star streams announced by the Dutch astronomer **Jacobus Kapteyn (1851–1922)** in 1905.

American physicist **Robert Millikan (1868–1953)** compares the intensity of ionization in two lakes at different altitudes, proving that the ionizing rays, discovered by **Victor F. Hess (1883–1964)** in 1912, come from outer space and giving them the name cosmic rays.

English-born American astronomer **Cecilia H. Payne (1900–1979)** publishes her monograph entitled *Stellar Atmospheres*, where she uses stellar spectral lines to demonstrate that the abundances of the elements are similar in the atmospheres of virtually all bright stars. Because hydrogen is terrestrially rare, she supposed that its intense spectral line in the Sun and other stars was due to some abnormality, but within three years the German astrophysicist **Albrecht Unsöld (1905–1995)** had shown that hydrogen is in fact the most abundant element in the outer atmosphere of the Sun.

Austrian physicist **Wolfgang Pauli (1900–1958)** proposes that no two electrons in the same atom can occupy the same energy state, or have the same values for all four of their quantum numbers. Pauli was awarded the 1945 Nobel Prize in Physics for his discovery of this Exclusion Principle, also known as the Pauli Principle.

1926

English astronomer **Arthur Eddington (1882–1944)** demonstrates that energetic radiation from very luminous stars will heat nearby interstellar material to a temperature of about 10,000 degrees kelvin, ionizing the atoms. In the same year he published his book *Internal Constitution of the Stars*, which collects and extends his major papers on the physical nature of stars and provides a foundation for much of our understanding of the structure, constitution and evolution of the stars.

English astronomer **Arthur Eddington (1882–1944)** shows that there is a maximum luminosity that can be produced by stellar accretion. The radiation of the in-falling material can become so intense, and its pressure so pronounced, that it blows the incoming material away and stops the accretion. We now realize that this results in a maximum X-ray luminosity, called the Eddington limit, which is equivalent to the visible luminosity of about 100 thousand stars like the Sun.

English physicist **Ralph H. Fowler (1889–1944)** shows that a white dwarf star, with about the size of the Earth and the mass of the Sun, is supported by electrons, which exert an outward push called degeneracy pressure owing to their inability to occupy exactly the same energy state. This is a consequence of the Exclusion Principle proposed by the Austrian physicist **Wolfgang Pauli (1900–1958)** the previous year. The degenerate electron pressure persists when the gas is cooled, even to zero degrees kelvin, and it can support a white dwarf indefinitely, without any need of nuclear reactions. Electron degeneracy pressure is nevertheless unable to halt the collapse of stars with masses greater than 1.46 solar masses.

American physicist **Robert H. Goddard (1882–1945)** launches the first liquid-propelled rocket on 16 March 1926 at his Aunt Effie's farm in Auburn, Massachusetts.

The Japanese astronomer **Kiyotsugu Hirayama (1874–1943)** discovers groups of asteroids that share very similar orbits, with common orbital inclinations and distances from the Sun, and suggests that each group is the broken fragments, or family, of a larger parent object. He named the Hirayama families, as they are now called, after the largest member asteroid, such as the Eos, Flora, Koronis and Themis families.

American astronomer **Edwin Hubble (1889–1953)** introduces a morphological classification of galaxies into three broad types: ellipticals, spirals and irregulars, with subdivisions that he interpreted as an evolutionary sequence.

American astronomer **Donald Menzel (1901–1976)** proposes that the hydrogen emission features, known as the Balmer lines, detected in planetary nebulae are the result of photoionization of hydrogen atoms by ultraviolet starlight, followed by recombination of the free electrons and protons.

American physicist **Albert Michelson (1852–1931)** measures the velocity of light over a 35-kilometer path between two mountains in California, and obtains a precise value of 299,796 kilometers per second with an uncertainty of 4 in the last digit.

Austrian physicist **Erwin Schrödinger (1887–1961)** proposes a wave-mechanical theory of the atom, using his formulation of a wave equation to describe the behavior of electrons in atoms. Schrödinger was awarded the 1933 Nobel Prize in Physics, together with the English theoretical physicist **Paul A.M. Dirac (1902–1984)** for their discovery of new productive forms of atomic theory.

1927

German physicist **Werner Heisenberg (1901–1976)** proposes the uncertainty principle, which states that it is impossible to specify precisely both the position and the simultaneous momentum (mass multiplied by velocity) of a particle. It explains why Newtonian mechanics is inapplicable at the atomic level, where quantum physics is invoked, and maintains that the result of an action can be expressed only in terms of the probability that a certain effect will occur. Heisenberg was awarded the 1932 Nobel Prize in Physics for the creation of quantum mechanics.

Belgian astrophysicist and Catholic priest **Georges Lemaître (1894–1966)** proposes that the receding velocities of extra-galactic nebulae, observed by the American astronomer **Vesto Slipher (1875–1969),** are the cosmic effect of the expansion of the Universe, in accordance with Einstein's *General Theory of Relativity*, and derives a theoretical expression for the linear increase of velocity with distance. He used observations to derive a Hubble constant of 625 kilometers per second per Megaparsec.

Dutch astronomer **Jan H. Oort (1900–1992)** uses the distribution of stellar radial velocities to show that the stars in our Milky Way move around a distant massive center in the direction of the constellation Sagittarius, with the inner stars orbiting faster than the outer ones.

The Dutch astronomer **Herman Zanstra (1894–1972)** uses quantum theory to explain the hydrogen lines of diffuse nebulae, including the planetary nebulae, whose hydrogen atoms are ionized by the ultraviolet light of bright stars followed by the recombination of the free electrons and protons to make hydrogen atoms.

1928

Danish physicist **Niels Bohr (1885–1962)** introduces the Principle of Complementarity, based on the wave-particle duality of light and matter (electrons). According to this Principle, the wave and particle interpretations on the quantum level are equally valid and indispensable, and a complete description requires them both with an application that depends on the experimental context. When combined with the Uncertainty Principle of the German physicist **Werner Heisenberg (1901–1976),** the Principle of Complementarity led to the Copenhagen Interpretation of Quantum Mechanics.

American astronomer **Ira S. Bowen (1898–1973)** shows that the spectral lines of planetary nebulae, previously attributed to a hypothetical element nebulium and first observed in 1864 by the English astronomer **William Huggins (1824–1910),** arise from ionized atoms of oxygen and nitrogen undergoing "forbidden" transitions in the low-density nebulae; such transitions are very improbable in the higher-density laboratory situation. The American astronomer **Donald Menzel (1901–1976)** and the Dutch astronomer **Herman Zanstra (1894–1972)** had set the stage for this work by showing, in 1926–27, that the intense ultraviolet light of hot stars embedded in planetary nebulae ionize the surrounding material and account for their hydrogen line emission.

English theoretical physicist **Paul A. M. Dirac (1902–1984)** formulates the relativistic theory of the electron from which he predicted the existence of the positron, or positive electron, the anti-matter particle of the electron.

Russian physicist **George Gamow (1904–1968)** uses quantum mechanics to show how energetic charged particles can escape radioactive atomic nuclei, tunneling through the nuclear electrical barrier that constrains them. The tunneling process involves a quantum waviness, uncertainty, or spread out character, which makes it possible for a sub-atomic particle to occasionally penetrate classical barriers. It makes nuclear fusion possible at temperatures that exist inside the Sun and other stars.

The German astrophysicist **Albrecht Unsöld (1905–1995)** shows that spectral lines observed in the Sun's visible light demonstrate that hydrogen is a million times more abundant than any other element in the Sun.

1929

The German physicist **Wilhelm Anderson (1880–1940)** and the English scientist **Edmund C. Stoner (1899–1968)** independently show that electron degeneracy pressure cannot support a star with a mass greater than abut 1.4 solar masses, the upper limit for the mass of a white dwarf star.

The Welsh astronomer **Robert d'Escourt Atkinson (1898–1982)** and the German physicist **Fritz Georg Houtermans (1903–1966)** provide the first attempt at a theory of nuclear energy generation in the hot central portions of stars, showing that the most likely nuclear reactions will involve light nuclei with low electrical charge, and that only a few, rare, high-velocity nuclei will be able to penetrate the electrical barrier between them, explaining why nuclear reactions proceed slowly inside stars. In 1931 and 1936 Atkinson developed the idea in greater detail, laying the foundation for **Hans Bethe's (1906–2005)** explanation, in 1938 and 1939, of the proton-proton chain which makes the Sun shine.

American astronomer **Edwin Hubble (1889–1953)** publishes his seminal article entitled *A Relation between Distance and Radial Velocity among Extra-Galactic Nebulae*, finding that the more distant an extra-galactic nebula is, the greater is its speed of recession as determined from its spectroscopic redshift. Hubble was using the 100-inch (2.5-meter) **Hooker telescope** on Mount Wilson to detect individual stars and determine the large distances of the extra-galactic nebulae, now known as galaxies. The American astronomer **Vesto Slipher (1875–1969)** had determined many of the radial velocities as early as 1917 using the 0.61-meter (24-inch) refractor at the Lowell Observatory, in Arizona, with additional crucial velocities of the most distant objects obtained by Hubble's colleague **Milton Humason (1891–1972)** using the 100-inch telescope. A linear increase of the recession speed with distance is now known as the Hubble law, and attributed to the expansion of the Universe. Also in 1929, Hubble published an exhaustive study of the stellar content of M31 based on approximately 350 photographs taken with the 60-inch and 100-inch reflectors over an interval of eighteen years, firmly establishing its remote distance of about 930,000 light-years and ending all doubts that the spiral nebulae are remote objects ablaze with stars, galaxies in their own right and separated from the Milky Way by wide gulfs of apparently empty space.

American physicist **Ernest O. Lawrence (1901–1958)** invents a circular particle accelerator, known as the cyclotron, capable of accelerating sub-atomic particles to high energy, and leading to the production of hundreds of radioactive isotopes, including carbon-14, iodine-131 and uranium-233.

FIG. 15   Discovery diagram of the expanding Universe A plot of the distance of extra-galactic nebulae, or galaxies, versus the radial velocity at which each galaxy is receding

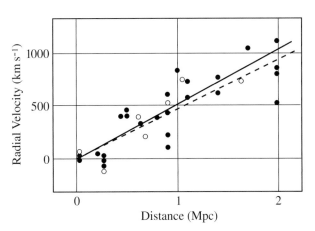

from Earth, published in 1929 by the American astronomer **Edwin Hubble (1889–1953).** The linear relationship between the distance and radial velocity of extragalactic nebulae, or galaxies, indicates that the Universe is expanding. Here the velocity is in units of kilometers per second, abbreviated km s^{-1}, and the distance is in units of millions of parsecs, or Mpc, where 1 Mpc is equivalent to 3.26 million light-years. Hubble underestimated the distances of the spiral nebulae, so the distance scale for modern versions of this diagram is about seven times larger (Fig. 30). The filled circles and solid line represent the solution for individual nebulae; the open circles and dashed line are for groups of them, the cross represents the mean distance of 22 nebulae whose distances could not be determined individually.

Japanese geophysicist **Motonori Matuyama (1884–1958)** provides evidence that reversely magnetized rocks may have originated during long periods in the past when the polarity of the Earth's magnetic poles was the opposite of what it is now.

1930–1939

1930

The Turkish-born French astronomer **Eugène M. Antoniadi (1870–1944)** publishes his monumental book *La Planète Mars*, which contains a complete description of the planet's changing surface features as observed with a telescope. He also supported the idea that the canals on Mars are an optical illusion, produced when the eye combines and integrates many small details.

English geophysicist **Sydney Chapman (1888–1970)** and his colleague **Vincent Ferraro (1907–1974)** postulate that electrically neutral plasma clouds ejected from the Sun during a solar flare cause geomagnetic storms and auroras, and that the Earth's magnetic field carves out a cavity in nearby space that deflects much of the cloud.

Worldwide photographic observations during the close approach of the asteroid Eros, in 1930–31, are used to estimate a value for the mean Earth-Sun distance, the astronomical unit, of between 149.4 million and 149.7 million kilometers.

American astronomer **Donald Menzel (1901–1976)** attributes the low mass density of the giant planets to an abundance of hydrogen, the lightest element. He supposed that the Sun and planets were formed from the same primeval gas, noted that the Sun was mainly composed of hydrogen, and used the dynamic arguments of **George Johnstone Stoney (1826–1911),** made in 1898, to show that the cold, massive giant planets would retain hydrogen.

Austrian physicist **Wolfgang Pauli (1900–1958)** proposes that an invisible particle, with no charge and little or no mass, is removing energy during a kind of radioactivity called beta decay. In this process, the nucleus of a radioactive atom emits an energetic electron, or beta particle, whose energy is less than that lost by the nucleus, and an unseen particle was postulated to carry away the remaining energy. Four years later, the Italian physicist **Enrico Fermi (1901–1954)** provided further evidence for Pauli's conjecture and called the particle the *neutrino*, Italian for "little neutral one".

American astronomer **Clyde William Tombaugh (1906–1997)** discovers a new planet, subsequently named Pluto, on 18 February 1930 during a systematic photographic search using a 0.33-meter (13-inch) telescope at the Lowell Observatory in Arizona. His meticulous search used a blink comparator to look at two photographs of the same part of the sky taken on different days, each photograph containing between 50,000 and 500,000 stars. Just one of the points of light exhibited a change of position, or motion, between the two times, a ninth planet. Pluto moves in a highly elongated orbit that carries it between 29.7 and 49.3 AU from the Sun, within and outside the orbit of Neptune. The American astronomer **William Henry Pickering (1858–1938)** predicted an unknown planet beyond Neptune in 1909, and the wealthy American **Percival Lowell (1855–1916),** the founder of the Lowell Observatory, also predicted its existence in 1915. Both Pickering and Lowell argued that such a planet was producing perturbations in the apparent motion of Uranus beyond those caused by Neptune. However, after accounting for the gravitational effects of Neptune, using a precise mass obtained when *Voyager 2* encountered the planet in 1989, the unexplained motions in Uranus disappeared and astronomers concluded that Neptune is the outermost major planet in the Solar System. Pluto is now known to be one of several small bodies that orbit the Sun beyond the orbit of Neptune.

Swiss-born, American astronomer **Robert J. Trumpler (1886–1956)** uses observations of the distances, sizes and reddening of open star clusters to convincingly demonstrate the interstellar absorption of starlight throughout the plane of our Milky Way Galaxy, even in regions that contain no dark nebulae.

1931

In an article entitled *Atomic Synthesis and Stellar Energy*, the Welsh astronomer **Robert d'Escourt Atkinson (1898–1982)** argues that the most effective nuclear reactions in stars involve protons, the nuclei of hydrogen atoms, and proposes that the observed relative abundances of the elements might be explained by the synthesis of heavy elements from hydrogen and helium by successive proton captures.

The Indian astrophysicist **Subrahmanyan Chandrasekhar (1910–1995)** derives the detailed equilibrium configurations in which degenerate electron gases support their own gravity. Chandrasekhar was awarded the 1983 Nobel Prize in Physics for his theoretical studies of the structure and evolution of stars.

After the discovery of the expanding Universe, German physicist **Albert Einstein (1879–1955)** retracts the cosmological constant, introduced by him in 1917 to keep a stationary Universe from collapsing.

American astronomers **Edwin Hubble (1889–1953)** and **Milton Humason (1891–1972)** confirm the linear correlation of the distance and radial velocity of extragalactic nebulae, or galaxies, and derive a Hubble constant of 558 kilometers per second per Megaparsec. The approximate expansion age of the Universe, inferred from the reciprocal of this constant, is 1.8 billion years, less than the age of some terrestrial rocks.

Belgian astrophysicist and Catholic priest **Georges Lemaître (1894–1966)** proposes that the observed systematic outward motion of the galaxies might be explained by the explosion of an incredibly small and dense "primeval atom" or "cosmic egg" which sent its contents out in all directions, just before the beginning of space and time. Such an explosive initiation of the expanding Universe has subsequently become known as the Big Bang.

French astronomer **Bernard Lyot (1897–1952)** invents the coronagraph, a telescope with an occulting disk that blocks out the intense light of the solar disk, permitting the observation of the solar corona.

Estonian lens- and mirror-maker **Bernhard Schmidt (1879–1935)** uses a spherical mirror and correcting plate to create a telescope with a wide angular field of view and high speed. These advantages made it possible to survey large areas of the sky in a reasonable time, something impossible with the narrow fields of conventional reflecting telescopes.

American chemist **Harold C. Urey (1893–1981)** and his colleagues discover heavy water and deuterium, the heavy isotope of hydrogen of mass 2, and show that the observed relative abundance of isotopes cannot have resulted from nuclear equilibrium at any single temperature. Urey received the 1934 Nobel Prize in Chemistry for his discovery of heavy hydrogen.

The German-born American astronomer **Rupert Wildt (1905–1976)** interprets infrared spectra of the giant planets, taken by the American astronomer **Vesto Slipher (1875–1969)** in 1909, as due to absorption by ammonia and methane. The American astronomer **Theodore Dunham Jr. (1897–1984)** verified these identifications in 1933.

1932

The American astronomers **Walter Adams (1876–1956)** and **Theodore Dunham Jr. (1897–1984)** identify the absorption lines of carbon dioxide in the atmosphere of Venus.

American physicist **Carl Anderson (1905–1991)** reports his discovery of the positron, or "positive electron", the first antimatter particle to be found, which was created when cosmic rays from outer space interacted with particles in the Earth's atmosphere. At the time of his discovery, Anderson was unaware of Dirac's theoretical prediction of the positron in 1928. Anderson was awarded the 1936 Nobel Prize in Physics for his discovery of the positron, sharing it with the Austrian physicist **Victor F. Hess (1883–1964)** for his 1912 discovery of cosmic radiation, now called cosmic rays.

English physicist **James Chadwick (1891–1974)** discovers the neutron when bombarding atoms with energetic particles. Chadwick was awarded the 1935 Nobel Prize in Physics for his discovery of the neutron.

English physicist **John Cockcroft (1897–1967)** and Irish physicist **Ernest Walton (1903–1995)** use a particle accelerator to create high-speed protons. They bombarded lithium with the protons, thereby producing helium nuclei, for which they shared the 1951 Nobel Prize in Physics.

German physicist **Albert Einstein (1879–1955)** and Dutch astronomer **Willem de Sitter (1872–1934)** abandon the cosmological constant and use the equations of the *General Theory of Relativity* to propose a homogeneous, isotropic expanding Universe with no space curvature, in which the gravitational pull of all the matter in the Universe precisely balances its expansion. The space in this ever-expanding Universe is described by Euclidean geometry.

Russian-Azerbaijani physicist **Lev Landau (1908–1968)** shows that the collapse of massive stars, which have depleted their nuclear fuel, cannot be arrested at the white dwarf stage, and that those stars whose mass exceeds the critical mass of about 1.5 solar masses must, because of increasing gravitational forces, collapse without limit, forming what is now known as a black hole.

1933

American physicist **Arthur H. Compton (1892–1962)** demonstrates that the Earth's magnetic field is deflecting cosmic rays incident from outer space, resulting in an increase of their intensity with latitude and showing that cosmic rays are very energetic charged particles.

English astronomer **Arthur Eddington (1882–1944)** publishes his book *The Expanding Universe*, which includes a description of an accelerating expansion driven by the cosmic repulsion of the cosmological constant.

American radio engineer **Karl Jansky (1905–1950)** announces his discovery of cosmic radio emission, which he found while investigating interference with radio communications equipment at the Bell Telephone Laboratories. His paper on these results was delivered at the 27 April 1933 Washington, D.C. meeting of the Union Radio Scientifique International, or URSI for short. On 27 June 1933 Jansky presented his classical paper entitled *Electrical Disturbances Apparently of Extraterrestrial Origin* in Chicago, Illinois at the annual convention of the Institute of Radio Engineers, abbreviated IRE, who published it in their *Proceedings* in October 1933. Within two years Jansky showed that the most intense radio emission is located in the direction of the constellation

Sagittarius, which is also the direction of the center of our Galaxy, and that weaker radio waves are coming from all directions in the Milky Way. Although this marked the beginning of radio astronomy, there was no follow-up to Jansky's discovery until the American radio engineer **Grote Reber (1911–2002)** confirmed it in 1940.

American physicist **Howard Robertson (1903–1961)** summarizes relativistic cosmology in the *Reviews of Modern Physics*, introducing the general metrics for homogeneous and isotropic Universes that Robertson had previously derived.

1934

The German-born American astronomer **Walter Baade (1893–1960)** and the Swiss astronomer **Fritz Zwicky (1898–1974)** propose that exploding stars, known as supernovae, release an energy comparable to the complete annihilation of a star's mass, produce and accelerate energetic cosmic ray particles, and represent the transition of an ordinary star to a neutron star.

Italian physicist **Enrico Fermi (1901–1954)** bombards uranium with neutrons and produces a radioactive substance, which he cannot identify. In 1938 the German radio chemists **Otto Hahn (1879–1968)** and **Fritz Strassman (1902–1980)** proposed that the uranium atoms were being broken apart, producing a much lighter element, radioactive barium. In the following year in Copenhagen, **Lise Meitner (1878–1968)** and **Otto Frish (1904–1979)** confirmed that uranium nuclei were being broken apart by neutron bombardment, introducing the term fission to describe this splitting of a nucleus.

The Italian physicist **Enrico Fermi (1901–1954)** proposes the name *neutrino*, Italian for "little neutral one", for the energy-carrying, uncharged particle first proposed by the Austrian physicist **Wolfgang Pauli (1900–1958)** in 1930 to explain radioactive beta decay, and provides further evidence for the existence of the neutrino. A radioactive element emits an electron, or beta particle, when a neutron is converted into a proton, an electron (a beta particle), and an antineutrino.

American astronomers **Edwin Hubble (1889–1953)** and **Milton Humason (1891–1972)** announce that there are at least as many galaxies in the Universe as there are stars in the Milky Way. They also reformulate the velocity-distance relation of galaxies, now known as Hubble's law, in a form that relates the radial velocity, or redshift, of a galaxy to its apparent luminosity, or apparent magnitude, and make a plot of this data that is now known as a Hubble diagram. It was later shown that the observed slope of 0.2 in the Hubble diagram is that expected for a homogeneous, isotropic expanding Universe that obeys **Albert Einstein's (1879–1955)** *General Theory of Relativity*.

The English astrophysicist **Edward A. Milne (1896–1950)** introduces the cosmological principle, which assumes that the Universe is both homogeneous and isotropic, or the same in all directions, on a sufficiently large scale.

1936

German physicist **Albert Einstein (1879–1955)** proposes that the gravitation of a foreground star can bend and focus the light of a background star. Although he explored these aspects of gravitational lensing by stars as early as 1912, Einstein did not publish the result until 1936, as the result of prodding by an amateur scientist.

American astronomer **Edwin Hubble (1889–1953)** publishes his book *The Realm of the Nebulae* that summarizes his investigations of extragalactic nebulae, or galaxies, including their average mass density, morphological classification and the linear correlation between their distance and radial velocity, now known as Hubble's law.

The Danish seismologist **Inge Lehmann (1888–1993)** discovers the Earth's solid inner core, with a radius of 1,225 kilometers.

The Russian biochemist **Alexandr Oparin (1894–1980)** publishes his book *The Origin of Life on Earth*, which suggests that the first living organisms could have been generated spontaneously if sufficient quantities of organic compounds were present in the shallow seas of the primitive Earth.

1937

Swiss astronomer **Fritz Zwicky (1898–1974)** measures the velocity dispersion of galaxies in the Coma cluster, concluding that it must contain ten times more invisible dark matter than the mass in visible galaxies to keep the cluster's galaxies from flying apart. Zwicky also proposes that the dark matter in galaxies or clusters of galaxies will act as a gravitational lens, diverting and focusing the light of more distant galaxies, and that the effect could be used to estimate the amount of dark matter. At about this time he also began using a Schmidt telescope to discover supernovae in galaxies.

1938

German-born American physicist **Hans Bethe (1906–2005)** and **Charles L. Critchfield (1910–1994)** demonstrate that the fusion of hydrogen nuclei, or protons, into helium nuclei by the proton-proton reaction provides the energy which makes the Sun shine. In the following year Bethe shows that the synthesis of helium nuclei from hydrogen nuclei, or protons, by the carbon-nitrogen-oxygen, or CNO, chain of nuclear reactions provides the temperature-dependent energy generation that accounts for the luminosity of main-sequence stars more massive than the Sun, and reasons that less massive, main-sequence stars shine by the direct fusion of protons to make helium nuclei. Bethe was awarded the 1967 Nobel Prize in Physics for his work on the production of energy within stars.

The German radio chemist **Otto Hann (1879–1968)**, who had been working with **Lise Meitner (1878–1968)**, showed that when uranium is bombarded with neutrons it could be split into two nearly equal fragments, releasing large amounts of energy. The result was confirmed by Meitner, who was in exile in Copenhagen, and it ultimately resulted in the United States Manhattan Project to build the first atomic bombs.

The Estonian astronomer **Ernst Öpik (1893–1985)** proposes that a giant star is formed when a star like the Sun depletes its central hydrogen. The core collapses, the outer envelope expands, and the compressed, heated material in the core undergoes new nuclear fusion processes that make the giant star shine.

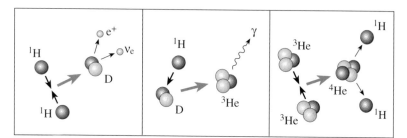

FIG. 16   Proton-proton chain Hydrogen nuclei, or protons, are fused together to form helium nuclei within the solar core, providing the Sun's energy. In 1939 the German-born American physicist **Hans Bethe (1906–2005)** described the detailed sequence of nuclear fusion reactions, called the proton-proton chain. It begins when two protons, here designated by the letter ^{1}H, combine to form the nucleus of a deuterium atom, the deuteron that is denoted by D, together with the emission of a positron, e^+, and an electron neutrino, v_e. Another proton collides with the deuteron to make a nuclear isotope of helium, ^{3}He, and then a nucleus of helium, ^{4}He, is formed by the fusion of two ^{3}He nuclei, returning two protons to the gas. Overall, this chain successively fuses four protons together to make one helium nucleus. Even in the hot, dense core of the Sun, only rare, fast-moving particles are able to take advantage of the tunnel effect and fuse in this way.

German physicist **Carl Friedrich von Weizsäcker (1912–)** discovers the carbon-nitrogen-oxygen, abbreviated CNO, chain of nuclear reactions in which carbon acts as a catalyst in the synthesis of helium from hydrogen.

1939

Danish physicist **Niels Bohr (1885–1962)** proposes his liquid-droplet model for the nucleus that describes how a heavy nucleus could undergo fission following the capture of a neutron and explains why uranium-235 could undergo fission with slow neutrons.

French chemists **Frédéric Joliot (1900–1958)** and **Irène Joliot-Curie (1897–1956)** demonstrate the possibility of a chain reaction when they split uranium nuclei.

In a paper entitled *On Continued Gravitational Contraction*, American physicists **J. Robert Oppenheimer (1904–1967)** and **Hartland S. Snyder (1913–1962)** use Einstein's *General Theory of Relativity* to show how a black hole will form as the result of the collapse of a sufficiently massive star.

In a paper entitled *On Massive Neutron Cores*, American physicist **J. Robert Oppenheimer (1904–1967)** and Canadian physicist **George M. Volkoff (1914–2000)** derive the first reliable estimates for the mass and size of neutron stars.

Swedish astronomer **Bengt Strömgren (1908–1987)** shows that the ultraviolet radiation from hot, massive blue stars creates spheres of ionized gas that envelop them, with a size, now called the Strömgren radius, that depends on the temperature of the star and the density of the surrounding material. For a typical interstellar hydrogen density of one atom per cubic centimeter, the Strömgren radius ranges between 30 and 300 light-years for central stars with disk temperatures between 26,000 and 48,000 degrees kelvin.

The German-born American astronomer **Rupert Wildt (1905–1976)** shows that the negative ion of hydrogen is an important source of opacity in stars, including the Sun.

1940–1949

1940

American radio engineer and amateur radio operator **Grote Reber (1911–2002)** uses a parabolic radio antenna, built in his back yard, to confirm the American radio engineer **Karl Jansky's (1905–1950)** discovery, in 1932, of cosmic radio static, showing that much of it comes from the Milky Way.

The German-born American astronomer **Rupert Wildt (1905–1976)** notices that carbon dioxide is so opaque to surface radiation that it will produce a substantial greenhouse effect on Venus, increasing its surface temperature to at least 400 degrees kelvin.

1941

American astronomer **Walter S. Adams (1876–1956)** and Canadian astronomer **Andrew McKellar (1910–1960)** independently infer the presence of radiation with a temperature of a few degrees from the rotational excitation of interstellar diatomic molecules detected in the visible-light spectra of bright stars; but the cosmological implications of the radiation was not realized until after the discovery of the three-degree cosmic background radiation in 1965.

Swedish astronomer **Bengt Edlén (1906–1993)** and German astronomer **Walter Grotrian (1890–1954)** show that emission lines from the solar corona, which were previously attributed to a hypothetical element coronium, are due to the forbidden transitions of known elements, which are highly ionized by an unexpectedly high, million-degree temperature in the corona. The conspicuous, bright green emission line of the corona is identified with iron missing 13 electrons, denoted Fe XIV.

Russian-born American physicist **George Gamow (1904–1968)** and the Brazilian astrophysicist **Mario Schönberg (1914–1990)** call attention to energy losses by neutrino processes during the final stages of stellar evolution and the onset of catastrophic stellar explosions known as supernovae.

The Serbian mathematician **Milutin Milankovitch (1879–1958)** publishes his most detailed account of how the major ice ages might be produced by periodic fluctuations in the wobble and tilt of the Earth's rotational axis and the shape of the Earth's orbit with periods of 23,000, 41,000 and 100,000 years. His idea that the ice ages are caused by a variation in the intensity and distribution of solar energy arriving at Earth was supported in 1976 by an analysis of deep-sea sediments, and by ice cores drilled in Greenland and Antarctica for more than a decade after 1985.

1942

A team led by American physicist **Arthur H. Compton (1892–1962)** and Italian-born American physicist **Enrico Fermi (1901–1954)** build the first self-sustaining nuclear reactor, using neutrons to initiate the fission of uranium. The first atomic bombs, exploded in 1945, used the same reaction but without control, resulting in a nuclear explosion.

English physicist **James Stanley Hey (1909–1990)** shows that severe jamming of terrestrial radar receivers is due to intense radio radiation emanated from the Sun during periods of solar activity.

1943

German-born American astronomer **Walter Baade (1893–1960)** uses the 100-inch (2.5-meter) **Hooker telescope** on Mount Wilson during wartime blackouts of nearby Los Angeles to discover red stars in the nuclear region of the nearby spiral galaxy Andromeda, or M31. Until that time, only the blue giant stars in the spiral arms had been resolved. Baade proposed that there are two groups of stars with differing structures and origins. They are now known as the young bluish Population I stars found in spiral arms, and the old, reddish Population II stars created in the nucleus.

The American astronomer **Carl K. Seyfert (1911–1960)** discovers evidence for high-speed motions in the centers of some spiral galaxies, subsequently named Seyfert galaxies in his honor.

1944

The English geologist **Arthur Holmes (1890–1965)** publishes his important textbook *Principles of Physical Geology*, arguing that heat inside the Earth could drive internal churning motions that might propel the moving continents from below.

The Dutch-born American astronomer **Gerard Kuiper (1905–1973)** discovers an atmosphere on Saturn's largest moon, Titan. Instruments aboard the *Voyager 1* spacecraft showed in 1980 that the dominant gas surrounding Titan is molecular nitrogen, with a surface pressure comparable to the Earth's atmosphere.

The American radio engineer and amateur radio operator **Grote Reber (1911–2002)** publishes the first contour maps of cosmic radio emission, and detects discrete sources of radio emission that lie in the direction of the galactic center and the constellations Cygnus and Cassiopeia.

The Austrian physicist **Erwin Schrödinger (1887–1961)** publishes his book *What is Life*, which discusses the physical aspects of the living cell and anticipates molecular biology and the subsequent discovery of deoxyribonucleic acid, abbreviated DNA.

1945

The Dutch astronomer **Hendrik C. "Henk" van de Hulst (1918–2000)** publishes his prediction of the possible detection of a radio frequency spectral line from neutral hydrogen, or H I, regions in interstellar space. In his paper, entitled *Hermomst der radiogolven uit het wereldruim*, or *Radio Waves from Space: Origin of Radiowaves*, he realized that these regions would be cold and that most of the hydrogen atoms would be in their lowest energy ground state. The electron of each hydrogen atom might then undergo a transition between the two possibilities of its spin, giving rise to emission or absorption at a wavelength of 21 centimeters. In his prophetic 1945 paper, van de Hulst also discussed the radio frequency recombination lines emitted when free electrons recombine with free protons in ionized hydrogen, or H II regions, and discussed the tremendous potential importance of radio astronomy to cosmology.

The American astronomer **Alfred Joy (1882–1973)** provides a detailed description of the young T Tauri stars, which exhibit irregular and unpredictable changes of light and appear to be still in gravitational contraction or just settling down on the main sequence of the Hertzsprung-Russell diagram.

On 6 and 9 August 1945 the cities of Hiroshima and Nagasaki, Japan, were destroyed by the first atomic bombs used in warfare. The bombs were produced at the Los Alamos Scientific Laboratories in New Mexico, under the direction of American physicist **J. Robert Oppenheimer (1904–1967)** with the participation of the most eminent physicists of the time, such as **Hans Bethe (1906–2005)** and **Richard Feynman (1918–1988)**.

1946

English physicists **Edward V. Appleton (1892–1965)** and **James Stanley Hey (1909–1990)** demonstrate that meter-wavelength solar radio noise originates in sunspot-associated active regions, and that sudden large increases in the Sun's radio output are associated with solar flares.

Working at the Radiation Laboratory, Massachusetts Institute of Technology, the American physicist **Robert Dicke (1916–1997)** invents a radio receiver that switches between a source and reference noise, greatly improving the sensitivity of radio telescopes.

American physicist **Scott E. Forbush (1904–1984)** and his colleagues note that solar flares produce transient increases in the amount of energetic charged particles arriving at the Earth.

Russian-born American physicist **George Gamow (1904–1968)** proposes that the observed relative abundances of the elements were determined by non-equilibrium nucleosynthesis during the early hot, dense stages of the expansion of the Universe.

Russian astrophysicist **Vitalii L. Ginzburg (1916–)** and Australian radio astronomers **David F. Martyn (1906–1970)** and **Joseph L. Pawsey (1908–1962)** independently use observations of the Sun's radio radiation to confirm the existence of a million-degree solar corona.

Belgian astrophysicist and Catholic priest **Georges Lemaître (1894–1966)** publishes his book *L'hypothese de l'Atome Primitif: Essai de cosmogonie* in 1946; the English translation *The Primeval Atom: An Essay in Cosmogony* was published in 1950.

The Russian physicist **Evgenij M. Lifshitz (1915–1985)** derives mathematical formulae to show that the expansion of the Universe pulls mass density concentrations apart, prohibiting the formation of galaxies.

Richard Tousey (1908–1997) and his colleagues at the United States Naval Research Laboratory use instruments aboard a captured German V-2 rocket to make the first extreme ultraviolet observations of the Sun.

1948

The American physicists **Ralph A. Alpher (1921–)** and **Robert C. Herman (1914–)** predict the existence of the pervasive relic cosmic background radiation, which originated in the early stages of the Big Bang, and estimate that it would have cooled to a present temperature of about 5 degrees kelvin.

In a paper entitled *The Origin of Chemical Elements*, the American physicist **Ralph A. Alpher (1921–)** and Russian-born American physicist **George Gamow (1904–1968)** carry out the first quantitative work on the nucleosynthesis of elements in the early Universe, following Gamow's 1946 proposal that thermonuclear reactions would occur in the early, hot dense stages of the Universe. Alpher and Gamow included the name of the German-born American physicist **Hans Bethe (1906–2005)** as a co-author of this paper, in order to make a pun on the first three letters of the Greek alphabet. Their α-β-γ theory assumed that the material Universe initially consisted of a cosmic "ylem", named after an ancient Greek word for the primeval substance out of which all matter was formed. The ylem was supposed to consist solely of neutrons at high temperature, about 10 billion or 10^{10}, degrees kelvin. The decay of neutrons into protons and successive neutron capture led to the formation of the elements. The buildup of elements by neutron capture during the early stages of the expansion of the Universe was subsequently developed in greater detail by Alpher and the American physicist **Robert C. Herman (1914–).**

Though not yet completely operational, the 200-inch (5.0-meter) **Hale telescope** on Palomar Mountain in southern California is dedicated and formally named the **Hale Telescope** after the American astronomer **George Ellery Hale (1868–1938).** The nearby 48-inch (1.25-meter) Schmidt telescope is also completed.

English cosmologist and astrophysicist **Fred Hoyle (1915–2001)** and two Austrian-born scientists, then at Cambridge University, **Hermann Bondi (1919–2005)** and **Thomas Gold (1920–2004)** propose the Steady State cosmology in which the Universe expands but does not evolve, remaining unchanged in time. Bondi and Gold proposed the name "perfect cosmological principle" for the notion that the Universe has both spatial and temporal homogeneity when viewed on sufficiently large scales.

Dutch-born American astronomer **Gerard Kuiper (1905–1973)** detects carbon dioxide in the Martian atmosphere, and predicts that it might be one of the main constituents. Instruments aboard the *Mariner 4* spacecraft confirmed this prediction in 1965.

The French astronomer **Evry Schatzman (1920–)** considers the possibility of thermonuclear reactions on white dwarf stars.

1949

Armenian astronomer **Viktor Ambartsumian (1908–1996)** notes that associations of bright, hot O and B stars cannot be bound by their own gravitation, and must be expanding and relatively young. The Dutch astronomer **Adriaan Blaauw (1914–)** observed the expansion of some of these stellar associations in 1952.

Australian radio astronomers **John G. Bolton (1922–1993), Gordon J. Stanley (1922–)** and **Owen Bruce Slee (1924–)** use a sea radio interferometer to obtain accurate enough positions for the identification of three discrete radio sources with optical objects seen in visible light. They are Taurus A, identified with the Crab Nebula, an expanding supernova remnant in our Galaxy; Virgo A identified with M87, an elliptical galaxy; and Centaurus A associated with another galaxy NGC 5128.

The Dutch astronomer **Hendrik C. "Henk" van de Hulst (1918–2000)** shows that the observed extinction of visible starlight is due to solid dust particles in interstellar space, with a size of about ten millionths, or 10^{-7}, meters. American astronomers **William A. Hiltner (1914–1991)** and **John Scoville Hall (1908–1991)** independently observe highly polarized light from reddened stars, attributed to elongated interstellar dust particles aligned by an interstellar magnetic field.

1950–1959

1950

The Swedish astrophysicist **Hannes Alfvén (1908–1995)** and his colleague **Nicolai Herlofson (1935–)** suggest that optically invisible radio stars, located in interstellar space, might emit non-thermal synchrotron radiation of high-speed electrons, accounting for the radio emission of the Milky Way.

Italian-born American physicist **Enrico Fermi (1901–1954)** and American physicist **Anthony Turkevich (1916–2002)** consider nuclear reactions during the early stages of the expanding Universe, and show that the lack of stable nuclei with atomic weights of 5 and 8 prevent significant production of heavier nuclei in the Big Bang.

Japanese physicist **Chushiro Hayashi (1920–)** notes that the temperature will exceed 10 billion degrees kelvin during the first few seconds after the Big Bang, which initiated the expansion of the Universe, and that sufficient energy was then available to form neutrinos and electron-positron pairs. As a result, the relative abundance of neutrons and protons is determined by conditions of thermal equilibrium between all forms of matter and radiation, instead of by neutron decay.

German astronomer **Karl O. Kiepenheuer (1910–1975)** suggests that the radio emission of the Milky Way is due to the non-thermal synchrotron radiation of high-velocity, cosmic ray electrons spiraling about an interstellar magnetic field. The Russian astrophysicist **Vitalii L. Ginzburg (1916–)** developed this idea in 1956, additionally proposing that the cosmic ray electrons are accelerated in stellar explosions known as supernovae, and extending the synchrotron radiation hypothesis to other galaxies.

Dutch astronomer **Jan Oort (1900–1992)** uses the size and orientation of the trajectories of long-period comets, with orbital periods greater than 200 Earth years, to show that they come from a remote, spherical shell located roughly a quarter the way to the nearest star and containing up to a million million, or 10^{12}, invisible comets orbiting the Sun at distances of about one light-year. This comet reservoir is now known as the Oort comet cloud.

In a paper entitled *A Comet Model I: The Acceleration of Comet Encke*, the American astronomer **Fred L. Whipple (1906–2004)** proposes the icy conglomerate, or dirty snowball, comet model, in which the solid nucleus of a comet is just a gigantic ball of ice, dust and rock. When the comet's orbit brings it close to the Sun, solar radiation causes the frozen surface material to evaporate, producing the dust and ion tails of a comet. Whipple also recognized that if the comet were rotating as it neared the Sun, the ejection of gases resulting from the sublimation of ices would act as a jet engine, pushing a comet ahead or delaying its journey. He used these non-gravitational forces to explain the apparent acceleration of Comet Encke, which appeared near the Sun earlier than predicted on the basis of gravitational theory alone.

Australian radio astronomer **John Paul Wild (1923–)** and colleagues begin nearly a decade of observations of radio bursts, or flares, on the Sun, obtaining evidence for outward moving shock waves and outward streams of high-energy electrons.

1951

German astronomer **Ludwig Biermann (1907–1986)** proposes that the straight ion tails of comets, which always point away from the Sun, are accelerated by moving, charged particles of the same type as those causing geomagnetic storms, and proposes that the Sun emits a continuous electrified flow of these solar corpuscles, which stream radially out in all directions from the Sun.

Harvard graduate student, **Harold I. "Doc" Ewen (1922–)** and his advisor, American physicist **Edward M. Purcell (1912–1997)** detect the 21-centimeter transition of interstellar neutral hydrogen, using a novel "switched-frequency" mode suggested by Purcell. The existence of this spectral feature was predicted by the Dutch astronomer **Hendrik C. "Henk" van de Hulst (1918–2000)** in 1945, and investigated in greater detail by the Russian astrophysicist **Iosif S. Shklovskii (1916–1985)** in 1949. Ewen and Purcell delayed the announcement of their discovery until it was fully confirmed by the Dutch astronomers **Karl Müller (1927–)** and **Jan Oort (1900–1992)** and in a cable from Australian radio astronomers.

Herbert Friedman (1916–2000) and his colleagues at the United States Naval Research Laboratory, abbreviated NRL, use instruments aboard Aerobee sounding rockets to show that the Sun emits enough X-rays to ionize most of the Earth's upper atmosphere. Similar experiments by the group subsequently indicated that the intensity of solar X-rays rises and falls with the 11-year sunspot cycle, and that the X-ray emission during a solar flare can outshine the entire Sun at these wavelengths. The NRL group used a captured German V-2 rocket to first detect the Sun's X-rays in 1948.

Austrian-born astrophysicist **Thomas Gold (1920–2004),** then at Cambridge University, proposes that intense, discrete radio sources lie outside the Milky Way, that only collapsed stars with amplified magnetic fields would emit intense radio emission, and that violent action at the centers of some visible galaxies might produce high-energy particles that emit synchrotron radiation at radio wavelengths.

Dutch-born, American astronomer **Gerard P. Kuiper (1905–1973)** proposes the existence of a flat, distant ring of small, icy bodies at the outer edge of the planetary realm, orbiting the Sun just beyond the orbit of Neptune and thought to be the source of short-period comets. It is now known as the Kuiper belt, but the discussion of such a belt by the less prominent Irish astronomer **Kenneth E. Edgeworth (1880–1972)** is sometimes acknowledged by using the name Edgeworth-Kuiper belt.

American astronomer **William W. Morgan (1906–1994)** and two graduate students **Donald E. Osterbrock (1924–)** and **Stewart Sharpless (1926–)** use the space distribution of emission nebulae, or H II regions, to delineate segments of spiral arms in the Milky Way.

Estonian astronomer **Ernst Öpik (1893–1985)** first realizes that when the temperature in the contracting core of a giant star rises above 400 million degrees kelvin, its helium will be converted into carbon by triple collisions of helium nuclei. In the following year, the American astrophysicist **Edwin E. Salpeter (1924–),** unaware of Öpik's work, presented in greater detail the arguments for the formation of carbon by helium burning in the cores of giant stars.

English radio astronomer **Martin Ryle (1918-1984)** presents the initial results of his interferometric radio surveys, arguing that a new kind of radio star has been discovered in the Milky Way.

1952

The German-born American astronomer **Walter Baade (1893–1960)** shows that there are two kinds of variable stars, respectively found in globular clusters in our Galaxy and in the arms of other spiral galaxies, and uses the difference to obtain a downward revision of Hubble's constant and an increase in the estimated age of the expanding Universe.

The Dutch astronomer **Adriann Blaauw (1914–)** measures the expansion velocities of some OB associations, inferring that it took about a million years for them to disperse to their present size. Since this is comparable to the age of the individual stars, the associations and their component stars must have formed together.

In a paper entitled *Nuclear Reactions in Stars without Hydrogen*, the American astrophysicist **Edwin E. Salpeter (1924–)** demonstrates how carbon nuclei can be synthesized by triple collisions of helium nuclei in the hot, dense cores of giant stars. The beryllium 8 nucleus formed momentarily by the collision of two helium 4 nuclei can capture a third one before breaking up into two helium nuclei again, and thus carbon 12 can be synthesized.

American astronomers **Allan R. Sandage (1926–)** and **Martin Schwarzschild (1912–1997)** present theoretical calculations of the evolution of a star after it has depleted its core hydrogen, showing that the star evolves from a main-sequence star to a red giant star and describes tracks in the observed Hertzsprung-Russell diagrams of globular star clusters.

1953

American physicists **Ralph A. Alpher (1921–)**, **James W. Follin (1919–)** and **Robert C. Herman (1914–)** delineate physical conditions in the initial stages of the expanding Universe, during the hot Big Bang. They also recognize the horizon problem in cosmology, in which widely separated regions could not have communicated with each other, since they are separated by more than the maximum distance that a light signal could have traveled since the regions originated.

English astronomer **Robert Hanbury Brown (1916–2002)** shows how a total intensity radio interferometer could be developed using correlation after detection at independent, separated radio telescopes.

English and American biophysicists **Francis Crick (1916–2004)** and **James Watson (1928–)** publish their discovery of the three-dimensional, double helix structure of deoxyribonucleic acid, or DNA for short, explaining how genetic information could be coded. New-Zealand born British biophysicist **Maurice Wilkins (1916–2004)** contributed to the discovery of the structure of DNA, and the three scientists shared the 1962 Nobel Prize in Medicine for their discoveries concerning the molecular structure of nucleic acids and its significance for information transfer in living material. In 1968 Wilson wrote the story of this discovery in *The Double Helix*, and in 1981 Crick wrote *Life Itself: Its Origin and Nature*, which suggests that the seed of life on Earth could have come from another planet.

American geologists **William Maurice Ewing (1906–1974)** and **Bruce C. Heezen (1924–1977)** discover deep canyons or rifts that run along a submerged mid-ocean ridge, which was later found to be part of a global network of underwater ridges some 60,000 kilometers long. The rifts mark the place where heat and molten rock, or magma, come out from inside the Earth to replenish the spreading seafloor.

English radio astronomers **Roger C. Jennison (1922–)** and **Mrinal Das Gupta (1923–1963)** construct a new type of intensity interferometer at Jodrell Bank, England, consisting of two completely independent radio telescopes and receivers, and use it to show that the intense radio source Cygnus A consists of two components of roughly equal intensity, separated by about 1.5 minutes of arc. At Cambridge, England, **F. Graham Smith (1923–)** used a conventional radio interferometer, with interconnected radio telescopes, to also refine the position of this remarkable object to about a minute or arc.

The American chemist **Stanley Miller (1930–)** produces amino acids, one of the basic building blocks of living organisms, using an electrical discharge in a simulation of the Earth's primitive atmosphere containing methane, ammonia and sterilized water.

This suggests that the carbon and hydrogen in methane, the nitrogen and hydrogen in ammonia, and the hydrogen and oxygen in water might be liberated by radiation or other external energy sources to produce the complex molecules that could have subsequently formed living organisms. The possibility that life might have originated spontaneously from organic compounds in the Earth's oceans was suggested in 1936 by the Russian biochemist **Alexandr Oparin (1894–1980)**.

Russian astrophysicist **Iosif S. Shklovskii (1916–1985)** argues that both the optical and radio emission of the Crab Nebula come from the synchrotron radiation of high-energy electrons spiraling about magnetic fields, accounting for the fact that the radio emission is a thousand times more intense than the visible-light emission.

American physicist **Charles H. Townes (1915–)** constructs the first working maser. Townes received the 1964 Nobel Prize in Physics for this invention; he shared the prize with **Nicolay Basov (1922–2001)** and **Aleksandr Prokhorov (1916–2002)**, two Russian scientists who independently constructed oscillators and amplifiers based on the maser-laser principle in 1955.

1954

Guided by accurate radio interferometric positions of the intense source Cygnus A, obtained by English radio astronomers in 1953, two German-born American astronomers **Walter Baade (1893–1960)** and **Rudolph Minkowski (1895–1976)** use the 200-inch (5.0-meter) **Hale telescope** on Mount Palomar, California to identify the discrete radio source with a distant galaxy about a billion light-years away, discovering the prototype of radio galaxies, whose radio luminosities are comparable to their optical, or visible-light, ones.

French astronomer **Gérard H. De Vaucouleurs (1918–1995)** presents evidence for a Local Supergalaxy, or Local Supercluster, of bright galaxies which contains several of the nearest clusters of galaxies, including the Virgo cluster, and is concentrated in a plane that runs approximately perpendicular to the Milky Way.

American physicist **Scott E. Forbush (1904–1984)** demonstrates the inverse correlation between the intensity of cosmic rays arriving at Earth and the number of sunspots over two 11-year solar activity cycles. As explained in 1956, the interplanetary magnetic field, which is anchored in the Sun, is enhanced at times of greater solar activity, blocking more cosmic rays from reaching the Erath.

The English astrophysicist **Fred Hoyle (1915–2001)** shows that carbon can only be produced in substantial amounts by triple helium collisions within giant stars if a resonant reaction is involved. The predicted excited state of carbon was observed a few years later by the American astrophysicist **William A. "Willy" Fowler (1911–1995)** and his colleagues in the terrestrial laboratory.

Dutch astronomers **Jan Oort (1900–1992)** and **Gart Westerhout (1927–)**, and Australian radio astronomer **Frank Kerr (1918–2000)** use observations of the 21-centimeter spectral line of neutral hydrogen, or H I regions, to show that the interstellar hydrogen in our Milky Way Galaxy is distributed in elongated spiral arms.

Russian astrophysicist **Iosif S. Shklovskii (1916–1985)** predicts that the optical jet of M87 is due to synchrotron emission associated with violent activity in the galaxy's nucleus. The synchrotron hypothesis was confirmed within a year, when **Walter Baade (1893–1960)** observed the polarization of the jet's visible light. Within another few decades, radio astronomers traced out a stream of high-speed electrons shooting out of M87 along its jet and into the lobes of a radio galaxy, much as Shklovskii had suggested.

1955

American radio astronomers **Bernard F. Burke (1928–)** and **Kenneth L. Franklin (1923–)** unexpectedly discover intense radio bursts from Jupiter while observing other cosmic radio sources. The periodic radio emission was subsequently attributed to energetic electrons trapped in the strong, rotating magnetic field of Jupiter.

American physicist **Leverett Davis Jr. (1914–2003)** proposes that solar corpuscular emission will carve out a cavity in the interstellar medium, now known as the heliosphere, accounting for some observed properties of low-energy cosmic rays.

1956

English astronomer **Robert Hanbury Brown (1916–2002)** builds a mirror intensity interferometer near Narrabri, Australia, and uses it to measure the angular diameter of the star Sirius. The intensity interferometer was used to measure the angular diameters of several main-sequence stars between 1958 and 1976.

American physicists **Clyde L. Cowan (1919–1974)** and **Frederick Reines (1918–1998)** detect electron antineutrinos emitted from the Savannah River nuclear reactor in South Carolina. Reines was awarded the 1995 Nobel Prize in Physics for the detection of the neutrino.

American astronomers **Milton Humason (1891–1972), Nicholas U. Mayall (1906–1993)** and **Allan R. Sandage (1926–)** publish the radial velocity and apparent brightness, or redshift and apparent magnitude, data for 474 extra-galactic nebulae out to a velocity of one-third the velocity of light. The redshift-magnitude plot, known as a Hubble diagram, indicates a linear relation, within the accuracy of the data, with a slope that is expected in a homogeneous, isotropic expanding Universe.

American scientist **Clair C. Patterson (1922–1995)** uses radioactive dating to obtain a definitive age for meteorites of 4.55 billion years, with an uncertainty of just 70 million years. A similar age has subsequently been obtained for the oldest rocks returned from the Moon, and the age of the Earth and other solid bodies in the Solar System is now assumed to be 4.6 billion years, with an uncertainty of 100 million years.

American physicists **Peter Meyer (1920–2002), Eugene N. Parker (1927–)** and **John A. Simpson (1916–2000)** propose that enhanced interplanetary magnetism at the peak of the solar activity cycle deflects cosmic rays from their Earth-bound paths, accounting for the inverse correlation between the intensity of cosmic rays arriving at Earth and the number of sunspots, which was noted by **Scott E. Forbush (1904–1984)** in 1954.

Austrian-born, American cosmologist **Wolfgang Rindler (1924–)** clarifies the horizon problem in which distances between similarly uniform regions are large compared with the horizon distance, the maximum distance that a light signal could have traveled since the regions originated.

The American chemists **Hans E. Suess (1909–1993)** and **Harold C. Urey (1893–1981)** provide a detailed description of the elemental isotopic abundances in the Sun, meteorites and stars, serving as an inspiration for explaining these abundances by nuclear fusion processes in stars.

1957

The English husband-and-wife team, **Geoffrey R. Burbidge (1925–)** and **E. Margaret Burbidge (1919–)** team up with American astrophysicist **William A. "Willy" Fowler (1911–1995)** and the English astrophysicist **Fred Hoyle (1915–2001)** to describe how most elements were formed within the centers of stars, during their long evolution, and as the result of the explosive supernova death of massive stars. The Canadian-born American astrophysicist **Alastair Cameron (1925–)** described many of the same processes in the same year. Fowler was awarded the 1983 Nobel Prize in Physics for his theoretical and experimental studies of the nuclear reactions of importance in the formation of the chemical elements in the Universe.

British scientist **Gordon Miller Bourne (G.M.B.) Dobson (1889–1976)** begins measurements of the ozone content in the atmosphere above Antarctica, leading to the discovery of the ozone hole.

The Soviet Union launches the first artificial Earth satellite, *Prosteyshij Sputnik,* the simplest satellite, on 4 October 1957, marking the start of the American–Soviet Union Space Race.

Two American chemists, **Roger Revelle (1909–1991)** and **Hans E. Suess (1909–1993),** predict that an increase in atmospheric carbon dioxide by the burning of fossil fuels – coal, oil and natural gas – will have far-reaching consequences for our weather and climate.

Swiss astronomer **Max Waldmeimer (1912–2000)** calls attention to the absence of intense coronal emission lines in vacant places that he called coronal holes.

1958

American astronomer **George Abell (1927–1983)** compiles a catalogue of thousands of rich clusters of galaxies.

Armenian astronomer **Viktor Ambartsumian (1908–1996)** provides a wide range of evidence for mass loss and other departures from equilibrium both in individual galaxies and in groups of galaxies, arguing that violent activity is common in extragalactic objects.

American scientist **Charles Keeling (1928–2005)** begins measurements of the carbon dioxide concentration in the atmosphere above Mauna Loa Observatory in Hawaii,

FIG. 17   Abundance and origin of the elements in the Sun The relative abundance of the elements in the solar photosphere, plotted as a function of their atomic number, Z. The abundance is

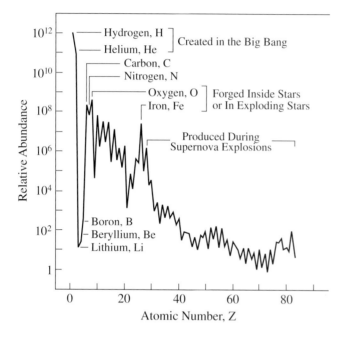

plotted on a logarithmic scale and normalized to a value a million, million, or 1.0×10^{12}, for hydrogen. Hydrogen, the lightest and most abundant element in the Sun, was formed about 14 billion years ago in the immediate aftermath of the Big Bang that led to the expanding Universe. Most of the helium now in the Sun was also created then. All the elements heavier than helium were synthesized in the interiors of stars that no longer shine, and then subsequently wafted or blasted into interstellar space where the Sun originated. Carbon, nitrogen, oxygen and iron, were created over long time intervals during successive nuclear burning stages in former stars, and also during the explosive death of massive stars. Elements heavier than iron were produced by neutron capture reactions during the supernova explosions of massive stars that lived and died before the Sun was born. The atomic number, Z, is the number of protons in the nucleus, or roughly half the atomic weight. The elements shown, He, C, N, O and Fe, have Z = 2, 6, 7, 8 and 26, with atomic weights of 4, 12, 14, 16, and 56, since each nucleus contains as many neutrons as protons with about the same weight. Hydrogen has one proton and no neutrons in its nucleus. The exponential decline of abundance with increasing atomic number and weight can be explained by the rarity of stars that have evolved to later stages of life. (Data courtesy of Nicolas Grevesse.)

which continue over subsequent decades and demonstrate the exponential rise in the amount of this heat-trapping, greenhouse gas.

American astronomer **Cornell H. Mayer (1921–)** and his colleagues announce that microwave observations indicate that Venus may have a very hot surface of at least 600 degrees kelvin. In 1971 the Russian probe *Venera 7* landed on the surface of Venus, measuring a temperature of 735 degrees kelvin.

Dutch astronomers **Jan Oort (1900–1992)** and **Gart Westerhout (1927–)**, and Australian radio astronomer **Frank Kerr (1918–2000)** map the intensity and velocities of the 21-centimeter line of neutral hydrogen in different directions in our Galaxy, from the northern and southern hemispheres of the Earth. The results permit the detection of spiral arms in our Galaxy, the determination of its rotation velocities at different distances from the galactic center, and evidence for recent violent activity within the nucleus of our Galaxy.

American physicist **Eugene N. Parker (1927–)** uses hydrodynamic equations to show that the million-degree electrons and protons in the outer solar atmosphere, or corona, will overcome the Sun's gravity and accelerate to supersonic speeds, naming the resultant radial outflow the solar wind. He also proposes that the solar wind will carry the Sun's magnetic field with it, forming a spiral magnetic pattern in interplanetary space as the Sun rotates.

English radio astronomer **Martin Ryle (1918–1984)** abandons his radio star hypothesis, and proposes that radio galaxies can be used to test cosmological models, interpreting his observations as an evolutionary effect in which the more distant, younger radio galaxies are either more luminous or more concentrated than the nearby older ones. Such evolutionary effects are inconsistent with the Steady State cosmology. Ryle was awarded the 1974 Nobel Prize in Physics for his pioneering observations and inventions in radio astrophysics, particularly the aperture synthesis technique.

The American astronomer **Allan R. Sandage (1926–)** obtains a value of Hubble's constant of 75 kilometers per second per Megaparsec, close to the presently accepted value, with an estimated age of the expanding Universe, given by the reciprocal of the Hubble constant, of about 13 billion years.

1959

American astronomer **George Field (1929–)**, and independently American radio astronomer **Frank Drake (1930–)** and visiting Norwegian scientist **Hein Hvatum (1923–)**, suggest that Jupiter's radio emission, discovered in 1955, is caused by the synchrotron radiation of magnetically trapped electrons surrounding Jupiter. Because of its fast rotation and large size, Jupiter has a stronger magnetic field and larger radiation belts than the Earth.

Austrian-born American astrophysicist **Thomas Gold (1920–2004)** proposes that geomagnetic storms are caused by a shock front associated with magnetic clouds ejected from the Sun, and coins the term magnetosphere for the region in the vicinity of the Earth in which our planet's magnetic field dominates all dynamical processes involving charged particles.

The Soviet Union's *Luna 2* spacecraft, launched toward the Moon on 12 September 1959, directly samples the solar wind for the first time, obtaining a particle volume density of about 5 million protons per cubic meter just outside the Earth.

The Soviet Union launches the space rocket *Luna 3* on 4 October 1959. It obtained the first photographs of the invisible, backside of the Moon, showing that it has fewer maria than the visible front side.

A 120-inch (3.0-meter) optical telescope is completed at Lick Observatory on the summit of Mount Hamilton near San Jose, California; it was named the **Shane telescope** in 1977, in honor of the astronomer **Charles Donald Shane (1895–1983)** who led the effort to acquire the necessary funds from the California Legislature.

In a paper entitled *Radiation Observations with Satellite 1958ε*. American scientists **James A. Van Allen (1914–), Carl E. McIlwain (1931–)** and **George H. Ludwig (1927–)** publish the first conclusive evidence for geomagnetically trapped, high-energy electrons and protons, found by the first American satellite, *Explorer 1* launched on 1 February 1958, in a belt far above the Earth's atmosphere. Also in 1959, Van Allen and his colleagues announce the results of a second satellite, *Explorer 3* launched on 26 March 1958, which confirmed the result and discovered a second belt of energetic electrons and protons higher above the Earth. These belts, which encircle the Earth's equator but do not touch it, are sometimes called the inner and outer Van Allen radiation belts, but the nomenclature does not imply either electromagnetic radiation or radioactivity.

1960–1969

1960

Austrian-born American astrophysicist **Thomas Gold (1920–2004)** and English astrophysicist **Fred Hoyle (1915–2001)** show that magnetic energy must power solar flares, and argue that flares are triggered by the interaction and reconnection of magnetic loops.

American mathematician **Martin D. Kruskal (1925–)** and Hungarian-born Australian mathematician **George Szekeres (1911–2005)** independently define a space-time in which the black-hole singularity at the Schwarzschild, or gravitational, radius is removed to the center of the black hole, using what is now known as the Kruskal-Szekeres coordinate transformation.

American astronomer and geologist **Eugene M. Shoemaker (1928–1997)** studies the structure and mechanics of meteorite impact. He demonstrated the similarity between the Meteor Crater, near Flagstaff, Arizona, and the craters produced during nuclear test explosions in Nevada, demonstrating its origin by external impact, and proposed that the Moon's craters had a similar origin.

1961

Dutch astronomer **Adrian Blaauw (1914–)** proposes that high-velocity "runaway" stars, of the hot O and B spectral class, are escaped members of binary star systems in which one star has become a exploding star, or supernova. He noticed that the runaway stars are massive stars whose high velocities are comparable to the orbital velocities expected from massive binary star systems, and that the more massive component would quickly evolve to the supernova stage and thereby release the other member as a high-velocity star.

The cosmonaut **Yuri A. Gagarin (1934–1968)** becomes the first human in space, orbiting the Earth on 12 April 1961 in the Soviet Union's *Vostok 1* capsule.

Japanese astrophysicist **Chushiro Hayashi (1920–)** describes the theory of stellar evolution in the early phases of gravitational contraction, accounting for the

Hertzsprung-Russell diagrams of extremely young star clusters observed by the American astronomer **Merle F. Walker (1926–)** in 1956 to 1961.

On 25 May 1961 American President **John F. Kennedy (1917–1963)** delivers an address to a joint session of Congress, declaring that "I believe that this nation should commit itself to achieving the goal, before the decade is out, of landing a man on the Moon and returning him safely to Earth".

The American physicist and astronomer **Robert B. Leighton (1919–1997)** publishes his discovery of five minute vertical velocity oscillations in the solar photosphere.

1962

Italian-born American astronomer **Riccardo Giacconi (1931–)** and his colleagues at the American Science and Engineering Company discover cosmic X-ray sources with an instrument on a five-minute rocket flight. The group subsequently used their X-ray telescope aboard NASA's *Uhuru* satellite, launched on 12 December 1970, to discover X-ray pulsars and stellar black hole companions of ordinary visible stars. Giacconi was awarded the 2002 Nobel Prize in Physics for his discoveries of cosmic X-ray sources.

The astronaut **John H. Glenn, Jr. (1921–)** becomes the first American to orbit the Earth, on 20 February 1962 in the *Friendship 7* capsule.

In a paper entitled *History of Ocean Basins*, the American geologist **Harry H. Hess (1906–1969)** elaborates his concept of sea-floor spreading, first suggested by him in 1960. The sea floor is formed in a volcanic rift in a mid-ocean ridge, spreading sideways and turning cold and heavy as it moves away from its source in two directions; the sea floor eventually sinks and disappears in a deep–ocean trench, where it is consumed. The spreading sea floor is thought to be the cause of continental drift.

1963

The world's largest radio telescope is inaugurated, in the hills near Arecibo, Puerto Rico. The metal reflecting surface of the **Arecibo Observatory** has a spherical shape with a diameter of 305 meters (1,000 feet). The telescope was resurfaced in 1974 to permit observations at shorter wavelengths.

American radio astronomer **Alan Barrett (1927–1991)** and his colleagues detect interstellar OH molecules at radio wavelengths.

NASA launches *Explorer 18,* the first *Interplanetary Monitoring Platform,* abbreviated **IMP 1,** which first mapped the Earth's magnetotail.

The American scientist **Herbert Friedman (1916–2000)** and his colleagues discover X-rays from the Crab Nebula during a brief rocket flight that carried the X-ray detectors above the atmosphere.

The first quasi-stellar object, abbreviated quasar, is discovered after English radio astronomer **Cyril Hazard (1928–)** and his colleagues establish a precise position of the radio source 3C 273 using its occultation by the Moon in 1962, and showing that the radio source coincides with a blue object at visible wavelengths. The Dutch-born American astronomer **Maarten Schmidt (1929–)** uses the 200-inch (5.0-meter) **Hale**

telescope on Mount Palomar to confirm the optical identification and obtain its spectrum, which he interprets in terms of a redshift of 0.16. The large redshift implies a remote distance and a brightness of hundreds of times that of nearby spiral galaxies for an object with a star-like image. Perhaps as a result of Schmidt's discovery, the American astronomers **Jesse L. Greenstein (1909–)** and **Thomas Matthews (–)** reexamine the optical spectrum of the star-like object coinciding with the radio source 3C 48, showing that this quasar has an even larger redshift of 0.37. The articles reporting these major discoveries appeared together in 1963 within an issue of the journal *Nature*.

The New Zealand mathematician **Roy Kerr (1934–)** finds analytic solutions to the field equations of the *General Theory of Relativity* that describe the curved space outside a rotating black hole.

American scientists **Marcia Neugebauer (1932–)** and **Conway W. Snyder (1918–)** announce that NASA's *Mariner 2* spacecraft, launched on 7 August 1962, has measured the density and velocity of the solar wind, with an estimated volume density of 5 million protons and 5 million electrons per cubic meter near the Earth and an average speed of about 500 kilometers per second.

British geophysicists **Frederick Vine (1939–1988)** and **Drummond Matthews (1931–1997)** analyze the magnetism of rocks at the floor of the Atlantic Ocean, discovering a symmetric pattern of magnetic field reversals on either side of the mid-ocean ridge. This discovery provided concrete evidence of sea-floor spreading, and when compared with radioactive dating of volcanic rocks on land was used to determine the rate of spreading.

1964

American astrophysicist **John N. Bahcall (1934–2005)** performs detailed calculations of the amount of solar neutrinos emitted by the Sun, and refines these theoretical predictions over subsequent decades.

American astrophysicist **William A. "Willy" Fowler (1911–1995)** and his English colleague **Fred Hoyle (1915–2001)** propose a new kind of supernova explosion, now known as type Ia, in which a close, orbiting normal star spills over onto its white-dwarf companion, triggering an explosion that obliterates the white dwarf star.

The American physicist **Murray Gell-Mann (1929–)** proposes that the proton and neutron are not themselves fundamental particles, but are instead composed of quarks which have fractional charges, occur in pairs and trios, and can never be detected singly. The quarks are bound together by exchanging gluons. The name quark is a whimsical designation taken from a passage in **James Joyce's (1882–1942)** *Finnegan's Wake*, which includes "Three quarks for Muster Mark."

English astronomers and cosmologists **Fred Hoyle (1915–2001)** and **Roger J. Tayler (1929–1997)** propose that helium was produced during the very hot, early stages of the expanding Universe, in the amounts now observed in nebulae and stars.

The American astronomer **Robert P. Kraft (1927–)** announces that many ex-novae occur in close, short-period binary star systems in which one component is a white dwarf star. His conclusions were extensions of observations by the American astronomers **Merle F. Walker (1926–)** and **Alfred H. Joy (1882–1973)** who respectively showed, in 1954 and 1956, that two old novae were components of binary stars in tight orbit about each other, where the atmosphere of one component might overflow to the other one.

The American astrophysicist **Edwin E. Salpeter (1924–)** and the Russian astrophysicist **Yakov B. Zeldovich (1914–1987)** independently propose that quasars are powered by supermassive black holes at their centers.

American astrophysicist **Irwin I. Shapiro (1929–)** announces that radar observations of Venus from the Arecibo Observatory indicate that Venus rotates with a period of 243 Earth days, in the retrograde direction, longer than the planet's orbital period of 224.7 Earth days.

1965

American astronomers **James E. Gunn (1938–)** and **Bruce Peterson (1941–)** show how the absence of atomic hydrogen absorption in the light of distant quasars places stringent limits on the amount of intergalactic unionized hydrogen.

Instruments aboard the *Mariner 4* spacecraft show that carbon dioxide is the main ingredient of the Martian atmosphere, but that it has a relatively low surface pressure.

English astrophysicist **Roger Penrose (1931–)** publishes his discovery that black holes must have a singularity.

German-born American radio engineer **Arno A. Penzias (1933–)** and American radio astronomer **Robert W. Wilson (1936–)** report their discovery of the cosmic microwave background radiation, while determining the sources of excess noise in a horn antenna at the Bell Telephone Laboratories. The radiation was coming from all directions at a wavelength of 7.35 centimeters with a temperature of three degrees kelvin, In a companion article in the same issue of the *Astrophysical Journal*, the American physicist **Robert H. Dicke (1916–1997)** and his colleagues explain the observed three-degree-kelvin radiation as the faint, cooled relic glow of the primeval Big Bang that initiated the expansion of the Universe. The American physicists **Ralph A. Alpher (1921–)** and **Robert C. Herman (1914–)** had already shown in 1948 that a current background temperature of about 5 degrees kelvin results from the expansion of the 10-billion-degree Big Bang. Penzias and Wilson were awarded the 1978 Nobel Prize in Physics for their discovery of the cosmic background radiation.

American radar astronomers **Gordon H. Pettengill (1926–)** and **Rolf B. Dyce (1929–)** announce that radar observations of Mercury from the Arecibo Observatory indicated that Mercury rotates with a period of 58.646 Earth days, just 2/3 of its orbital period of 87.969 Earth days.

The Nobel Prize in Physics is awarded to Japanese physicist **Sin-Itiro Tomonaga (1906–1979)** and American physicists **Richard P. Feynman (1918–1988)** and **Julian Schwinger (1918–1994)** for their fundamental work in quantum electrodynamics with important implications for the physics of elementary particles. They proposed that supposedly empty, sub-atomic space seethes with vacuum energy in the form of virtual particles that spontaneously pop in and out of existence; in so short a time that one cannot measure them directly. But the unseen "virtual" particles do produce predictable effects on "real" elementary particles, and these effects have been detected in the laboratory.

Canadian geologist **John Tuzo Wilson (1908–1993)** formulates a new concept of plate tectonics in a paper entitled *A New Class of Faults and Their Bearing on Continental Drift*. According to this theory, the Earth's outer shell is subdivided into a mosaic of large rigid plates that are in continual motion, producing earthquakes and building mountains at their boundaries by *tectonics*, from the Greek word for "carpenter" or "building". The moving plates carry continents on their backs, accounting for continental drift, and are pushed along by the spreading sea floor.

1966

American astronomers **Stirling A. Colgate (1925–)** and **Richard H. White (1934–)** use calculations developed for nuclear bomb studies during World War II to describe the explosive shocks developed by exploding stars called supernovae, and propose that just a small fraction of the neutrinos produced during the implosion of the stellar core would lead to its explosion.

Russian astrophysicists **Igor D. Novikov (1935–)** and **Yakov B. Zeldovich (1914–1987)** propose that a black hole might be detected by the X-ray radiation of material falling into it from a nearby star that is orbiting the black hole. The black hole uses its powerful gravity to suck in the outer layers of its companion, and as the in-falling material is compressed to fit into the hole, it is heated to millions of degrees kelvin, emitting almost all of its radiation at X-ray wavelengths.

American cosmologist **P. James E. Peebles (1935–)** uses the observed cosmic microwave background radiation to extrapolate backward into time and compute the amount of helium produced in the Big Bang, showing that is about 27 percent by mass, comparable to the amount now observed in stars and nebulae.

American astronomers **Peter G. Roll (1935–)** and **David T. Wilkinson (1935–2002)** confirm the existence of the three-degree cosmic microwave background radiation, at a wavelength of 3.2 centimeters.

1967

American physicist **Raymond Davis Jr. (1914–)** detects neutrinos emitted from the Sun using a massive container of cleaning fluid buried deep underground in the Homestake Gold Mine near Lead, South Dakota. This pioneering experiment was continued for more than a quarter-century, always detecting about one third of the expected amount of solar neutrinos. This discrepancy between the observed and predicted amounts, known as the solar neutrino problem, was eventually resolved when it was realized that some neutrinos were changing form on their way from the Sun to the subterranean experiment, thereby eluding detection. Davis and the Japanese physicist **Masatoshi Koshiba (1926–)** shared the 2002 Nobel Prize in Physics for their discovery of cosmic neutrinos. The American astronomer **Riccardo Giacconi (1931–)** also shared the 2002 Nobel Prize for his discovery of cosmic X-rays.

While investigating scintillations of cosmic radio sources produced by the solar wind, English radio astronomer **Antony Hewish (1924–)**, and his Irish graduate student **Jocelyn Bell (1943–)** discover the rapid periodic bursts of the first known pulsar, designated CP 1919, whose radio bursts repeated at precisely 1.3372795 seconds, with each burst lasting only milliseconds. Hewish was awarded the 1974 Nobel Prize in Physics for his decisive role in the discovery of pulsars.

Italian astrophysicist **Franco Pacini (1939–)** notes that the gravitational energy released during the collapse of a normal star into a neutron star will be converted into rotational energy, and that the stellar magnetic field will be amplified during the collapse. As a result, a neutron star has a strong, rapidly rotating magnetic field that can generate radiation, providing the luminosity of the Crab Nebula supernova remnant over its entire lifetime.

American radio astronomers **Bruce Partridge (1940–)** and **David T. Wilkinson (1935–2002)** show that the cosmic microwave background radiation is remarkably uniform, with the same temperature measured in every direction to within 0.1 percent.

Theoretical astrophysicists **Frank H. Shu (1943–)** and **Chia C. Lin (1916–)** show how density waves temporarily concentrate stars in spiral arms.

American astrophysicists **Robert V. Wagoner (1938–)** and **William A. "Willy" Fowler (1911–1995)** team up with English astrophysicist **Fred Hoyle (1915–2001)** to make the first detailed calculations of all the light elements – deuterium, helium and lithium – produced during the early stages of the expanding Universe, showing that they are created in amounts comparable to those observed now if the present radiation temperature of the cosmic microwave background is about three degrees kelvin.

1968

The Advanced Research Project Agency, abbreviated ARPA and funded by the United States Department of Defense, begins construction of ARPANET, the prototype of a computer network that led to the Internet.

Austrian-born American astrophysicist **Thomas Gold (1920–2004)** proposes that radio pulsars are highly magnetized, rapidly rotating neutron stars, successfully predicting that the pulsar periods will lengthen as the star loses rotational energy and that pulsars with much shorter periods will be found.

English radio astronomer **Antony Hewish (1924–)**, his Irish graduate student **Jocelyn Bell (1943–)** and their colleagues publish the discovery of the first radio pulsar, designated CP 1919, and three weeks later publish their discovery of three additional pulsars. Hewish was awarded the 1974 Nobel Prize in Physics for his decisive role in the discovery of pulsars.

The American astrophysicist **Lyman Spitzer, Jr. (1914–1997)** publishes his book *Diffuse Matter in Space*, which summarizes the various processes that determine the temperature and physical state of interstellar hydrogen, delineated in the preceding decades by Spitzer and others.

American physicist **Charles H. Townes (1915–)** and his colleagues detect the radio signatures of interstellar water and ammonia molecules.

The Italian particle physicist **Gabriele Veneziano (1942–)** develops a theoretical description of the strong nuclear force, which was subsequently interpreted by other particle theorists in terms of little, vibrating, and one-dimensional strings. Although a more convincing theory, known as quantum chromodynamics, was derived around 1973 to describe the strong nuclear force, the string theory was eventually developed as a candidate theory for uniting quantum mechanics, particle physics and gravity, and applied to the very early stages of the expanding Universe before a hypothetical inflation.

1969

Radar measurements of the distance to Venus establish a precise value of 149.597870 million kilometers for the mean distance between the Earth and the Sun, the Astronomical Unit, abbreviated AU.

The Charge-Coupled Device, abbreviated CCD, is invented at the Bell Telephone Laboratories. It has been widely used to improve the light-gathering power of telescopes, both astronomical and military.

American astronomers **Donald D. Clayton (1935–)**, **Gerald J. Fishman (1943–)** and **Sterling A. Colgate (1925–)** propose that newly synthesized radioactive nickel will power the brilliant light output of a supernova, and that the nickel will eventually decay into abundant iron, emitting gamma rays in the process. Supporting evidence for these ideas came from observations of the supernova SN 1987A.

NASA's *Lunar Module Eagle* carries two American *Apollo 11* astronauts to the lunar surface on 20 July 1969, and **Neil A. Armstrong (1930–)** becomes the first human to set foot on the Moon with **Edwin E. "Buzz" Aldrin Jr. (1930–)** at his heels.

English astronomer **Donald Lynden-Bell (1935–)** proposes that the nuclei of nearby galaxies are the aged, quiescent counterparts of younger, active quasars.

English theoretical astronomer **Roger Penrose (1931–)** shows that one can inject matter into a rotating black hole and extract rotational energy from it.

1970–1979

1970

The British physicist **Brandon Carter (1942–)**, working in France, announces an Anthropic Principle, which asserts that the Universe we observe is one that is compatible with our existence as observers.

Australian radio astronomer **Kenneth C. Freeman (1940–)** notices that the outer clouds of hydrogen atoms in spiral galaxies, detected by the 21-centimeter line, must be moving under the gravitational attraction of dark matter that is at least as massive as the mass of the visible stars in the galaxy. Several radio astronomers soon showed that the outer parts of spiral galaxies move unexpectedly fast, suggesting that each galaxy is enveloped by a halo of dark matter at least 10 times more massive than its visible components.

English theoretical astronomer **Stephen W. Hawking (1942–)** shows that the area of the horizon of a black hole can never decrease and must always increase, like entropy in thermodynamics.

The American astronomer **James C. Kemp (1927–)** and his colleagues measure the intense, million-Gauss magnetic fields of white dwarf stars, confirming their compact, Earth-sized radii. The magnetic flux is conserved in gravitational collapse, so the magnetic field strength is amplified by the inverse square of the radius.

American astronomers **Vera C. Rubin (1928–)** and **W. Kent Ford (1931–)** show that the outer, visible parts of the nearest spiral galaxy, Andromeda or M31, orbit its center as fast as the inner visible parts do, suggesting the presence of substantial amounts of non-luminous matter that holds the outer, fast-moving regions within the galaxy. Rubin and her colleagues subsequently obtained similar evidence for dark matter in other nearby spiral galaxies.

The *Uhuru* satellite was launched from Kenya on 12 December 1970, on the seventh anniversary of Kenyan independence, when it was named *Uhuru*, the Swahili word for "freedom". It was equipped with a sensitive X-ray telescope that was used to discover both periodic and irregularly variable X-ray sources, subsequently identified with rotating neutron stars and stellar black holes in close orbits around ordinary visible companion stars. Instruments aboard *Uhuru* also discovered the diffuse X-ray emission from clusters of galaxies, which accounts for some of the so-called dark matter in the Universe. The mission ended in March 1973.

The Russian probe *Venera 7* achieves a soft landing on Venus, obtaining direct measurements of the temperature and pressure of the planet's hot, dense atmosphere all the way to the surface, where the temperature is 735 degrees kelvin and the pressure 92 times the atmospheric pressure at sea level on Earth.

1971

NASA launched the *Mariner 9* spacecraft to Mars on 30 May 1971. It orbited the red planet for nearly a year, from 13 November 1971 to 27 October 1972, obtaining images of towering volcanoes, vast canyons, such as the Valles Marineris named for the spacecraft, and deep, wide outflow channels that have been attributed to catastrophic floods of running water at previous epochs.

The first good, space-based observation of a coronal mass ejection is obtained on 14 December 1971 using a coronagraph aboard NASA's *Seventh Orbiting Solar Observatory,* abbreviated *OSO 7.*

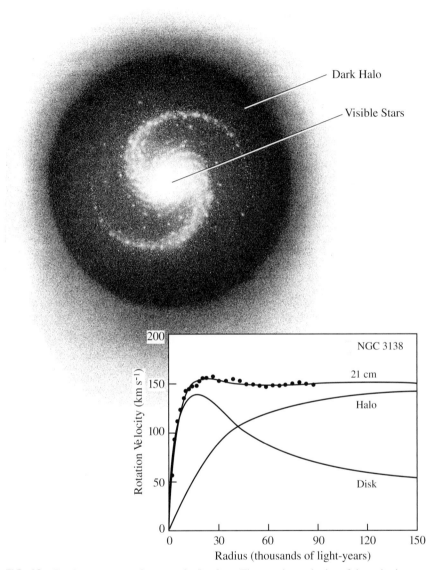

FIG. 18   Dark matter envelops a spiral galaxy The rotation velocity of the spinning spiral galaxy NGC 3198 plotted as a function of radius from its center *(bottom)*. The observed neutral hydrogen 21-centimeter data are attributed to an optically luminous disk, containing all the visible stars, and a dark halo that contributes most of the mass at distant regions from the center. They are illustrated in a hypothetical drawing of a spiral galaxy seen from above with its visible stars and surrounding halo of dark matter *(top)*. The fact that the rotational speed of the cool hydrogen gas remains high even at the largest distances indicates that the outermost gas must be constrained and held in by the gravitational pull of extra matter far outside the visible part of the galaxy.

American radar astronomer **Irwin I. Shapiro (1929–)** and his colleagues measure the travel time for a radar signal from Earth to Venus and return, with enough precision to determine the extra time delay caused by the Sun's curvature of nearby space, confirming the *General Theory of Relativity* with greater accuracy than the visible-light solar eclipse results in 1919. Radio astronomers substantiated the theory with even greater accuracy in the 1990s by observing the change in the positions of radio sources, and the delay of radio signals from spacecraft, when their line of sight passed near the Sun.

1972

Swiss space scientist **Johannes Geiss (1926–)** and Canadian-born French astronomer **Hubert Reeves (1932–)** publish measurements of the solar wind helium abundance using foil collectors unfurled on the Moon by American astronauts.

English astronomers **B. Louise Webster (1941–1990)** and **Paul Murdin (1942–)** use observations of the visible stellar companion of the X-ray source Cygnus X-1 to show that the invisible companion is more massive than eight solar masses, and most likely a stellar black hole. The Canadian astronomer **Charles Bolton (1943–)** independently confirmed the observations in the same year.

1973

American scientists **Ray W. Klebesadel (1932–), Ian B. Strong (1930–)** and **Roy A. Olson (1924–)** announce the discovery of cosmic gamma-ray bursts that were unexpectedly detected several years before using military *Vela* satellites designed to monitor Soviet compliance with an international treaty banning the tests of nuclear warheads in the Earth's atmosphere or in outer space.

American astrophysicists **Jeremiah P. Ostriker (1937–)** and **P. James E. Peebles (1935–)** show that the stellar disks of spiral galaxies cannot remain stable unless they are surrounded by huge, roughly spherical halos of dark, invisible matter that constrain and mold the spiral galaxies from outside.

Canadian-born French astronomer **Hubert Reeves (1932–)**, French astronomer **Jean Audouze (1941–)** and American astronomers **William A. "Willy" Fowler (1911–1995)** and **David N. Schramm (1945–1997)** show that deuterium cannot be made in stars, but must be made in the Big Bang, using comparisons of the observed deuterium abundance with primordial nucleosynthesis calculations to obtain accurate measurements of the baryon density of the Universe, in both visible and invisible form. It is well below the critical density needed to stop the expansion of the Universe in the future.

American astronomers **John B. Rogerson (1922–)** and **Donald G. York (1944–)** announce their determination of the interstellar deuterium abundance using ultraviolet spectroscopy toward the bright star Beta Centauri from NASA's *Copernicus* satellite, the *Third Orbiting Solar Observatory,* with an abundance close to that determined for the solar wind from the *Apollo 11* mission by the Swiss space scientist **Johannes Geiss**

(1926–) and the Canadian-born French astronomer **Hubert Reeves (1932–)** reported the following year.

NASA's manned orbiting solar observatory, *Skylab,* is launched on 14 May 1973, and manned by three-person crews until 8 February 1974. The Apollo Telescope Mount on *Skylab* was used to spatially resolve solar flares at soft X-ray and ultraviolet wavelengths, to fully confirm coronal holes, coronal loops and X-ray bright points from X-ray photographs of the Sun, and to observe coronal mass ejections with a coronagraph, finding that many coronal mass ejections are not associated with solar flares and that some move fast enough to produce interplanetary shocks.

In a paper entitled *Is the Universe a Vacuum Fluctuation*, the American physicist **Edward Tryon (1940–)** proposes that our Universe originated as a quantum fluctuation in the vacuum, providing the first scientific mechanism by which the Universe might spontaneously have arisen from "nothing".

The Italian-born American astronomer **Giuseppe S. Vaiana (1935–1991)** and his colleagues at the American Science and Engineering Company use high-resolution, solar X-ray observations, taken from rockets, to discover loops, holes and bright points in the million-degree solar corona.

1974

American astronomer **Roger D. Blandford (1949–)** and English astronomer **Martin Rees (1942–)** propose that a rotating, supermassive black hole generates magnetic fields and propels high-energy electrons along its rotation axis, creating the radio jets and lobes found in radio galaxies and quasars.

American radio astronomers **Russell A. Hulse (1950–)** and **Joseph H. Taylor Jr. (1941–)** discover the radio pulsar PSR 1913+16, colloquially known as the binary pulsar, which consists of two neutron stars, each with a mass of about 1.4 solar masses, in orbit about a common center of mass. Precise timing observations of the radio pulses indicated that the radio pulsar is slowly approaching its silent, invisible companion at the rate expected if their orbital energy is being radiated away in the form of gravitational waves. Hulse and Taylor received the 1993 Nobel Prize in Physics for their discovery of this new type of pulsar, which opened up new possibilities for the study of gravitation.

Two American chemists, **Mario J. Molina (1943–)** and **F. Sherwood Rowland (1927–)**, show that the chlorine in man-made chemicals called chlorofluorocarbons will destroy enormous amounts of ozone in the atmosphere. Molina and Rowland were awarded the 1995 Nobel Prize in Chemistry, together with the German chemist, **Paul J. Crutzen (1933–)**, for their work in atmospheric chemistry, particularly concerning the formation and decomposition of ozone.

American astronomers **Jeremiah Ostriker (1937–)**, **P. James E. Peebles (1935–)** and **Amos Yahil (1943–)** show that galaxy masses increase linearly with radius out to a million light-years and a mass of a trillion, or 10^{12}, times that of the Sun, indicating that about 90 percent of the mass of spiral galaxies is dark and invisible, enveloping their smaller, visible stellar component. The Estonian astronomers **Jaan Einasto (1929–)**,

Ants Kaasik (–) and **Enn Saar** (–) presented dynamical evidence for such massive unseen haloes of galaxies at about the same time.

1975

The first observations are obtained from the **Very Large Array,** abbreviated **VLA,** a radio interferometer located near Socorro, New Mexico. It became fully operational in 1980.

NASA's *Viking 1* and *Viking 2* spacecraft are launched, on 20 August 1975 and 9 September 1975 respectively, each consisting of an orbiter and a lander. The orbiters imaged the entire surface of Mars at a resolution of 150 to 300 meters, showing volcanoes, lava plains, immense canyons, cratered areas, wind-formed features and evidence for catastrophic floods of water across the Martian surface in the past. The landers found no detectable organic molecules, implying that this part of the surface now contains no cells, living, dormant or dead. The landers also provided years of information about local Martian weather.

1976

American astronomers **Sandra M. Faber (1944–)** and **Robert E. Jackson (–)** discover a new method of determining the distance of an elliptical galaxy through a correlation between its intrinsic luminosity and its velocity dispersion, which is now known as the Faber-Jackson relation.

American climate scientists **James D. Hays (1933–)** and **John Imbrie (1925–)** and English geophysicist **Nicholas J. "Nick" Shackleton (1937–)** use an analysis of oxygen isotopes in deep-sea sediments to show that the timing of the major ice ages is controlled by variations in the Earth's orbit around the Sun, as suggested by the Serbian mathematician **Milutin Milankovitch (1879–1958)** in 1920 to 1941. Analysis of ice cores drilled in Antarctica and Greenland for two decades after 1985 confirmed that the major ice ages are initiated every 100 thousand years by orbital-induced changes in the intensity and distribution of sunlight arriving at Earth.

Measurements from a U-2 spy plane, under the supervision of Principal Investigator **Richard A. Muller (1944–),** conclusively demonstrate a dipolar anisotropy in the cosmic microwave background radiation, attributed to the Doppler effect of the Earth's motion through the Universe.

1977

American oceanographers, diving underwater in the deep-sea submersible vessel named Alvin, discover that living creatures such as tube worms proliferate under dark, high-pressure conditions at the bottom of the ocean, feeding on materials emerging from deep-sea volcanic vents.

American astronomer **James L. Elliot (1943–)** unexpectedly discovers the thin, widely spaced rings of Uranus, just before and after observing a star pass behind the planet.

American astrophysicist **David N. Schramm (1945–1997)** and his colleagues use computations of the synthesis of helium in the early moments of the Big Bang to limit the number of families of elementary particles in the Universe.

American astronomers **R. Brent Tully (1943–)** and **J. Richard Fisher (1935–)** discover a new method of determining distances to spiral galaxies using a correlation between their intrinsic luminosity and their velocity of rotation, as measured from the width of the 21-centimeter hydrogen line. This correlation is now known as the Tully-Fisher relation.

1978

American astronomer **James W. Christy (1938–)** notices a varying elongation on images taken to refine the orbit of Pluto, realizing that the changing orientation of the elongation implied that Pluto has a satellite and announcing the discovery on 7 July 1978. His colleague, **Robert S. Harrington (1942–1993),** calculated that Pluto and its moon would undergo a series of mutual eclipses and occultation, beginning in early 1985. The first successful observation of one of these events, on 17 February 1985, fully confirmed the existence of the satellite, which was officially named Charon. It moves around Pluto at a distance of 19,640 kilometers, once every 6.387 Earth days, implying that Pluto has a mass of only 0.2 percent (0.002) of the Earth's mass.

NASA launched the ***Einstein Observatory,*** the first fully imaging X-ray telescope, on 12 November 1978. It orbited Earth and obtained data until April 1981, with a sensitivity several hundred times greater than any X-ray mission before it. The ***Einstein Observatory*** provided the first high-resolution X-ray studies of supernova remnants, discovered unexpectedly intense coronal emission from main-sequence stars, resolved numerous X-ray sources in nearby galaxies, and detected X-ray jets from Centaurus A and M87, aligned with radio jets. The observatory is named after the German-born American physicist **Albert Einstein (1879–1955)** who revolutionized our understanding of space, matter and time.

Stephen A. Gregory (1948–) and **Laird A. Thompson (1947–)** find that nearby galaxy clusters are linked together by bridges of galaxies, and speculate that all rich clusters are located in such superclusters.

NASA's ***International Ultraviolet Explorer,*** abbreviated ***IUE,*** satellite obtained over 104,000 ultraviolet spectra of cosmic objects between 26 January 1978 and 30 September 1996.

English astrophysicist **Martin Rees (1942–)** shows how a relativistic jet, moving toward an observer at nearly the velocity of light, can seem to move at superluminal speed, and also account for the one-sided radio jets observed for some radio galaxies and quasars.

English-born American astronomer **Wallace Sargent (1935–)** and his collaborators obtain dynamical evidence for a very massive core in the giant elliptical galaxy M87, most likely a supermassive black hole containing billions of solar masses within a radius of 150 light-years.

1979

American astrophysicists **Robert H. Dicke (1916–1997)** and **P. James E. Peebles (1935–)** review various enigmas and nostrums of Big Bang cosmology, including the flatness and horizon problems.

American astronomer **Bruce Margon (1948–)** and his colleagues announce the discovery of varying wavelength changes in the optical emission lines of SS 433, with simultaneous redshifts and blueshifts. It is listed as the 433rd object in the catalogue of emission line stars created by **Charles Bruce Stephenson (1929–)** and **Nicholas Sanduleak (1933–)** in 1977. If the wavelength changes were attributed to the Doppler effect of source motion, then the star seemed to be both coming and going with speeds of up to 50,000 kilometers per second. But the star itself wasn't moving; matter was being ejected in two narrow, collimated jets of matter from a central neutron star. A bright massive companion star, weighing in at 10 to 20 solar masses is losing its outer layers in a stellar wind that flows down toward the neutron star, overloading an accretion disk that cannot digest all of the in-falling matter and therefore hurls some of it out in opposite directions along the tightly collimated jets.

Russian astrophysicist **Alexei A. Starobinsky (1948–)** proposes a new kind of cosmological model, based on quantum gravity, in which the early Universe went through a stage of exponential expansion.

NASA's *Voyager 1* and *Voyager 2* spacecraft fly by Jupiter, measuring the planet's winds and excess heat, discovering active volcanoes on the giant planet's satellite, Io, and finding a ring around Jupiter.

1980–1989

1980

American geologist **Walter Alvarez (1940–),** his physicist father **Luis Walter Alvarez (1911–1988),** and their colleagues discover a thin layer of clay that is rich in the rare element iridium, deposited across the Earth 65 million years ago when the dinosaurs disappeared. They propose that the iridum was delivered when a huge asteroid struck the Earth 65 million years ago, destroying most of the planet's life forms, including the dinosaurs.

NASA launches the *Solar Maximum Mission,* abbreviated *SMM,* satellite on 14 February 1980, with an in-orbit repair from the *Space Shuttle Challenger* on 6 April 1984 and a mission end on 17 November 1989. It excelled in X-ray and gamma ray spectroscopy of solar flares, and in coronagraph observations of coronal mass ejections, during a maximum in the 11-year solar activity cycle. Radiometric data taken from *SMM* during the decade after launch were used to show that the total solar irradiance of Earth, known as the solar constant, varies in step with the 11-year cycle of solar activity. The so-called solar constant has a total decline and rise of about 0.1 percent.

The radio interferometric **Very Large Array,** or **VLA** for short, became fully operational. Located near Socorro, New Mexico, the **VLA** consists of twenty-seven antennas, each 25 meters in diameter, which are moveable along the arms of a giant Y with antenna separations of up to 34 kilometers. Also in 1980, the **Multi-Element Radio-Linked Interferometer Network,** abbreviated **MERLIN,** came into operation in England, consisting of seven linked radio telescopes with a maximum separation of 217 kilometers.

NASA's *Voyager 1* spacecraft flies by Saturn, providing new insights to the planet's rings of water ice and showing that Saturn's largest satellite, Titan, has a dense atmosphere that is mainly composed of molecular nitrogen, with a surface pressure comparable to that of the Earth's atmosphere, whose dominant gas is also nitrogen.

Russian physicist **Yakov B. Zeldovich (1914–1987)** and his colleagues suggest that hot dark matter induces the formation of galaxies by the fragmentation of larger structures, from the top down, but computer simulations using hot dark matter produced too many large objects.

1981

NASA launches the *Dynamics Explorer 1* and *2* satellites on 3 August 1981. American physicist **Louis A. Frank (1938–)** and his colleagues use the *Dynamics Explorer 1* satellite to obtain more than a decade of images of the entire aurora oval from space.

American elementary-particle physicist **Alan H. Guth (1947–)** announces an inflationary Universe as a possible solution to vexing aspects of the observed Universe, known as the flatness and horizon problems. Guth's version of the inflation theory proposes a very rapid and extensive growth in the size of the Universe in the first moments after the Big Bang, before any matter was around, and that all the mass and energy now in the observable Universe were made during inflation from seemingly empty space. Some of the difficulties with this initial theory were overcome in 1983 when the Russian astrophysicist **Andrei D. Linde (1948–)** proposed a chaotic inflation theory.

American astronomer **Robert Kirshner (1949–)** and his colleagues find evidence for a giant void or empty volume spanning nearly 300 million light-years across, containing practically no galaxies and lying in the direction of the Böotes constellation.

NASA's launches the *Space Shuttle Columbia,* the world's first reusable manned space vehicle, on 12 April 1981.

1982

American radio astronomer **Donald C. Backer (1943–)** discovers the first millisecond radio pulsar, designated PSR 1937+21, with a period of 1.56 milliseconds, or 0.0156 seconds.

1983

NASA launches the *InfraRed Astronomical Satellite,* abbreviated *IRAS,* on 26 January 1983. It scanned most of the sky and made observations of 10,000 selected objects at infrared wavelengths, including protoplanetary disks that surround young stars. In 1983 the American astronomers **Hartmut "George" Aumann (1940–)** and **Frederick Gillett (1937–2001)** reported the *IRAS* discovery of infrared radiation from a circumstellar, protoplanetary disk around Vega.

Russian astrophysicist **Andrei D. Linde (1948–)** proposes a chaotic inflation theory in which inflation is a natural, and perhaps inevitable, consequence of chaotic conditions in the early Universe.

1984

American astronomers **Bradford A. Smith (1931–)** and **Richard J. Terrile (1951–)** detect a circumstellar, protoplanetary disk around the star Beta Pictoris, seen edge-on in reflected visible light.

1985

Twenty-seven years of continuous measurements of the ozone content above Halley Bay, Antarctica, begun by the British scientist **Gordon Miller Bourne "G. M. B." Dobson (1889–1976)** in 1957–8, conclusively show exceptionally low ozone concentrations, called the ozone hole, that forms above the South Pole in the local winter. Since ozone absorbs energetic solar ultraviolet radiation, its depletion meant that more of the dangerous ultraviolet would reach the Earth's surface, potentially causing widespread skin cancer and eye cataracts in humans.

The gang of four, two English astronomers **Carlos S. Frenk (1951–)** and **George Efstathiou (1955–)** and two American astronomers **Marc Davis (1947–)** and **Simon White (1951–)** use computer simulations to suggest that cold dark matter could assist the formation of galaxies out of smaller building blocks, from the bottom up. Although the calculations seemed to reproduce most of the available observations, too many small galaxies were produced and the simulations could not account for the subsequent discovery of huge massive structures, such as the Great Walls of galaxies, which could not be assembled in the available time.

ESA launches its *Giotto* spacecraft on 2 July 1985 on its way to encounter Comet Halley, on 13 March 1986, when instruments aboard *Giotto* obtained images of the comet's dark, irregular jet-spewing nucleus and measured the amount of water and dust emitted. The mission is named after the Florentine painter **Giotto di Bondone (c. 1267–c. 1337)**, whose frescoes in the Arena Chapel in Padua, Italy illustrate the lives of Jesus Christ and of the Virgin Mary, including the birth of Jesus with the star of Bethlehem portrayed as a comet.

Ice cores drilled in Greenland and Antarctica from 1985 to 2000 are used to determine the local temperature and atmospheric carbon dioxide and methane content in the Polar Regions for up to 420,000 years ago, confirming that the major ice ages are triggered every 100,000 years by variations in the Earth's orbit around the Sun, and showing that the climate repeatedly warms and cools when the amounts of carbon dioxide and methane increase or decrease, respectively.

1986

University of Paris graduate student **Valérie de Lapparent (1962–)** and American astronomers **Margaret Geller (1947–)** and **John Huchra (1948–)** publish a three-dimensional distribution of galaxies in different directions and depths, or redshifts, within a pie-shaped slice of the sky, showing great sheet-like walls of hundreds of thousands of galaxies distributed along the peripheries of gigantic, bubble-like voids, typically 300 million light-years across.

NASA's *Space Shuttle Challenger* explodes 73 seconds after it takes off 28 January 1986, apparently because a rocket seal failed after being degraded by freezing temperatures.

1987

The American astronomer **Alan Dressler (1948–)** and his colleagues show that all the galaxies within 100 million light-years are being pulled thorough space by "The Great Attractor" located about 150 million light-years away and with an estimated mass equivalent to about 50,000 galaxies like our Milky Way Galaxy.

Representatives of 24 nations, including the United States, sign the *Montreal Protocol on Substances That Deplete the Ozone Layer,* agreeing to a 50 percent reduction of important destroyers of the ozone in the Earth's upper atmosphere below 1986 levels by mid-1998. The treaty was subsequently strengthened by amendments that resulted in an outright ban of the ozone-destroying substances.

The supernova explosion of a blue supergiant star in the Large Magellanic Cloud is detected on 23 February 1987; it is the first supernova to be visible to the unaided, or naked, eye since 1604. Dubbed SN 1987A, the supernova was discovered by **Ian Shelten (1957–)**. The Japanese **Kamiokande II** and the American **Irvine-Michigan-Brookhaven,** abbreviated **IMB,** underground neutrino detectors recorded neutrinos released during the collapse of the stellar core, with a neutrino luminosity that briefly equaled the visible luminous output of all the stars in the Universe. Both detectors recorded the neutrinos hours before the first visible sighting of the supernova.

1989

The English computer scientist **Timothy "Tim" Berners-Lee (1955–)** implements a hypertext system for information access at the Conseil Européen pour la Recherche Nucléaire, or CERN for short, releasing it to the public and naming it the World Wide Web in 1992. It is a language and protocol for marking up documents to be sent over the Internet.

NASA launches the *COsmic Background Explorer,* abbreviated *COBE,* on 18 November 1989; instrument operations were terminated on 23 December 1993. *COBE* instruments measured the angular distribution and spectrum of the three-degree comic microwave background radiation, showing that it has a faint anisotropy and a blackbody spectrum with a temperature of 2.726 ± 0.010 degrees kelvin.

NASA launches the *Galileo* spacecraft from the *Space Shuttle Atlantis* on 18 October 1989. Galileo arrived at Jupiter on 7 December 1995, orbiting the giant planet and vastly increasing our understanding of Jupiter and the four Galilean satellites, Io, Europa, Ganymede and Callisto. The mission is named after the Italian astronomer **Galileo Galilei (1564–1642)** who discovered the four large moons of Jupiter known as the Galilean satellites. The mission was terminated on 21 September 2003.

American astronomers **Margaret Geller (1947–)** and **John Huchra (1948–)** discover the Great Wall, a sheet of galaxies more than 500 million light-years long and 200 million light-years wide, but only 15 million light-years thick.

ESA launches its *HIgh Precision PARallax COllecting Satellite,* abbreviated *HIPPAR-COS,* on 8 August 1989 and operates it until August 1993. It was the first space mission dedicated to astrometry, measuring the accurate positions, distances, motions, brightness and colors of stars. *HIPPARCOS* measured the parallax and proper motions of 120,000 stars with a precision of 0.001 to 0.002 seconds of arc, up to 100 times better than possible from the ground. The acronym also alludes to the Greek astronomer **Hipparchus of Nicea (c. 190–c. 120 BC),** who recorded accurate star positions more than two millennia previously.

American astronomers **Roger Lynds (1928–)** and **Vahé Petrosian (1938–)** obtain observations that indicate that a cluster of galaxies acts as a gravitational lens, spreading the light of remote background galaxies into an array of luminous arcs and providing information about dark matter in the cluster of galaxies.

NASA launches its *Magellan* spacecraft to Venus on 4 May 1989. It arrived at the planet on 10 August 1990 and orbited the planet for four years, using radar to map 98 percent of the planet's surface with an unprecedented detail of 100-meter vertical resolution and 50-kilometer horizontal resolution. The *Magellan* radar images revealed rugged highlands, plains smoothed out by global lava flows, ubiquitous volcanoes, and sparse, pristine impact craters with debris modified by the thick atmosphere on Venus.

NASA's *Voyager 2* spacecraft flies by Neptune, measuring its excess heat and dynamic weather, and discovering its thin, dark rings and inclined magnetic field.

1990–1999

1990

Spectral measurements from NASA's *COsmic Background Explorer,* abbreviated *COBE,* reported by American scientist **John C. Mather (1946–),** demonstrate the blackbody spectrum of the cosmic microwave background radiation with high accuracy, and establish a temperature of 2.726 ± 0.010 degrees kelvin for the radiation.

NASA launches the world's first space-based optical telescope, the *Hubble Space Telescope,* abbreviated *HST,* on 24 April 1990 from the *Space Shuttle Discovery,* but with a flawed mirror that was fixed by shuttle astronauts in December 1993. Other *HST* servicing missions by shuttle astronauts occurred in February 1997, December 1999, and February 2002. The *HST* has a 2.4-meter (94.5-inch) primary mirror. It has observed newly formed galaxies when the Universe was half its present age, contributing to our understanding of the age and evolution of the Cosmos, watched supermassive black holes consuming the material around them, and helped astronomers determine how a mysterious "dark energy" has taken over the expansion of the Universe. The telescope was named for the American astronomer **Edwin P. Hubble (1889–1953)** who demonstrated that spiral nebulae are galaxies like our own and found the first evidence for the expansion of the Universe.

NASA launches the *RÖntgen SATellite,* abbreviated *ROSAT,* an X-ray observatory developed through a cooperative program between Germany, the United States, and the United Kingdom, on 1 June 1990. The mission ended on 12 February 1999. It carried out an all-sky survey of cosmic X-rays, and obtained detailed information about the X-ray emission from comets, T Tauri stars, globular clusters, neutron stars, supernova remnants, planetary nebulae, nearby normal galaxies, active galactic nuclei, and clusters of galaxies. The mission was named after **Wilhelm Röntgen (1845–1923),** the German physicist who discovered X-rays in 1895.

NASA launches the ***Ulysses*** spacecraft, a joint undertaking of ESA and NASA, from the ***Space Shuttle Discovery*** on 6 October 1990. During an encounter with Jupiter on 8 February 1992, the giant planet's gravitational field was used to accelerate the spacecraft and hurl it above the Sun's poles, but never at a distance closer to the Sun than the Earth is. ***Ulysses*** passed over the solar south pole in September 1994 and the north pole in July 1995, and then again over the south pole in November 2000 and the north pole in December 2001. Comparisons of ***Ulysses*** observations with those from other solar spacecraft, such as ***Yohkoh*** and **SOHO,** showed that much, if not all, of the high-speed solar wind comes from open magnetic fields in polar coronal holes, at least during the minimum of the Sun's 11-year cycle of magnetic activity, and that the slow wind is narrowly confined to low latitudes near an equatorial streamer belt.

1991

NASA launches the ***Compton Gamma Ray Observatory,*** abbreviated **CGRO,** from the ***Space Shuttle Atlantis*** on 5 April 1991; **CGRO** re-entered the Earth's atmosphere on 4 June 2000. It was the second of NASA's four great observatories in space, designed to study the Universe in an invisible, high-energy form of radiation known as gamma rays. It detected more than 2,600 gamma-ray bursts, as well as gamma rays from black holes, pulsars, quasars and supernovae. The gamma ray observatory was named after the American physicist **Arthur H. Compton (1892–1962).**

The Japanese Institute of Space and Astronautical Science, abbreviated ISAS, launched its ***Solar-A*** spacecraft on 30 August 1991, renaming it ***Yohkoh,*** which means "sunbeam" in English. ***Yohkoh*** collected high-energy radiation from the Sun at soft X-ray and hard X-ray wavelengths, with high angular and spectral resolution, for more than a decade, providing new insights to the mechanisms of solar flare energy release by the magnetic reconnection of coronal loops, and the heating of the million-degree solar corona. It was a collaborative effort of Japan, the United States and the United Kingdom.

1992

Measurements from NASA's ***COsmic Background Explorer,*** abbreviated **COBE,** reported by **George Smoot (1945–)** and his colleagues, reveal very small temperature fluctuations in the three-degree cosmic microwave background radiation.

Austrian-born American astronomer **Thomas Gold (1920–2004)** proposes that early life began in the deep, hot biosphere, kilometers below the Earth's surface, and that such life fed on chemical sources and was not dependent on solar energy or photosynthesis for its primary energy supply.

American astronomers **Jane X. Luu (1963–)** and **David C. Jewitt (1958–)** discover a small body revolving around the Sun beyond the orbit of Neptune, using an electronic detector with the University of Hawaii's 2.2-meter (87-inch) reflector on Mauna Kea. It was the first of several such objects that have now been found within a flat, distant ring at the outer edge of the Solar System, named the Kuiper belt in recognition of its prediction by the Dutch-born American astronomer **Gerard P. Kuiper (1905–1973)** in 1951.

American astronomers **Charles A. Meegan (1944–)** and **Gerald J. Fishman (1943–)** announce that the numerous gamma-ray bursts detected from NASA's *Compton Gamma Ray Observatory,* abbreviated *CGRO,* are equally distributed in all directions of the sky. The lack of any concentration of the burst sources in the Milky Way suggests that gamma-ray bursts lie at remote distances outside our Galaxy.

American radio astronomers **Aleksander Wolszczan (1946–)** and **Dale A. Frail (1961–)** discover two Earth-sized planets orbiting the pulsar PSR 1257+12.

1993

The first **Keck telescope,** with a mirror 10-meters (396-inches) in diameter, begins operation in May 1993, on Mauna Kea in Hawaii. The second **Keck telescope,** a nearby twin of the first one, was inaugurated in October 1996.

The crew of the *Space Shuttle Endeavor* outfits the $1.6 billion *Hubble Space Telescope* with corrective lenses to fix an embarrassing flaw in its mirror.

The **Very Long Baseline Array,** abbreviated **VLBA,** is completed. It consists of ten radio telescopes spread across the United States with a maximum baseline of 8,000 kilometers.

1994

The *Clementine* spacecraft, a joint project of the Strategic Defense Initiative and NASA, was launched on 25 January 1994, entering lunar orbit on 21 February 1994. Instruments aboard *Clementine* mapped the global surface composition and topography of the Moon. The spacecraft was named after the miner's darling daughter in the old Gold Rush ballad.

Instruments aboard the *Hubble Space Telescope* obtain confirming evidence for a supermassive black hole at the center of the giant elliptical galaxy M87.

Fragments of Comet Shoemaker-Levy-9, discovered by American astronomers **Eugene M. Shoemaker (1928–1997),** his wife **Carolyn Shoemaker (1929–),** and **David H. Levy (1948–),** collide with the planet Jupiter between 16 and 22 July 1994.

1995

American astronomer **Arthur F. Davidsen (1944–2001)** announces the discovery of intergalactic absorption of ultraviolet quasar radiation by singly ionized helium, designated He II, at a distance of about 10 billion light-years. This confirmed that helium was synthesized in the first moments of the Big Bang, in amounts comparable to those found in our Galaxy today.

Images of quasars taken with the *Hubble Space Telescope* confirm that quasars are located in the cores of large galaxies.

Swiss astronomers **Michel Mayor (1942–)** and **Didier Queloz (1966–)** discover the first planet around a Sun-like star; it is a Jupiter-sized planet in close orbit around the star 51 Pegasi.

American astronomer **C. Robert O'Dell (1937–)** and his colleagues publish panoramic, high-resolution images of the Orion Nebula using the *Hubble Space Telescope,* detecting flattened disks of dust swirling around many young stars in the nebula.

NASA launches the *Rossi X-ray Timing Explorer,* abbreviated *RXTE,* on 30 December 1995, with about a decade of subsequent observations of variable cosmic X-ray sources at high time resolution, providing new insights to black holes, gamma ray bursts, neutron stars and pulsars. The spacecraft is named for Italian-born, American physicist **Bruno B. Rossi (1905–1993),** a pioneer in studies of cosmic rays, X-ray astronomy and space plasma physics.

NASA launches the *SOlar and Heliospheric Observatory,* abbreviated *SOHO,* on 2 December 1995 with operations continuing into 2006. It is a cooperative project between ESA and NASA to study the Sun. *SOHO* has obtained important new information about the structure and dynamics of the solar interior, the heating mechanism of the Sun's million-degree outer atmosphere, the solar corona, and the origin and acceleration of the solar wind.

1996

American astronomers **Geoffrey W. Marcy (1955–)** and **R. Paul Butler (1962–)** announce the discovery of Jupiter-sized planets in close orbits about the Sun-like stars 70 Virginis and 47 Ursae Majoris.

NASA launches its *Mars Global Surveyor* on 7 November 1996, placing it into orbit around the red planet on 12 September 1997. High-resolution images, beginning on 4 April 1999, recorded abundant evidence for past and recent water flow, dust storms and volcanic activity on Mars. Global topography was determined with a laser altimeter, and a magnetometer established the global distribution of remnant magnetic fields preserved in the ancient highland crust.

NASA launches the *Mars Pathfinder* lander and its roving vehicle, *Sojourner,* on 4 December 1996. They landed on Chryse Planitia, near the mouth of an outflow channel, on 4 July 1997, finding evidence that liquid water once flowed there. This suggested that the planet's atmosphere was once thicker, warmer and wetter than at present, but perhaps several billion years ago.

1997

The Italian-Dutch *BeppoSAX* satellite is used to pinpoint the exact location of a gamma-ray burst emitted on February 28, 1997 and designated GRB 970228, enabling an international team led by the Dutch astronomer **Jan van Paradijs (1946–1999)** to detect the visible-light component using the 4.2-meter (165-inch) **William Herschel telescope** at La Palma in the Canary islands. An underlying host galaxy came into view as the afterglow faded. In the same year, the 10-meter (396-inch) **Keck II telescope** was used to obtain the optical spectra and redshifts of the host galaxies of two other gamma-ray bursts. A redshift obtained by American astronomers **Shrinivas R. Kulkarni (1956–)**, **S. George Djorgovski (1956–)**, and their colleagues indicated that one gamma-ray burst, designated GRB 971214, is located at a distance of 12 billion light-years, moving at almost the velocity of light near the apparent edge of the observable Universe. For just one second it emitted about as much gamma-ray energy as the visible-light output of all the stars of all the galaxies of the visible Universe.

NASA launches the *Cassini-Huygens* spacecraft on 15 October 1997. Instruments aboard the spacecraft, which reached Saturn in July 2004, improved our understanding of the planet Saturn, its rings, its magnetosphere, its moon Titan and its other moons. NASA's *Cassini Orbiter* will orbit Saturn and its moons for four years. ESA's *Huygens Probe* was parachuted into the thick, hazy atmosphere of Saturn's satellite Titan, landing on its surface on 14 January 2005. It revealed a hilly terrain riddled with channels or riverbeds carved by a liquid even though the surface is more than 200 degrees below the freezing point of water. The *Cassini Orbiter* is named after **Gian (Giovanni) Domenico (Jean Dominique) Cassini (1625–1712),** the Italian-born, French astronomer who in 1675 was the first to distinguish two zones within what was thought to be a single ring around Saturn; it is divided by a dark gap now known as Cassini's division. The *Huygens Probe* is named after **Christiaan Huygens (1629–1695),** a Dutch astronomer who in 1655 discovered Titan, Saturn's largest moon.

English astrophysicist **Martin Rees (1942–)** and American astrophysicist **Peter Mészáros (1943–)** propose a relativistic, expanding fireball model for gamma-ray bursts, the American astronomer **Stanford E. "Stan" Woosley (1944–)** proposes that the bursts occur during the collapse of a massive, rotating giant star into a black hole, and **Bohdan Paczynski (1940–)**, **David Eichler (1951–)** and **Tsvi Piran (1949–)** independently speculate that the gamma-ray bursts occur when two neutron stars merge.

1998

Launched on 24 October 1998 and retired on 18 December 2001, the *Deep Space 1* spacecraft flight tested an ion engine and other advanced technologies, and was redirected to an encounter with Comet Borrelly on 22 September 2001, taking images of the comet nucleus and showing that it is an irregular chunk of ice and rock covered by dark slag that reflects only about four percent of the incident sunlight.

NASA launches the *Lunar Prospector* spacecraft on 6 January 1998. It orbited the Moon from 15 January 1998 to 31 July 1999, determining the global distribution of elements and magnetic fields on the lunar surface, measuring the detailed gravity of the Moon, and detecting the lunar core.

The Supernova Cosmology Project, headed by American physicist **Saul Perlmutter (1959–)**, and the High-Z Supernova Search, involving Australian astronomer **Brian P. Schmidt (1967–)**, American astronomers **Adam G. Riess (1969–)**, **Robert P. Kirshner**

(1949–) and others, report that the Universe is accelerating. Both groups observed type Ia supernovae located at different distances and measured their velocities of recession, announcing that the older galaxies are receding faster than predicted by a uniform expansion of the Universe. Some sort of mysterious dark energy is counteracting the pull of gravity and speeding the expansion up as time goes on.

Measurements with the Japanese **Super-Kamiokande** neutrino detector indicate that atmospheric neutrinos change form, or oscillate, from muon neutrinos into tau neutrinos, indicating that neutrinos have a very small mass.

The **Transition Region And Coronal Explorer,** abbreviated **TRACE,** spacecraft was launched in April 1998. It provided images of solar plasma at temperatures from 10 thousand to 10 million degrees kelvin with high temporal resolution and second-of-arc angular resolution using observations at ultraviolet and extreme ultraviolet wavelengths. **TRACE** showed that the million-degree corona is comprised of thin magnetic loops that are naturally dynamic and continuously evolving.

1999

NASA deploys the **Chandra X-ray Observatory,** abbreviated **CXO,** on 23 July 1999, from the **Space Shuttle Columbia.** The CXO studies the high-energy, X-ray radiation from objects such as active galactic nuclei, black holes, clusters of galaxies, dark matter, galaxies, neutron stars, pulsars, quasars, supernova remnants, supernovae, and white dwarfs. The X-ray observatory was named after the Indian-born American astrophysicist **Subrahmanyan Chandrasekhar (1910–1995).**

More than 100 nations sign the *Kyoto Protocol,* agreeing to reduce the emissions of heat-trapping gases that can warm the planet. The treaty was ratified by Russia in 2004, but has yet to be agreed to by the United States and many other industrial nations.

NASA launches the **Stardust** spacecraft on 7 February 1999. After nearly four years of space travel, it encountered Comet Wild 2, on 02 January 2004, where it obtained close-up images of the comet nucleus and collected dust and carbon-based samples using a substance called aerogel. The cargo was parachuted in a re-entry capsule to the Earth's surface on 15 January 2006.

ESA launches its **X-ray Multi-Mirror – Newton,** abbreviated **XMM–Newton,** spacecraft on 10 December 1999, with a designed 10-year lifetime. It provides X-ray images and spectroscopy of selected X-ray objects, such as accreting black holes, active galactic nuclei, clusters of galaxies, and supernovae. The **XMM–Newton** is named after the English mathematician and astronomer **Isaac Newton (1643–1727)** who invented spectroscopy and is best remembered for his laws of motion and gravity.

2000–2010

2000

NASA launches its **Near Earth Asteroid Rendezvous,** abbreviated **NEAR,** mission on 17 February 1996; it arrived at the near-Earth asteroid 433 Eros on Valentine's Day

14 February 2000, becoming the first spacecraft to orbit an asteroid, and the first to land on one. The spacecraft was renamed *NEAR Shoemaker* after launch, in honor of the astronomer-geologist **Eugene M. Shoemaker (1928–1997)**, a pioneering expert on asteroid and comet impacts. *NEAR Shoemaker* circled the asteroid 433 Eros for a year, determining its detailed size, shape, mass and mass density, and showing that it has a battered, homogeneous structure.

2001

American astronomer **Wendy L. Freedman (1957–)** and her colleagues publish observations of Cepheid variable stars made with the *Hubble Space Telescope* to directly measure galaxy distances out to 65 million light-years, combining their results with other observations, such as those of type Ia supernovae, to establish a refined Hubble diagram out to a billion light-years and infer a Hubble constant of 72 kilometers per second per Megaparsec and an age of about 14 billion years for the expanding Universe.

Canadian physicist **Arthur "Art" McDonald (1943–)** announces that observations with the underground **Sudbury Neutrino Observatory,** abbreviated **SNO,** demonstrate that solar electron neutrinos change to muon- or tau-type neutrinos before reaching Earth, indicating that neutrinos have mass. These and subsequent **SNO** observations show that the total number of electron neutrinos produced by nuclear reactions in the core of the Sun is equal to that predicted by detailed solar models, but that roughly two-thirds of them change type in transit to Earth.

NASA launches its *2001 Mars Odyssey* spacecraft on 7 April 2001, placing it into orbit around the red planet on 24 October 2001. It mapped the amount and distribution of chemical elements and minerals on the Martian surface, and provided evidence for substantial subsurface water ice.

NASA launches its *Wilkinson Microwave Anisotropy Probe,* abbreviated *WMAP,* on 30 June 2001, publishing definitive full sky maps of the faint anisotropy, or temperature variations, of the three-degree cosmic microwave background radiation in early 2003. When combined with the results of other cosmic measurements, the *WMAP* data suggest that the Universe is 13.7 billion years old, that the first stars began to shine about 200 million years after the Big Bang, that the radiation in the *WMAP* images originated 379,000 years after the Big Bang, that the Hubble constant has a value of 71 kilometers per second per Megaparsec, and that the Universe is composed of 4.4 percent atoms, 22.6 percent cold dark matter and 73 percent dark energy. The probe is named in honor of **David T. Wilkinson (1935–2002)** of Princeton University, a pioneer in cosmic microwave background research and the *MAP* instrument scientist.

2002

ESA launches its *INTErnational Gamma Ray Astrophysics Laboratory,* abbreviated *INTEGRAL,* on 17 October 2002, dedicated to the detailed spectroscopy and imaging of celestial gamma ray sources.

NASA launches the ***Ramaty High Energy Solar Spectroscopic Imager,*** abbreviated ***RHESSI,*** on 5 February 2002. It obtained high-resolution imaging spectroscopy of solar flares at hard X-ray and gamma ray wavelengths, in order to study solar flare particle acceleration and flare energy release. Instruments aboard ***RHESSI*** determined the frequency, location and evolution of impulsive flare energy release in the corona and located the sites of particle acceleration and energy deposition at all phases of solar flares. The spacecraft was named after **Reuven Ramaty (1937–2001),** a pioneer in the fields of solar physics, gamma ray astronomy, nuclear physics and cosmic rays.

The **Two Degree Field,** abbreviated **2dF, Galaxy Redshift Survey,** is used to demonstrate that invisible dark matter is located in the same place as visible matter, and that there is no bias in the distribution of dark matter.

The importance of the observed solar neutrino deficit was recognized by the award of the 2002 Nobel Prize in Physics to **Raymond Davis Jr. (1914–),** who pioneered the field of neutrino astrophysics, and to the Japanese physicist **Masatoshi Koshiba (1926–)** for the discovery of cosmic neutrinos, from both the Sun and the distant exploding star, or supernovae, SN 1987A. Another astrophysicist, the Italian-born American **Riccardo Giacconi (1931–)** shared the 2002 prize for his pioneering investigations of cosmic X-rays. And since the prize cannot be given to more than three scientists, other significant contributors to neutrino astronomy or the theory of neutrinos could not be recognized.

2003

NASA launches the ***GALaxy Evolution eXplorer,*** abbreviated ***GALEX,*** on 28 April 2003. It is designed to observe galaxies in ultraviolet light across 10 billion years of cosmic history, providing information on star formation and the evolution of galaxies.

NASA launches its twin ***Mars Exploration Rovers,*** named ***Spirit*** and ***Opportunity,*** on 10 June and 7 July 2003, respectively. ***Spirit*** landed in Gusev Crater on 4 January 2004 and ***Opportunity*** landed in Meridiani Planum on 25 January 2004, obtaining evidence that liquid water once flowed across Mars.

NASA's ***Space Shuttle Columbia*** breaks apart upon re-entry on 1 February 2003.

NASA launches the ***Spitzer Space Telescope,*** abbreviated ***SST,*** on 25 August 2003. It is an infrared telescope with an 0.85-meter (33.5-inch) primary mirror, obtaining images and spectra of the infrared energy, or heat, radiated by cosmic objects at wavelengths between 3 and 180 microns, where 1 micron is one-millionth of a meter, providing information on the formation, composition and evolution of planets, stars and galaxies. The mission was initially designated the ***Space Infra-Red Telescope Facility,*** or ***SIRTF*** for short, and renamed for the American astrophysicist **Lyman Spitzer, Jr. (1914–1997).**

Measurements from NASA's ***Wilkinson Microwave Anisotropy Probe,*** abbreviated ***WMAP,*** reported by **Charles L. "Chuck" Bennett (1956–),** conclusively determine the distribution and characteristic sizes of temperature fluctuations in the three-degree cosmic microwave background radiation. The angular power spectrum of this data indicates that ordinary baryonic matter constitutes about 4.4 percent of the critical mass density needed to stop the expansion of the Universe, and when combined with other astronomical measurements indicates a Hubble constant of 71 kilometers per second per Megaparsec, an age of 13.7 billion years for the expanding Universe, and a "flat" space described by Euclidean geometry, with 22.6 percent dark non-baryonic matter density and 73 percent dark energy density. The combined data also indicate that the first stars in the Universe began to shine when the expanding Universe was only about 200 million years old.

2004

The *Cassini-Huygens* spacecraft, launched by NASA on 15 October 1997, reaches Saturn in July 2004, and releases ESA's *Huygens probe* on 24 December 2004, beginning a three-week journey to the surface of Saturn's moon Titan.

NASA launches *Gravity Probe B* on 20 April 2004, a gyroscope experiment designed for precise tests of the *General Theory of Relativity*.

English theoretical astronomer **Stephen W. Hawking (1942–)** concludes that all the matter and energy consumed by a black hole is eventually sent back into the place they came from, but in a transformed unrecognizable state.

The first results from the **Sloan Digital Sky Survey** of galaxies in different directions and depths, or redshifts, reveal a wall of galaxies 1.37 billion light-years across, together with huge voids, or seemingly empty places with few galaxies.

2005

On 14 January 2005, ESA's *Huygens probe,* released from NASA's *Cassini orbiter,* lands on the surface of Saturn's satellite Titan. Although the surface is more than 200 degrees below the freezing point of water, it shows signs of river channels and flowing liquid, probably methane.

2006

On 15 January 2006, NASA's *Stardust* spacecraft sucessfully returned to Earth samples of Comet Wild 2, via a sample return capsule.

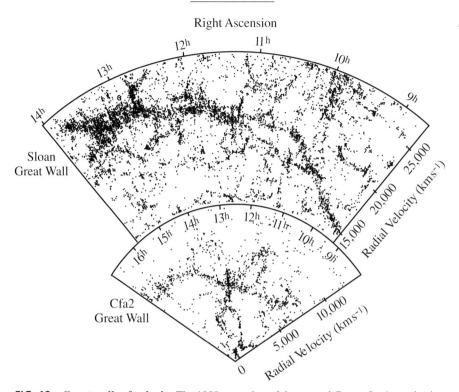

FIG. 19   Great walls of galaxies The 1989 extension of the second Center for Astrophysics survey, designated the CfA2 survey, revealed a Great Wall of galaxies (*bottom*) that stretches about 500 million light-years across the northern sky, between 8 hours and 17 hours right ascension at radial velocities between 5,000 and 10,000 kilometers per second, abbreviated km s[-1]. In 2004, the Sloan Digital Sky Survey extended the northern hemisphere survey to radial velocities of up to 28,000 km s[-1] and an estimated distance of 1.5 billion light-years, discovering a more remote and larger Sloan Great Wall of galaxies that is 1.37 billion light-years long (*top*). It is the largest structure yet observed in the Universe. The Earth is at the apex of both plots, and the radial coordinate, or depth, is the radial velocity, determined from the redshift of spectral lines. Each point on the CfA2 map corresponds to one of 1,732 galaxies, while each dot in the Sloan map corresponds to one of 11,243 galaxies.

GLOSSARY

with Data Tables

Aberration: A tilting of the apparent position of a star in the direction of the motion of the observer. The size of the effect depends on the velocity of the Earth, the velocity of light, and the angle between the direction of observation and the direction of motion, with a maximum value of 20.47 seconds of arc. The English astronomer **James Bradley (1693–1762)** discovered stellar aberration in 1728, proving that the Earth is not fixed in space, but moves through it, while also showing that light travels at a finite velocity, which he estimated as 308,300 kilometers per second using his observed value of aberration.

Absolute luminosity: The total radiant energy of a celestial object emitted per unit time, and a measure of its intrinsic brightness. Denoted by L, the amount of energy radiated per unit time, measured in units of watts. One Joule per second is equal to one watt of power, and to ten million erg per second. The absolute luminosity of the thermal radiation from a blackbody is given by the Stefan-Boltzmann law in which $L = 4\pi\sigma R^2 T_{eff}^4$, where $\pi = 3.14159$, the Stefan-Boltzmann constant, $\sigma = 5.67051 \times 10^{-8}$ Joule m^{-2} $^{\circ}$K^{-4} s^{-1}, the radius is R and the effective temperature is T_{eff}. *See* apparent luminosity, brightness, luminosity and Stefan-Boltzmann law.

Absolute magnitude: The absolute magnitude of a celestial object is the apparent magnitude the object would have at a standard distance of 10 parsecs from the Earth. The absolute magnitude is denoted by the symbol M, and the apparent magnitude by m. The absolute magnitude may be derived from the apparent magnitude and the parallax by the formula $M = m + 5 + 5 \log \pi$, where π is the annual parallax in seconds of arc. The annual parallax is the reciprocal of the distance, D, in units of parsecs, so $M = m + 5 - 5 \log D$, with D in parsecs. The absolute visual magnitude of the Sun is +4.82 magnitudes and its apparent visual magnitude is −26.72 magnitudes. The absolute magnitudes of most stars lie between −5 and + 15. *See* annual parallax, magnitude, parallax, stars, and visual magnitude.

Absolute temperature: The temperature as measured on a scale whose zero point is absolute zero, the point at which all motion at the molecular level ceases. The unit of absolute temperature is the degree kelvin, denoted by $^{\circ}$K or just K and writted kelvin without a capital K. The freezing temperature of water is 273°K and the boiling temperature of water is 373°K. Absolute zero is 0°K. The equivalent temperature in degrees Celsius is $^{\circ}C = {}^{\circ}K − 273$, and the equivalent temperature in degrees Fahrenheit is $^{\circ}F = (9^{\circ}K/5) − 459.4$. *See* Celsius, Fahrenheit and Kelvin.

Absolute zero: The lowest possible temperature, −273 degrees Celsius and 0 degrees on the Kelvin scale. *See* absolute temperature.

Absorption: The process by which the intensity of radiation decreases as it passes through a material medium. The energy lost by the radiation is transferred to the medium.

Absorption line: A "dark" line of decreased radiation intensity at a particular wavelength of the electromagnetic spectrum, produced when radiation from a distant source passes through a gas cloud closer to the observer. An absorption line can be formed when a cool, tenuous gas, between a hot, radiating source and an observer, absorbs electromagnetic radiation of that wavelength, as in the spectra of stars whose hot internal radiation is absorbed by the cooler, outer stellar layers. This spectral feature looks like a line when the radiation intensity is displayed as a function of wavelength; such a display is called a spectrum. Different atoms, ions and molecules produce characteristic patterns of absorption lines, and observations of these lines enable identification of the chemical ingredients of the gas between the light source and the observer. Absorption lines can also be used to determine the velocity of the source, using the Doppler effect. The absorption lines in the Sun's spectrum are often called Fraunhofer lines. *See* Doppler effect, Fraunhofer line, spectroheliograph and spectrum.

Acceleration: An increase in an object's speed and velocity over time.

Accretion: The formation of a large body as the result of collisions of small particles or the gravitational attraction of smaller bodies. The capture of gas or dust by a single body to form a larger one, like a moon, planet or star. Also the process by which a star accumulates matter as it moves through a cloud of interstellar gas.

Accretion disk: A flattened, rotating disk of gas in orbit around a black hole or a young star. The material in an accretion disk surrounding a black hole spirals into the black hole, emitting X-rays on the way in; the gas can be replenished by material pulled from a visible stellar companion. Planets may be formed in the accretion disk of a young star. These disks contain dust that is detected at infrared wavelengths.

Achernar: The ninth brightest star, located in the constellation Eridanus, and therefore also known as Alpha Eridani. Achernar is a blue main-sequence, or dwarf, star of spectral type B3 with an apparent visual magnitude of +0.45, a distance of 142 light-years, an absolute visual magnitude of −2.7, and a luminosity of about 1,150 times that of the Sun. *See* B star, stars and stellar classification.

Acrux: The brightest of the stars in the constellation Crux, popularly known as the Southern Cross. Acrux, alternately named Alpha Crucis, is located at a distance of 320 light-years. It is a visual double star with a combined apparent visual magnitude of 0.77 and a combined absolute visual magnitude of −4.19. One member of the double star system is a blue B0.5 subgiant with an absolute luminosity of 25,000 times that of the Sun, and the other member is a blue B1 dwarf star with an absolute luminosity of 16,000 times the solar value. *See* Crux.

Active galactic nucleus: A very bright, compact region found at the center of an active galaxy. The brightness of an active galactic nucleus is attributed to an accretion disk around a supermassive black hole. Quasars are the most luminous type of active galactic nucleus. *See* accretion disk, black hole, quasar, radio galaxy, Seyfert galaxy, and supermassive black hole.

Active region: A region in the solar atmosphere, from the photosphere to the corona, that develops when strong magnetic fields emerge from inside the Sun. The magnetized realm in, around and above sunspots is called an active region. Radiation from active regions is enhanced, when compared to neighboring areas in the chromosphere and corona, over the whole electromagnetic spectrum, from X-rays to radio waves. Active regions may last from several hours to a few months. They are the sites of intense explosions, called solar flares, which last a few minutes to hours. The number of active regions varies in step with the 11-year sunspot cycle. *See* flare, solar activity cycle, and sunspot cycle.

Adaptive optics: A method for obtaining sharp visual images of cosmic objects by correcting for the distortion of light as it travels through the Earth's atmosphere. This can be accomplished by rapidly adjusting the shape and alignment of a telescope mirror.

Advanced X-ray astrophysics facility: *See Chandra X-ray Observatory.*

Aether: An invisible medium that was thought to diffuse all space, providing something for light to propagate in. Also, the fifth, unchanging heavenly element in Aristotle's cosmology; the other four are earth, air, fire and water, which all change with time. *See* ether and luminiferous ether.

Afterglow: The fading fireball of a gamma-ray burst that is observable at less energetic, longer wavelengths, such as X-rays, optical radiation and radio radiation. After an initial, brief explosion, a gamma-ray burst generates the afterglow, which is visible for several weeks or months.

Albedo: A measure of the fraction of sunlight reflected by a body in the Solar System. Rocky objects have a low albedo, while clouds or water ice have a high albedo.

Aldebaran: Also known as Alpha Tauri, the brightest star in the constellation Taurus. Aldebaran is an orange K5 giant star that is 65 light-years away from the Earth, with an apparent visual magnitude of about 0.87, an absolute visual magnitude of -0.6, and an absolute luminosity about 160 times that of the Sun. *See* stars, and K star.

Algol: The first variable star to be discovered, and the brightest eclipsing binary star. It is located in the constellation Perseus, and also known as either the "Demon Star" or Beta Persei. The first recorded observations of Algol's variability were made in 1669 by the Italian astronomer **Geminiano Montanari (1633–1687)**, and in 1782 the English astronomer **John Goodricke (1764–1786)** first suggested that the variability resulted from a binary star system in which one of its stars passes in front of the other, brighter component. The pair of stars is located at a distance of 93 light-years with a combined apparent visual magnitude of 2.12 and absolute visual magnitude of –0.10. Every 2.867 days one star passes in front of the other, and eclipses it, and light that we observe from the binary star system dims to 30 percent of its original brightness. Algol is a prototype for a class of eclipsing binaries known as Algol-type variables. *See* eclipsing binary.

Almagest: A comprehensive collection of astronomical works by the Egyptian astronomer **Ptolemy (c. 87–c. 165)**, or **Claudius Ptolemaeus**, which he called *Mathematical Syntaxis;* awed Arabic translators of the ninth century called it *The Greatest Composition – Almageste,* or, as it became known, the *Almagest.* Ptolemy presented an Earth-centered, or geocentric, description of the motions of the planets that succeeded so well that it was still being used to predict the locations of the planets in the sky more than a thousand years after Ptolemy's death. The *Almagest* also included the celestial latitude and longitude and brightness of nearly 850 stars obtained much earlier by the Greek astronomer **Hipparchus (c. 190–120 BC)**. In 1718 the English astronomer **Edmond Halley (1656–1742)** compared his observed positions of bright stars with those found in the *Almagest,* concluding that three of them had changed positions by between 15 and 20 minutes of arc since Hipparchus' time, and were apparently moving. *See* proper motion.

Alpha Centauri: The Alpha Centauri system consists of three co-orbiting dwarf, or main sequence, stars – two bright and one faint. The two brightest stars, designated Alpha Centauri A and Alpha Centauri B, are located at a distance of 4.39 light-years, in the southern sky, with apparent visual magnitudes of –0.01 and 1.33 and absolute visual magnitudes of 4.34 and 5.68. The brightest member of the pair is Rigil Kentaurus, the "Centaur's Foot"; this star is also sometimes called Alpha Centauri A. It is the brightest star in the constellation Centaurus and the third brightest star in the night sky. Like the Sun, Alpha Centauri is a G2 dwarf, or main sequence, star. It appears bright in comparison with other stars because it is relatively near; many other exceptionally bright stars are farther away but shine with greater luminosity. The closest star to our Solar System is the faint red dwarf component of the Alpha Centauri system, known as Proxima Centauri, with a parallax of 0.772 seconds of arc and a distance of 4.22 light-years from the Earth, or about 271,000 times more distant than the Sun. *See* annual parallax, parallax, Proxima Centauri, and stars.

Alpha decay: The disintegration of an atomic nucleus, in which an alpha particle is emitted, carrying away two protons and two neutrons from the nucleus. *See* alpha particle.

Alpha particle: The nucleus of a helium atom, consisting of two protons and two neutrons. The helium nucleus has a charge twice that of a proton and a mass just 0.007, or 0.7 percent, less than the mass of four protons. The alpha particle is a helium ion. *See* helium and ion.

Altair: Also known as Alpha Aquilae, the brightest star in the constellation Aquila, with an apparent visual magnitude of 0.76, a distance of 16.8 light-years, an absolute visual magnitude of $+2.20$, and an intrinsic luminosity of 11 times that of the Sun. Altair is a dwarf, or main sequence, star of spectral type A7. *See* A star and stellar classification.

Altitude: The angular distance of a celestial object above the observer's horizon measured along the great circle passing through the object and the zenith.

Ammonia: An interstellar molecule and one of the three main "ices" that existed in the outer Solar System during its formation. One molecule of ammonia, designated NH_3, consists of one atom of nitrogen, N, and three atoms of hydrogen, H.

Amor asteroid: *See* near-Earth asteroid.

Andromeda: A large constellation of the Northern Hemisphere, adjoining the Square of Pegasus.

Andromeda galaxy: The nearest spiral galaxy, at a distance of 2.4 million light-years from Earth, also known as M31 or NGC 224 and the Andromeda Nebula. It is similar in shape and size to our own Milky Way Galaxy. The Andromeda galaxy is the most distant object visible to the unaided eye, seen from the Northern Hemisphere as a fuzzy, luminous patch in the constellation of Andromeda, and is about 1 by 4 degrees in angular extent. Two elliptical galaxies, M32 and NGC 205, accompany the Andromeda galaxy, and they all share membership with our own Galaxy in the Local Group. The Andromeda galaxy is currently approaching our Galaxy, the Milky Way. *See* galaxy, local group and spiral galaxy.

Anglo-Australian Observatory: Located at Siding Spring Mountain in New South Wales, an observatory jointly operated by the United Kingdom and the Australian governments. It contains a 3.9-meter (150-inch) Anglo-Australian Telescope and the 1.2-meter (48-inch) UK Schmidt telescope.

Ångström: A unit of wavelength designated by the symbol Å, and equal to 10^{-10} meters, or 0.00000001 of a centimeter and 0.1 nanometers. An Ångström is on the order of the size of an atom. Blue light has a wavelength of about 4400 Å, yellow light 5500 Å, and red light 6500 Å. The unit is named after the Swedish astronomer **Anders Jonas Ångström (1814–1874)** who in 1868 published *Researches of the Solar System*, in which he presented measurements of the wavelengths of more than a hundred dark absorption lines in the Sun's spectrum using this unit of measurement.

Angular momentum: The product of angular velocity and mass for an object in rotation.

Angular resolution: *See* resolution.

Anisotropy: The characteristic or quality of being dependent upon direction. *See* isotropic.

Annual parallax: Half the apparent angular displacement of a nearby star observed against the more distant stars at intervals of six months from opposite sides of the Earth's orbit. The annual parallax is also known as the trigonometric parallax. The distance of a nearby star in parsecs is given by the reciprocal of its annual parallax in seconds of arc, or arc seconds. The first star to have its distance measured by this method was 61 Cygni, in 1838 by the German astronomer **Friedrich Wilhelm Bessel (1784–1846)**. Accurate parallax measurements of 120,000 stars have been determined with a precision of 0.001 seconds of arc using instruments aboard the *HIPPARCOS* satellite. *See* Alpha Centauri, Barnard's Star, *HIPPARCOS*, Kapteyn's Star, parallax, parsec, Proxima Centauri, and 61 Cygni.

Antares: Also known as Alpha Scorpii, the brightest star in the constellation Scorpius. Antares is a bright, red M1.5 supergiant star, located at a distance of 604 light-years with an apparent visual magnitude of 0.90 and an absolute visual magnitude of –5.30, which is intrinsically 12,000 times brighter than the Sun at visible wavelengths. The star is about the same color and apparent brightness as Mars, resulting in the name *Ant Ares*, Greek for "like Mars". From the measured angular diameter and distance, a radius of over 800 times that of the Sun is obtained, which is almost 4 AU, or almost 4 times the mean distance between the Earth and the Sun.

Antennae: A pair of interacting galaxies, also known as NGC 4038 and NGC 4039, in the constellation Corvus.

Anthropic principle: First proposed in 1970 by the British physicist **Brandon Carter (1942–)**, working in France, the Anthropic Principle states that the Universe we observe is necessarily compatible with our existence as observers. In other words, the Universe we inhabit has properties that have made it possible for life to develop.

Antielectron: The antiparticle of an electron, also called a positron. *See* antimatter, antiparticle, and positron.

Antimatter: Matter made of particles with identical mass and spin as those of ordinary matter, but with opposite charge. The antimatter particle of the electron is the positron, or "positive electron". Antimatter has been produced by cosmic rays striking atoms in the Earth's atmosphere, in terrestrial particle accelerators, and during flares on the Sun. When a particle of antimatter collides with the corresponding particle of ordinary matter they annihilate each other, producing energetic radiation called gamma rays. *See* antiparticle, and positron.

Antineutrino: The antiparticle of a neutrino. *See* antimatter, antiparticle, and neutrino.

Antiparticle: A particle equal in mass and most other properties to an elementary particle, but opposite in charge. The antiparticle of an electron is the positron or "positive electron", and the antiparticle of the neutrino is an antineutrino. When a particle meets with its antiparticle, they annihilate each other, producing energetic radiation called gamma rays. *See* antimatter and positron.

Aperture: The diameter of the light-gathering lens or mirror of a telescope, a measure of its light-gathering ability.

Aperture synthesis: A technique, pioneered by the English radio astronomer **Martin Ryle (1918–1984)**, in which an array of radio dishes is connected electronically and used with the Earth's rotation to synthesize the resolution of a much larger telescope. *See* Australia Telescope, MERLIN, Mullard Radio Astronomy Observatory, National Radio Astronomy Observatory, Very Large Array, Very Long Baseline Array, and Westerbork Synthesis Radio Telescope.

Apollo asteroid: *See* near-Earth asteroid.

Apollo Missions: Six *Apollo* spacecraft transported twelve humans to the Moon and back between 1969 and 1972. They left seismometers to measure the lunar interior and laser corner reflectors used to determine the Moon's increasing distance from Earth, and returned a total of 382 kilograms of rocks that contained no water and had never been exposed to it, but were used to decipher the Moon's history and provide clues to its origin. The *Apollo 11* astronaut, **Neil A. Armstrong (1930–)** was the first human to step on the Moon, from the *Lunar Module Eagle* on 20 July 1969. *See* Moon.

Apparent luminosity: The total amount of energy per second per unit receiving area detected from a celestial object. The apparent luminosity is equal to the absolute luminosity divided by the square of the distance. *See* absolute luminosity, brightness, inverse square law, and luminosity.

Apparent magnitude: A measure of the relative brightness of a star, or other celestial object, as perceived by an observer on Earth. It is denoted by the symbol m. Apparent magnitude depends on both the absolute amount of light energy emitted, or reflected, and the distance to the object. The smallest apparent magnitudes correspond to the greatest brightness. The apparent magnitude of the Sun is –26.72, the apparent magnitude of the full Moon is –12.6, and the apparent magnitude of the planet Venus when it is brightest is –4.7. *See* absolute magnitude, magnitude, parallax and visual magnitude.

Arachnid: A volcanic feature on Venus caused by rising hot material that produces circular and radial fractures in the surface, nicknamed *arachnids,* from the Greek and modern Latin words for "spider."

Arc degree: A unit of angular measure of which there are 360 in a full circle. An arc degree is denoted by the symbol °.

Arc minute: A unit of angular measure equal to one sixtieth of a degree of angular measure, or to 1/60 of an arc degree. There are 60 arc minutes in 1 arc degree. An arc minute is denoted by the symbol ′, and it is also called a minute of arc. The Moon is 31 arc minutes across.

Arc second: A unit of angular measure equal to 1/3,600 of an arc degree and 1/60 of an arc minute. An arc second is denoted by the symbol ″, and it is also called a second of arc. There are 2.06265×10^5 seconds of arc in one radian, and 2π radians in a full circle of 360 degrees, where the constant $\pi = 3.14159$. Jupiter is 40 seconds of arc across.

Arcturus: The brightest star in the northern hemisphere and the fourth brightest star on the celestial sphere. Arcturus has been called the "Bear Watcher", since it follows Ursa Major, the Great Bear, around the northern pole, and the name Arcturus derives from *arktos,* the Greek word for "bear". Arcturus is also known as Alpha Boötis, the brightest star in the constellation Boötes. It is an orange K1 giant star at a distance of 37 light-years, with an apparent visual magnitude of –0.04 and an absolute visual magnitude of –0.30. *See* K star, stars and stellar classification.

Arecibo Observatory: The world's largest radio telescope located near Arecibo, Puerto Rico, dedicated in 1963 and resurfaced in 1974. Its metal reflecting surface has a spherical shape with a diameter of 305 meters (1,000 feet), and focuses incoming radio waves to a moveable structure positioned about 168 meters (550 feet) above. The transmission and reception of radar signals from the **Arecibo Observatory** in the 1960s resulted in an accurate determination of the astronomical unit and the first correct measurements of the rotation periods of Mercury and Venus. The **Arecibo Observatory** is used to study radio emission from all kinds of cosmic objects, such as pulsars, including the discovery in 1974 of the Binary Pulsar, PSR 1913+16, used to provide evidence for gravitational radiation, and the discovery in 1992 of the pulsar PSR 1257+12, which has at least two Earth-sized planets orbiting it. *See* binary pulsar, extrasolar planet, gravitational radiation, Mercury, millisecond pulsar, PSR 1257+12, PSR 1913+16, PSR 1937+21, radar astronomy, and Venus.

A star: A hot, white star of spectral type A, with a disk temperature much hotter than the Sun, such as Sirius, Altair, Fomalhaut, Vega, and Deneb. *See* Altair, Deneb, Fomalhaut, Hertzsprung-Russell diagram, Sirius, spectral classification, and Vega.

Asteroid: A relatively small, low mass, solid object of irregular shape that orbits the Sun and shines by reflected sunlight. The Sicilian astronomer **Giuseppe Piazzi (1749–1826)** discovered the first asteroid accidentally, on 1 January 1801, the first day of the 19th century. He named the tiny object Ceres, in honor of the patron goddess of Sicily. Another asteroid, named Pallas, was located the following year, and the third and fourth, designated Juno and Vesta, were found in 1804 and 1807. Ceres is the largest asteroid, with a diameter of 475 kilometers, about twice as large as Pallas and Vesta. Most asteroids are found in the asteroid belt located between the orbits of Mars and Jupiter, but a few of them, the near-Earth asteroids, come closer to the Earth's orbit. Although there are millions of asteroids, their total mass is less than that of the Earth's Moon. The name asteroid derives from their star-like appearance in telescopes, due to their small size rather than a large distance. Groups of asteroids, known as Hirayama families, have similar orbits, suggesting a common origin in the breakup of a larger body. Asteroids are also known as minor planets, but unlike the familiar major planets the gravity of an asteroid is too weak to hold onto an atmosphere or to pull it into a spherical shape. *See* Ceres, Eros, Hirayama families, near-Earth asteroid, **NEAR Shoemaker**, Pallas and Vesta.

Astrolabe: Sighting instrument employed from the time of the ancient Greeks down to the 17th century. It was used to show the appearance of the celestial sphere at a given moment and to determine the altitude above the horizon of the Sun, stars, or other celestial objects. Eventually replaced by the sextant. *See* sextant.

Astrometry: The branch of astronomy concerned with the measurement of positions and motions of celestial objects. *See* parallax and proper motion.

TABLE 1   Asteroids – largest[a]

Asteroid	Type	Radius (km)	Mass[b] (10^{21} kg)	Mass Density (kg m^{-3})	Rotation Period (hours)	Semi-major Axis (AU)
1 Ceres	C	475	0.95	2120 ± 40	9.1	2.77
2 Pallas	B	266[c]	0.214	2710 ± 110	7.8	2.77
4 Vesta	V	265	0.267	3440 ± 120	5.3	2.36

[a] The size is in units of kilometers, abbreviated km, the mass is in units of kilograms, abbreviated kg, and the mass density is in units of kilograms per cubic meter, abbreviated kg m^{-3}.

[b] The mass determinations are from the work of E. M. Standish at the Jet Propulsion Laboratory where he uses observations of Mars to solve for the masses of the larger asteroids.

[c] Pallas is not quite spherical. A tri-axial ellipsoid fit to occultation observations has diameters of 559 km, 525 km, and 532 km with a mean diameter of 538 ± 12 km.

Astronomical unit: Abbreviated AU, a unit of astronomical distances especially within the Solar System, equal to the mean distance between the Earth and the Sun, about 150 million kilometers or 92.8 million miles. The astronomical unit has a more exact value of 1.495 978 706 1 $\times$ 10^{11} meters, which is equal to 499.012 light-seconds. The astronomical unit is also defined as the distance from the Sun at which a mass-less particle in an unperturbed orbit would have an orbital period of 365.256 898 3 days. Mercury, the nearest planet to the Sun, lies at an average distance of 0.39 AU from the Sun. Jupiter's mean distance from the Sun is 5.2 AU. A light-year is a much larger unit of distance, equal to 63 240 AU.

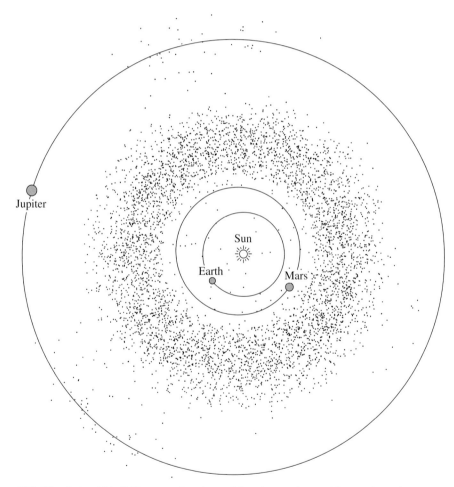

FIG. 20 Asteroid belt The exact locations of five thousand asteroids, or minor planets, whose orbits are accurately known. The vast majority of the asteroids orbit the Sun in the main belt located between the orbits of Mars and Jupiter. A few of them pass inside the orbit of Earth, while others move about 60 degrees ahead of and behind Jupiter in similar orbits.

Astronomy: The scientific study of planets, stars, galaxies and the Universe, including their motions, positions, sizes, composition, and the processes by which they formed and evolve.

Astrophysics: The study of the physical properties of celestial objects.

Aten asteroid: *See* near-Earth asteroid.

Atmosphere: The mixture of gases that surrounds a planet, natural satellite, and the outermost gaseous layers of a star, held near them by their gravity. Since gas has a natural tendency to expand into space, only bodies that have a sufficiently strong gravitational pull can retain atmospheres. The ability of a planet or satellite to retain an atmosphere depends on its mass, and on the gas temperature, determined by both its distance from the Sun and the atmosphere's greenhouse effect. Hotter, lighter gas molecules move faster, and are more likely to escape from a planet or satellite; less massive objects such as our Moon are less likely to retain an atmosphere. The term atmosphere is used for the tenuous outer material of the Sun because it is relatively transparent at visible wavelengths. The solar atmosphere includes, from the deepest layers outward, the photosphere, the chromosphere, and the corona. The Earth's atmosphere is transparent and consists mainly of molecular nitrogen (78 percent) and molecular oxygen (21 percent) with trace amounts of carbon dioxide (0.033 percent) and water vapor (variable, a few times 0.0001 percent). *See* Earth, escape velocity, Mars, and Venus.

Atmospheric window: A wavelength band in the electromagnetic spectrum for which radiation is able to pass through the Earth's atmosphere with relatively little attenuation by absorption, scattering or reflection. There are two main windows – the optical window of visible light and the radio window.

Atom: The smallest particle of a chemical element that still has the characteristics and properties of that element. An atom is composed of a dense and massive nucleus, containing protons and neutrons, surrounded by electrons. The hydrogen atom has one proton and no neutrons in its nucleus. A neutral atom has as many electrons as there are protons, and so is without net charge. An ionized atom has fewer electrons than protons, and therefore a positive charge. The atomic number is equal to the number

TABLE 2 Atmospheres – Venus, Mars and Earth

	Venus	Mars	Earth
Constituent[a]			
Carbon dioxide, CO_2	96	95	0.035
Nitrogen, N_2	3.5	2.7	77
Argon, Ar	0.007	1.6	0.93
Water vapor, H_2O	0.010	0.03 (variable)	1 (variable)
Oxygen, O_2	0.003	0.13	21
Surface pressure (bar)	92	0.007 to 0.010	1.0 (at sea level)
Surface temperature (K)	735	183 to 268	288 to 293

[a] Percentage composition of atmospheric constituent.

of protons in the atom's nucleus, while the atomic mass number is equal to the total number of protons and neutrons in the nucleus. The French physicist **Jean Baptiste Perrin (1870–1942)** demonstrated the existence of atoms by careful measurements of Brownian motion in 1909. *See* atomic number, Brownian motion, element, and ion.

Atomic mass number: The total number of protons and neutrons in the nucleus of an atom. The atomic mass number is designated by the capital letter A, and it is also known as the mass number of the atom. Hydrogen has a mass number of one, helium four, carbon twelve, and oxygen sixteen.

Atomic mass unit: A unit of mass that is equal to one-twelfth the mass of a carbon-12 atom, with a value of $1.660\,54 \times 10^{-27}$ kilograms. A neutron and proton have masses of about 1.008 atomic mass units.

Atomic number: The number of protons in the nucleus of an atom. The atomic number is designated by the capital letter Z. It is also equal to the element's position in the periodic table. Hydrogen has an atomic number of one, helium two, carbon six, and oxygen eight.

AU: Abbreviation for astronomical unit. *See* astronomical unit.

Aurora: A display of rapidly varying, colored light usually seen from magnetic polar regions of a planet. The light is given off by collisions between charged particles and the atoms, ions and molecules in a planet's upper atmosphere or ionosphere. Auroras are visible on Earth as the aurora borealis, or northern lights, near the North Pole, and the aurora australis, or southern lights, near the South Pole. They include the green and red emission lines from oxygen atoms and nitrogen molecules that have been excited by high-speed electrons. Terrestrial auroras occur at heights of around 100 to 250 thousand kilometers, which is within the Earth's ionosphere.

Auroral oval: One of the two oval-shaped zones around the Earth's geomagnetic poles in which the auroras are observed most often. They are located at latitudes of about 67 degrees north and south, and are about 6 degrees wide. Their position and width both vary with geomagnetic and solar activity. The auroral ovals are detected from Earth-orbiting satellites. Ground-based observers detect only a section of an oval, as shimmering ribbons and curtains of light.

TABLE 3   Atmospheres – giant planets and Sun[a]

Constituent	Sun	Jupiter	Saturn	Uranus	Neptune
Hydrogen, H_2	84	86.4	97	83	79
Helium, He (atom)	16	13.6	3	15	18
Water, H_2O	0.15	(0.1)	—	—	—
Methane, CH_4	0.07	0.21	0.2	2	3
Ammonia, NH_3	0.02	0.07	0.03	—	—

[a]The percentage abundance by number of molecules for the Sun, cooled to planetary temperatures so that the elements combine to form the compounds listed, and for the outer atmospheres of the giant planets below the clouds. Dashes indicate unobserved compounds. (Courtesy of Andrew P. Ingersoll.)

TABLE 4   Aurora – emission lines

Wavelength (nm)	Emitting atom, ion or molecule	Altitude (km)	Visual Color
391.4	N^+ (nitrogen ion)	1,000	violet-purple
427.8	N^+ (nitrogen ion)	1,000	violet-purple
557.7	O (oxygen atom)	90–150	green
630.0	O (oxygen atom)	>150	red
636.4	O (oxygen atom)	>150	red
661.1	N_2 (nitrogen molecule)	65–90	red
669.6	N_2 (nitrogen molecule)	65–90	red
676.8	N_2 (nitrogen molecule)	65–90	red
686.1	N_2 (nitrogen molecule)	65–90	red

Australia Telescope: An aperture synthesis radio telescope in New South Wales, consisting of six 22-meter (72-foot) dishes spread over a 6-kilometer baseline, another 22-meter (72-foot) dish at Mopra, about 100 kilometers to the south, and the 64-meter (210-feet) radio telescope at the Parkes Observatory located 200 kilometers farther south. *See* aperture synthesis, Australia Telescope National Facility, interferometer, Parkes Observatory, radio telescope, radio interferometer, and Very Long Baseline Interferometry.

Australia Telescope National Facility: Abbreviated ATNF, the main radio astronomy institute of Australia, and a division of the Commonwealth Scientific and Industrial Research Organization, or CSIRO for short. The Australia Telescope National Facility operates the Australia Telescope, which consists of its Compact Array at Narrabri and the Parkes and Mopra radio telescopes. These telescopes can be used together for aperture synthesis and Very Long Baseline Interferometry. *See* aperture synthesis, Parkes Observatory and Very Long Baseline Interferometry.

Autumnal equinox: The point where the Sun crosses the celestial equator in the autumn, on or near 23 September. *See* equinox.

AXAF: Acronym for *Advanced X-ray Astrophysics Facility*. *See Chandra X-ray Observatory*.

Background radiation: *See* cosmic background radiation.

Baily's beads: A string of bright lights observed at the extreme edge of the Sun's disk, seen during a total eclipse of the Sun just before or after totality. They are caused by sunlight shining through the valleys on the Moon's edge, or limb, while mountains on the limb block other rays of sunlight. The English astronomer **Francis Baily (1774–1844)** first described them in his account of the total eclipse of the Sun on 15 May 1836, and hence the name Baily's beads. *See* solar eclipse.

Balmer series: A mathematical formula devised in 1885 by the Swiss mathematics teacher **Johann Balmer (1825–1898)** to give the wavelengths, or frequencies, of the visible spectral lines of the hydrogen atom seen in the light of the Sun and other stars. The first Balmer line is called the red hydrogen alpha line, designated Hα, at a wavelength of 656.3 nanometers, which is used to detect the chromosphere across the face of the Sun. The series of wavelengths was explained in 1913 when the Danish physicist **Niels Bohr (1885–1962)** supposed that the lone electron in the hydrogen atom revolves about the nuclear proton in specific orbits with definite quantized values of energy. The Balmer series is radiated, or absorbed, when an electron jumps between these allowed orbits. *See* Bohr atom, chromosphere, Fraunhofer lines, hydrogen-alpha line, Lyman series, spectral line and spectroheliograph.

Barnard's Star: A fast-moving red M5 dwarf star with the largest proper motion, traveling across the sky at 10.36 seconds of arc per year, and the closest star to the Sun after the Alpha Centauri system. As its name implies, the American astronomer **Edward Barnard (1857–1923)** discovered the star in 1916. Barnard's Star has an apparent visual magnitude of 9.54, an absolute visual magnitude of 13.30, and an intrinsic luminosity only 0.0004 that of the Sun. It is the second closest star to the Earth, with a parallax of 0.549 seconds of arc and a distance of 5.96 light-years; only the Alpha Centauri triple star system, containing Proxima Centauri, is closer. At its distance of 5.96 light-years, the proper motion of Barnard's Star translates into a speed of 90 kilometers per second across the line of sight. Combined with its radial velocity of 108 kilometers per second, obtained from the Doppler effect of its spectral lines, Barnard's Star moves through space at 140 kilometers per second, a rate which is seven times that of the average star. *See* Doppler effect, parallax and proper motion.

Barringer Meteorite Crater: *See* Meteor Crater.

Barycenter: The center of mass of two or more celestial bodies – the point about which they revolve. *See* center of mass.

Baryon: A massive elementary particle with half-integral spin that responds to the strong nuclear force. Protons and neutrons, which comprise most of the mass of ordinary matter, are baryons. Each baryon is composed of three quarks.

Baryonic matter: Matter made out of baryons, including anything composed of elements in the periodic table, but perhaps accounting for only 4 percent of the matter-energy content of the Universe. The "ordinary" matter that we see around us, from people and trees to planets and stars, which are composed of baryons. Non-baryonic matter consists of subatomic particles that are not baryons, perhaps accounting for dark matter that may amount to 23 percent of the matter in the Universe. The other 73 percent is attributed to some sort of mysterious dark energy. *See* cold dark matter, dark energy, dark matter, hot dark matter, and non-baryonic matter.

Beta decay: The disintegration of a radioactive nucleus, in which a neutron turns into a proton by ejecting an electron, historically called a beta particle, and an antineutrino. The resulting nucleus has a charge one greater than the initial one.

Beta particle: An electron emitted from an atomic nucleus as a result of a nuclear reaction or in the course of radioactive decay. Such particles were originally called beta rays. *See* neutrino.

Beta Pictoris: An A-type star 63 light-years from the Earth, with a disk of dust surrounding it.

Betelgeuse: The apparently brightest star on the celestial sphere, a red M2 Supergiant star, located 427 light-years away in the constellation Orion, and also known as Alpha Orionis. It has an apparent visual magnitude of 0.45 and an absolute visual magnitude of –5.1. In 1921 the American physicist **Albert Michelson (1852–1931)** and the American astronomer **Francis G. Pease (1881–1938)** reported that they had used an optical interferometer to obtain an angular diameter of about 0.045 seconds of arc for Betelgeuse, which at a distance of 430 light-years corresponds to a physical radius of about 435 million kilometers and 625 times the radius of the Sun. In our Solar System, the star would extend to a radius of about 3 AU or three times the distance between the Sun and the Earth. When using up all its thermonuclear fuel, this massive star will explode as a supernova.

Biela's Comet: A former comet discovered in 1826 by the Austrian army officer and astronomer **Wilhelm von Biela (1782–1856)**. The comet had a period of 6.62 years, broke into two pieces when it came close to the Sun in 1846, and has not been seen since 1852. The Andromedid meteor shower of 1872 was produced when the Earth crossed the orbital path of the disintegrated comet, showing that meteor showers, or shooting stars, are produced when small pieces of comets enter the Earth's atmosphere.

Big Bang: The beginning of the observable Universe, and an event that propelled the Universe into expansion about 14 billion years ago, when all the mass of the Universe was compressed to a very high density and an extremely high temperature exceeding 10 billion kelvin. The cosmic microwave background radiation is considered to be the residual radiation of the Big Bang. *See* cosmic microwave background.

Big-Bang nucleosynthesis: The creation of light elements, such as hydrogen, deuterium and helium, in the first few minutes after the Big Bang. *See* nucleosynthesis.

Billion: One thousand million, written 1 000 000,000 or 10^9.

Binary pulsar: The radio pulsar designated PSR 1913 + 16, with a period of 53 milliseconds, and a companion star that is also a neutron star. The binary pulsar was discovered in 1974 by the American radio astronomers **Russell A. Hulse (1950–)** and **Joseph H. Taylor Jr. (1941–)** using the Arecibo Observatory. Precise timing observations of the radio pulses indicated that the radio pulsar is slowly approaching its silent, invisible companion at the rate expected if their orbital energy is being radiated away in the form of gravitational waves. *See* Arecibo Observatory, gravitational radiation and PSR 1913 + 16.

Binary star: A double star system, in which the two stars are bound together by their mutual gravitational force and orbit a common center of mass, revolving around each other. A star in such a double star system is also known as a binary star. Double stars are very common, and only a third of stars are solitary stars like the Sun.

Binding energy: The minimum energy needed to split an atomic nucleus into its constituents.

Bipolar outflow: The high-speed jetting of gas in opposite directions from disk-shaped systems, usually emitted from the rotational poles in directions perpendicular to the plane of the disk.

Blackbody: A heated body that absorbs the entire radiation incident on it, a perfect absorber of radiation. The intensity of the radiation emitted by a blackbody and the way it varies with wavelength depends only on the temperature of the body and can be predicted by quantum theory. The German physicist **Gustav Kirchhoff (1824–1887)** introduced the idea of a perfect absorber of radiation, the blackbody, in 1862. *See* blackbody radiation, and blackbody spectrum.

Blackbody radiation: The radiation that would be emitted by a blackbody, a heated object that absorbs all the energy that strikes it, with an intensity and spectrum determined by the temperature alone. Blackbody radiation is also called thermal radiation, and the intensity and spectra are determined by the criterion of thermal equilibrium. The spectrum of the radiation emitted by a blackbody is a continuous spectrum, and the wavelength of greatest emission is inversely proportional to the body's temperature. The total amount of energy emitted by a blackbody is proportional to the product of its area and the fourth power of its temperature. *See* blackbody, blackbody spectrum, continuous spectrum, effective temperature, Planck's constant, Stefan-Boltzmann law, and thermal radiation.

Blackbody spectrum: The radiation intensity at different wavelengths for a blackbody. The wavelength of greatest emission is inversely proportional to the body's temperature. The formula for the energy of the radiation from a blackbody at any given wavelength, or frequency, and temperature was derived in 1900 by the German physicist **Max Planck (1858–1947)**, who assumed that the energy consists of the sum of individual quanta particles, whose quantum energy is proportional to the frequency with a constant of proportionality now known as Planck's constant. *See* Planck's constant.

Black hole: An extremely dense object whose gravity is so intense that neither matter nor radiation can escape, or in other words the object's escape velocity exceeds the velocity of light, the fastest thing in the Universe. Stellar black holes form when the most massive stars, born with about forty times the mass of the Sun, run out of fuel and collapse. Cygnus X-1 is an example of such a stellar black hole. Supermassive black holes, with masses equivalent to millions of times the mass of the Sun or more, are found in the center of galaxies. A swirling disk of material, called an accretion disk, may surround a black hole, and jets of matter may arise from its vicinity. There is a singularity at the center of a black hole, where the matter is compressed into an infinitely small volume and an infinitely large mass density. *See* accretion disk, Cygnus X-1, escape velocity, event horizon, Schwarzschild radius, singularity, stellar black hole and supermassive black hole.

Blazar: Galaxies with exceptionally active nuclei. The name is derived from BL Lacertae object and quasar. *See* BL Lacertae object and quasar.

BL Lacertae object: A galaxy with a highly luminous, active nucleus, such as the prototype BL Lacertae. Their radiation, from radio waves to X-rays, is variable over time scales as short as a few hours.

Blue giant: A giant star with spectral type O or B.

Blueshift: The Doppler shift in the wavelength of a spectral line toward a shorter, bluer wavelength, caused when the source of radiation and its observer are moving toward each other. The greater the blueshift is, the faster an object is moving toward us. *See* Doppler shift.

Blue supergiant: A hot, young supergiant star with spectral type O or B, such as the star Rigel in the constellation Orion. *See* Rigel.

Bode's law: *See* Titius-Bode law.

Bohr atom: In 1913 the Danish physicist **Niels Bohr (1885–1962)** proposed a model, now known as the Bohr atom, in which the lone electron in a hydrogen atom revolves about the atomic nucleus, a proton, in specific orbits with definite, quantized values of energy, and an electron only emits or absorbs radiation when jumping between these allowed orbits. The model explains the wavelengths of the visible radiation emitted by the hydrogen atoms in the Sun, described in the mathematical formula derived in 1885 by the Swiss mathematics teacher **Johann Balmer (1825–1898)**. *See* Balmer series.

Bolshoi Teleskop Azimutalnyi: A 6-meter (236-inch) visible-light telescope located on Mount Pastukhov in the Caucasus range, in operation since 1976. It is part of the Zelenchukskaya Observatory, and operated by the Special Astrophysical Observatory of the Russian Academy of Sciences.

Boltzmann's constant: The constant of proportionality between energy per degree of freedom and temperature, denoted by the symbol k. Boltzmann's constant has the value $k = 1.380\,66 \times 10^{-23}$ Joules per degree Kelvin. The constant is named after the Austrian physicist **Ludwig Boltzmann (1844–1906)** who used it in 1868 to relate the mean total energy and the temperature of molecules in thermal equilibrium.

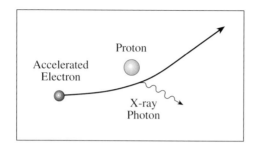

FIG. 21 **Bremsstahlung** When a hot electron moves rapidly and freely outside an atom, it inevitably moves near a proton in the ambient gas. There is an electrical attraction between the electron and proton because they have equal and opposite charge, and this pulls the electron toward the proton, bending the electron's trajectory and altering its speed. The electron emits electromagnetic radiation known as *bremsstrahlung* from the German word for "braking radiation".

Bremsstrahlung: Radiation that is emitted when an energetic electron is deflected and accelerated by an ion. It is also called a free-free transition because the electron is free both before and after the encounter, remaining in an unbound hyperbolic orbit without being captured by the ion. *Bremsstrahlung* is German for "braking radiation."

Brightness: The amount of light coming from an object, usually meaning the luminous intensity or amount of energy per second. The intrinsic brightness, or intrinsic luminosity, is the amount of light an object gives off, while the apparent brightness or apparent luminosity is the amount of light we see, which has been dimmed by the inverse square of the distance to the object. The intrinsic brightness is also known as the absolute luminosity. *See* absolute luminosity, apparent luminosity, inverse square law, and luminosity.

Brown dwarf: A star that forms like the Sun and other stars, by the collapse and fragmentation of an interstellar gas cloud, but is not massive enough to enable the hydrogen fusion required for a star to shine from inside, or to ignite its hydrogen burning. A brown dwarf has a mass of between 0.01 and 0.08 solar masses; less massive objects are giant planets and more massive ones are normal stars. A brown dwarf might be able to burn deuterium in its very early youth, making it shine by itself. *See* hydrogen burning and solar mass.

Brownian motion: The continual, random motion of small solid particles suspended in a liquid or a gas, discovered in 1827 by the Scottish botanist **Robert Brown (1773–1858)** when using a microscope to observe very fine pollen grains suspended in water. The German physicist **Albert Einstein (1879–1955)** explained this Brownian motion, as it is now called, in 1905, proposing that the molecules of water would move about at random and therefore randomly knock a sufficiently small particle into Brownian motion. In 1909 the French physicist **Jean Baptiste Perrin (1870–1942)** demonstrated the existence of atoms by careful measurements of Brownian motion.

B star: A hot, blue star of spectral type B, with a disk temperature hotter than the Sun, such as Achernar and Regulus, B-type dwarf, or main sequence, stars, and Rigel, a B-type supergiant star. Even hotter blue stars are designated spectral type O. *See* Achernar, Hertzsprung-Russell diagram, Regulus, Rigel, and stellar classification.

Bubble Nebula: A faint emission nebula in the constellation Cassiopeia, appearing as a bubble surrounding a seventh-magnitude star, but without the characteristics of a planetary nebula. The Bubble Nebula has an angular diameter of 3 minutes of arc. *See* emission nebula and planetary nebula.

Bulge: The spherical structure at the center of a spiral galaxy that is made up primarily of old stars, gas, and dust. The Milky Way's bulge is roughly 15,000 light-years across; it is old and dense.

C

Calcium H and K lines: Spectral lines of ionized calcium, denoted by Ca II, in the violet part of the spectrum at 396.8 nanometers (H) and 393.4 nanometers (K). They are conspicuous features in the spectra of many stars, including the Sun. The designations H and K were given by the German physicist **Joseph von Fraunhofer (1787–1826)** in 1814, and are still commonly used. The Sun's chromosphere and its magnetic chromospheric network are detected across the face of the Sun using the calcium H and K lines with a spectroheliograph. *See* chromosphere, chromospheric network, Fraunhofer lines, plage and spectroheliograph.

California Nebula: An emission nebula in the constellation Perseus, which is designated NGC 1499 and illuminated by the star ξ Persei. The California Nebula receives its name from its shape, which resembles the United States state of California. The nebula is about 0.66 × 2.5 degrees across. *See* emission nebula.

Callisto: The second largest and outermost of Jupiter's four largest moons, discovered in 1610 by the Italian astronomer **Galileo Galilei (1564–1642)** using the newly invented telescope. Callisto has a radius of 2410 kilometers and a mass density of 1834 kilograms per cubic meter. It is a primitive world whose surface of ice and rock is the most heavily cratered in the Solar System. Callisto has a variable magnetic field, apparently generated in an internal shell of liquid water as Jupiter's powerful magnetic field sweeps by. *See* Galilean satellites.

Caltech: Abbreviation for the California Institute of Technology, located in Pasadena, California.

Canopus: The second brightest star on the celestial sphere, visible from the Southern Hemisphere, located in the constellation Carina and also known as Alpha Carinae. The yellow-white F0 supergiant star is located at a distance of 326 light-years, with an apparent visual magnitude of –0.62 and an absolute visual magnitude of –5.63, corresponding to 14,800 times the absolute luminosity of the Sun.

Capella: The third brightest star seen from the Northern Hemisphere and the sixth brightest star on the celestial sphere. Capella is the brightest star in the constellation Auriga, with the designation Alpha Aurigae, lying 42 light-years away from the Earth and consisting of two yellow G8 and G1 giant stars in tight orbit, separated by 0.7 AU and with a 104-day period, a combined apparent visual magnitude of 0.08, and a combined absolute visual magnitude of –0.47. The brighter G8 giant is 82 times more luminous than the Sun, and the dimmer G1 giant has 51 times the Sun's luminosity. The name *Capella* means "the She-Goat" in old Latin, and the goat is carried in the constellation Auriga, the Charioteer.

Carbon: The element with atomic number six and atomic mass number twelve. Carbon is produced during helium burning in red giant stars and is ejected into interstellar space when these stars form planetary nebulae. Carbon also comes from high-mass stars that explode as supernovae. *See* helium burning, planetary nebula, red giant and supernova.

Carbonaceous chondrite meteorite: A rare class of meteorites that contain appreciable amounts of carbon and water, with unusually low mass densities in the range 2200 to 2900 kilograms per cubic meter. They are thought to be among the most primitive and least altered samples of solids in the Solar System. *See* meteorite.

Carbon dioxide: A molecule, symbolized by CO_2, composed of one atom of carbon, denoted by C, and two atoms of oxygen, each abbreviated by O. Carbon dioxide gas is the main constituent of the atmospheres of Venus and Mars, but only a fraction of the Earth's atmosphere. Carbon dioxide is a greenhouse gas that traps solar energy and warms the surfaces of the Earth and Venus. *See* greenhouse effect, Mars, and Venus.

Carbon monoxide: A molecule, symbolized by CO, composed of one atom of carbon, denoted by C, and one atom of oxygen, denoted by O. Carbon monoxide is the most abundant interstellar molecule after molecular hydrogen and is especially useful because it radiates at radio wavelengths and can be used to map the distribution of molecular hydrogen, which does not emit radio waves. *See* molecular cloud.

Carbon-nitrogen-oxygen cycle: A sequence of nuclear reactions, abbreviated as the CNO cycle, in which hydrogen nuclei are converted into helium nuclei inside main-sequence stars more massive than 1.5 solar masses and in giant and supergiant stars of all masses. Carbon, nitrogen and oxygen catalyze the nuclear reactions, their total number remaining unchanged at the end of the cycle. It occurs at temperatures above 15 million kelvin. The German physicist **Carl Friedrich von Weizsäcker (1912–)** published a description of the CNO cycle in 1938, and in the following year the American physicist **Hans Bethe (1906–2005)** showed that the CNO chain of nuclear reactions provides the temperature-dependent energy generation that accounts for the luminosity of main-sequence stars more massive than the Sun. *See* hydrogen burning.

Cassegrain telescope: An arrangement for diverting the light received by the parabolic primary mirror of a reflecting telescope by using a convex secondary mirror to reflect the light through a central hole in the primary to a focus, called the Cassegrain focus, on the axis of the telescope. The French priest **Laurent Cassegrain (1629–1693)** invented this arrangement in 1672. *See* reflecting telescope and telescope.

Cassini-Huygens Mission: NASA launched the ***Cassini-Huygens*** spacecraft on 15 October 1997. Instruments aboard the spacecraft, which reached Saturn in July 2004, have improved our understanding of the planet Saturn, its rings, its magnetosphere, its principal moon Titan and its other moons. NASA's ***Cassini Orbiter*** will orbit Saturn and its moons for four years, and ESA's ***Huygens Probe*** dove through the thick, hazy atmosphere of Saturn's moon Titan and landed on its surface on 14 January 2005, detecting signs of flowing, liquid methane. The ***Cassini Orbiter*** was named after **Gian (Giovanni) Domenico (Jean Dominique) Cassini (1625–1712)**, the Italian-born, French astronomer who in 1675 was the first to distinguish two zones within what was thought to be a single ring around Saturn, divided by a dark gap now known as Cassini's division. The *Huygens Probe* is named after **Christiaan Huygens (1629–1695)**, a Dutch astronomer who in 1655 discovered Titan, Saturn's largest moon.

Cataclysmic binary: A binary star system consisting of a degenerate white dwarf star and a companion that is usually a cool main-sequence star of late spectral type G, K or M. They have orbital periods of 1 to 15 hours. The mass of the companion star fills its Roche lobe, causing gas to flow from it onto the white dwarf, producing a nova outburst by the sudden nuclear fusion of hydrogen. *See* nova and white dwarf.

Cat's Eye Nebula: A planetary nebula, designated NGC 6543, located in the constellation Draco. It is about 20 seconds of arc in diameter and of ninth magnitude. The Cat's Eye Nebula is named for its oval shape and green color. In 1864 the English astronomer **William Huggins (1824–1910)** discovered three bright emission lines in the object, indicating a gaseous composition and proving that not all nebulae consist of stars. *See* planetary nebula.

CCD: Acronym for a Charge Coupled Device, an electronic imaging system invented at the Bell Telephone Laboratories in 1969. A CCD consists of a square array of closely spaced electrodes, which convert incident light into an electronic charge. The charges from all the electrodes are "coupled" together and read by computers to create digital images from them. Since a CCD produces electrically charged signals from most of the light falling on it, it has been used to greatly improve the light-gathering power of telescopes, both astronomical and military. The military now uses CCDs in high-flying aircraft and spy satellites, sending back images of enemy battlefields and installations. And since CCDs can hold a charge corresponding to variable shades of light, they are also used as imaging devices for digital scanners, fax machines, photocopiers, bar-code readers, and both digital and video cameras.

Celestial coordinates: The position of an object on the celestial sphere. The right ascension, symbolized by α or RA, is the celestial longitude measured eastward along the celestial equator from the Vernal Equinox, also called the First Point of Aries. The declination, symbolized by δ or Dec., is the celestial latitude, measured from the celestial equator along a great circle either to the north (positive) or to the south (negative). Owing to precession, the position of the Vernal Equinox moves westward along the ecliptic at a rate of about 50.3 seconds of arc, or 3.35 seconds of time each year, and as a result the celestial coordinates are continuously changing at a slow rate. A periodic oscillation of the Earth's rotational axis, known as nutation, also causes a change in celestial coordinates. They are therefore specified by the year or epoch they were determined, and corrected to the date of observation for precession and nutation. *See* declination, equinox, nutation, precession, right ascension, and Vernal Equinox.

Celestial equator: The projection of the Earth's equator on the celestial sphere.

Celestial sphere: An imaginary sphere with the Earth at the center, containing all celestial objects such as stars, galaxies, and quasars. The ecliptic is the projection of the Earth's orbital plane on the celestial sphere, inclined at an angle of 23.5 degrees to the celestial equator and intersecting it at the two equinoxes. The positions of an object

FIG. 22 Celestial coordinates Stars, galaxies and other cosmic objects are placed upon an imaginary celestial sphere. The celestial equator divides the sphere into northern and southern halves, and the ecliptic is the annual path of the Sun on the celestial sphere. The celestial equator intersects the ecliptic at the Vernal Equinox and the Autumnal Equinox. Every cosmic object has two celestial coordinates. They are the right ascension, designated by the angle alpha, α, or by R.A., and the declination, denoted by the angle delta, δ, or Dec. Right ascension is measured eastward along the celestial equator from the Vernal Equinox to the foot of the great circle that passes through the object. Declination is the angular distance from the celestial equator to the object along the great circle that passes through the object, positive to the north and negative to the south. Precession results in a slow motion of the Vernal Equinox, producing a steady change in the celestial coordinates.

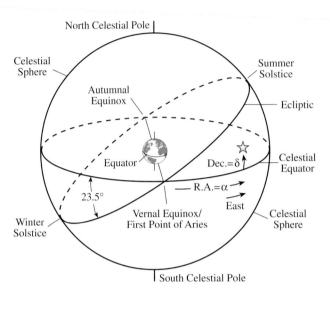

on the celestial sphere are called the right ascension and the declination. *See* equinox, declination, right ascension, and Vernal Equinox.

Celsius: A unit of temperature denoted by C or by °C for degrees Celsius. Absolute zero on the Celsius temperature scale is at -273°C, the freezing point of water is at 0°C, and the boiling point of water is at 100°C. The equivalent temperature in degrees kelvin is given by °K = °C + 273. The equivalent temperature in degrees Fahrenheit is given by °F = 9°C/5 + 32. The temperature unit is named for the Swedish astronomer **Anders Celsius (1701–1744)**, who proposed the temperature scale in 1742. *See* absolute temperature, Fahrenheit, and Kelvin.

Centaurus A: A strong radio source associated with the giant elliptical galaxy NGC 5128, located in the constellation Centaurus at a distance of about 10 million light-years.

Centaurus X-3: A bright X-ray source in the direction of the constellation Cygnus. Observations from the **Uhuru** satellite resulted in the discovery of X-ray pulsation from Centaurus X-3 in 1971, and the discovery of its binary nature the following year. It is an X-ray pulsar accreting material from a nearby companion and slowly speeding up as a result. *See* pulsar and **Uhuru Mission**.

Center for Astrophysics: The Harvard-Smithsonian Center for Astrophysics, abbreviated CfA, combines the resources and research facilities of the Harvard College Observatory and the Smithsonian Astrophysical Observatory. Headquartered at Cambridge, Massachusetts, the Center for Astrophysics has hundreds of scientists engaged in research in astronomy, astrophysics, the Earth, space sciences, and science education.

Center of mass: The point around which two gravitationally linked objects revolve. If the two objects are of equal mass, then the center of mass lies exactly halfway between them. The more unequal the masses, the closer the center of mass is to the more massive object. The center of mass is also known as the barycenter.

Centigrade: An older name for the Celsius unit of temperature, derived from the fact that there are 100 degrees between the freezing and boiling temperatures of water. *See* Celsius.

Cepheid: *See* Cepheid variable.

Cepheid variable: A very bright, pulsating star with a well-defined relation between its absolute luminosity and period of pulsation, useful for obtaining distances out to about 10 million light-years. It is a yellow, pulsating, variable supergiant star that periodically expands and contracts, producing a regular variation in its intrinsic brightness or absolute luminosity. The longer the star's period of variation the greater the star's mean intrinsic brightness or absolute luminosity. To determine a Cepheid variable's distance, one measures the variation period, infers the star's absolute luminosity, and compares it to the apparent luminosity. Cepheid variables are so bright that they can be seen in other galaxies, establishing distances to nearby galaxies outside our Milky Way Galaxy. Cepheid variable stars take their name from the first of the type to be discovered, Delta Cephei, the variability of which was noted by English astronomer **John Goodricke (1764–1786)** in 1784. The American astronomer **Henrietta Swan Leavitt (1868–1921)** discovered the period-luminosity relation of Cepheid variable stars in 1912. *See* absolute luminosity, apparent luminosity, inverse square law, and period-luminosity relation.

Ceres: The first asteroid to be discovered, on the night of 1 January 1801 by the Sicilian astronomer and monk **Giuseppe Piazzi (1746–1826)**. He named it Ceres after the patron goddess of Sicily. Ceres has a radius of 475 kilometers and a mass density of 2120 kilograms per cubic meter. *See* asteroid.

CERN: Acronym for Centre Européen pour la Recherche Nucléaire, the European Organization for Nuclear Research, based in Geneva, Switzerland.

Cerro Tololo Inter-American Observatory: Abbreviated **CTIO**, an observatory on Cerro Tololo peak about 80 kilometers east of La Serena, Chile. Its largest telescopes are the 4-meter (158-inch) reflector, named after the American astronomer **Victor M. Blanco (1918–)**, and since 2005 the 4.1-meter (162-inch) SOuthern Astrophysical Research, abbreviated SOAR, telescope.

CfA: Acronym for Center for Astrophysics. *See* Center for astrophysics.

CGRO: Acronym for the *Compton Gamma Ray Observatory*. *See* *Compton Gamma Ray Observatory*.

Chandra Mission: *See **Chandra X-ray Observatory.***

Chandrasekhar limit: The maximum mass of a white dwarf star, about 1.4 solar masses. A collapsing star of mass greater than this value will become a neutron star or a black hole. If a white dwarf receives material from a companion star and exceeds the Chandrasekhar limit, the white dwarf can explode as a type Ia supernova. This upper mass limit for white dwarfs was first derived by the German physicist **Wilhelm Anderson (1880–1940)** and the English physicist **Edmund C. Stoner (1899–1968)**, in 1929 and 1930 respectively, but it is often named for the Indian-born American astrophysicist **Subrahmanyan Chandrasekhar (1910–1995)**, who subsequently derived detailed equilibrium configurations in which degenerate electron gases support their own gravity. *See* black hole, neutron star, supernova and white dwarf.

Chandra X-ray Observatory: NASA launched the ***Chandra X-ray Observatory***, abbreviated ***CXO***, on 23 July 1999, from the ***Space Shuttle Columbia***. It investigates the high-energy, X-ray regions of the Universe from objects such as active galactic nuclei, black holes, clusters of galaxies, dark matter, galaxies, neutron stars, pulsars, quasars, supernova remnants, supernovae, and white dwarfs. The X-ray observatory was previously known as the ***Advanced X-ray Astrophysics Facility***, or ***AXAF*** for short, and was renamed after the Indian-American astrophysicist and Nobel laureate, **Subrahmanyan Chandrasekhar (1910–1995)**.

Charge-coupled device: *See* CCD.

Charged particles: Fundamental components of subatomic matter, such as protons, other ions and electrons, which have electrical charge. The charged subatomic particles are surrounded by electrical force fields that attract particles of opposite charge and repel those of like charge. Charged particles are either negative (electron) or positive (proton) and are responsible for all electrical phenomena.

Cherenkov radiation: Light or other forms of electromagnetic radiation emitted when a charged particle has a velocity that exceeds the velocity of light in a medium. A neutrino can be detected by the cone of blue Cherenkov light emitted when a neutrino enters a large underground tank of water and accelerates an electron by collision to a velocity larger than the velocity of light in water. The Cherenkov effect is named for the Russian physicist **Pavel A. Cherenkov (1904–1990)**, who first observed it in 1934. *See* Kamiokande, Sudbury Neutrino Observatory, and Super Kamiokande.

Chlorine experiment: The subterranean neutrino detector located in the Homestake Gold Mine near Lead, South Dakota, in which a neutrino strikes a chlorine nucleus in a huge tank of cleaning fluid to produce a nucleus of radioactive argon. The American physicist **Raymond Davis Jr. (1914–)** used the experiment to detect neutrinos from the Sun for more than a quarter of a century, beginning in 1967, always detecting about one-third of the expected amount of solar neutrinos. *See* Homestake, and solar neutrino problem.

Chromatic aberration: Introduction of spurious colors by a lens.

Chromosphere: The layer or region of the solar atmosphere lying above the photosphere and beneath the transition region and the corona. The Sun's temperature rises to about 10 000 kelvin in the chromosphere. The name literally means a "sphere of color". The chromosphere is normally invisible because of the glare of the photosphere shining through it, but it is briefly visible near the beginning or end of a total solar eclipse as a spiky red rim around the Moon's disk. At other times, the chromosphere can be studied by spectroscopy, observing it across the solar disk in the red light of the hydrogen-alpha line or the calcium H and K lines. Solar filaments, plage, prominences and spicules are all seen in the chromosphere. A thin transition region separates the chromosphere and the corona. *See* calcium H and K lines, chromospheric network, corona, filament, hydrogen-alpha line, photosphere, plage, prominence, spicule, and transition region.

Chromospheric network: A large-scale cellular pattern visible in spectroheliograms taken in the calcium H and K lines. The network appears at the boundaries of the photosphere supergranulation cells, and contains magnetic fields that have been swept to the edges of the cells by the flow of material in the cells. *See* calcium H and K lines, spectroheliograph, and supergranulation.

Chronometer: A very accurate clock designed for determining longitude at sea. This is done by comparing the local time observed astronomically with the corresponding time at the port of embarkation as given by the chronometer. Such a marine chronometer was first built by the English horologist and instrument maker **John Harrison (1693–1776)** in 1728, and subsequently used at sea to obtain longitude with an accuracy of about one minute of arc. *See* longitude, and Royal Greenwich Observatory.

Circle: A closed plane curve everywhere equidistant from its center. An ellipse possessing but one focus.

Circumstellar disk: *See* protoplanetary disk.

Clementine spacecraft: A joint project of the Strategic Defense Initiative and NASA, the *Clementine* spacecraft was launched on 25 January 1994, entering lunar orbit on 21 February 1994. Instruments aboard *Clementine* mapped the global surface composition and topography of the Moon, and obtained possible evidence for water ice at the lunar poles. The spacecraft was named after the miner's darling daughter in the old Gold Rush ballad. *See* Moon.

Climate: The average weather conditions of a place over a period of years.

Closed Universe: A cosmological model in which a homogeneous, isotropic Universe, without dark energy or a cosmological constant, eventually stops expanding and begins to collapse, most likely ending in a future Big Bang that initiates another expansion. Such a Universe is geometrically spherical. *See* critical mass density, dark energy, non-Euclidean geometry, and omega.

Cluster of galaxies: A collection of hundreds to thousands of galaxies, bound together by their mutual gravitation. A supercluster consists of several clusters of galaxies joined together. The nearest rich clusters of galaxies, containing about a thousand galaxies each, are the Virgo Cluster, 45 million light-years away, and the Coma Cluster, 300 million light-years away. *See* dark matter, supercluster, and Virgo Cluster.

CMB: Acronym for Cosmic Microwave Background.

CMBR: Acronym for Cosmic Microwave Background Radiation.

CME: Acronym for Coronal Mass Ejection. *See* coronal mass ejection.

CNO cycle: Acronym for Carbon-Nitrogen-Oxygen cycle. *See* carbon-nitrogen-oxygen cycle.

CO: *See* carbon monoxide.

CO₂: *See* carbon dioxide.

Coalstack: A large, prominent dark nebula in the constellation Crux, the Southern Cross. Its angular dimensions are 5 by 7 degrees. *See* dark nebula.

COBE: Acronym for *COsmic Background Explorer*. *See COsmic Background Explorer*.

Cold dark matter: Hypothetical subatomic particles that move slowly compared with the speed of light. *See* dark matter and hot dark matter.

Colliding galaxies: Two galaxies that pass close enough to gravitationally disrupt each other's shape. The collision rips streamers of stars from the galaxies, triggers star birth, and can ultimately result in both galaxies merging into one. *See* antennae.

Comet: An object made of ice, dust and rock, or dirty snowball, in our Solar System which is much smaller than a planet and about ten kilometers across, orbiting the Sun usually far beyond the planets and often in a highly eccentric orbit. In 1950 the Dutch astronomer **Jan Oort (1900–1992)** used the observed trajectories of long-period comets, with orbital periods greater than 200 years, to show they come from a remote, spherical shell, now known as the Oort comet cloud, located about a quarter of the way to the nearest star. In the same year, the American astronomer **Fred L. Whipple (1906–2004)** proposed the icy conglomerate, or dirty snowball, comet model, in which the solid nucleus of a comet is just a gigantic ball of ice, dust and rock, and close-up observations of Halley's Comet, by the *Giotto* spacecraft in 1986, confirmed this model. In July 1994, Comet Shoemaker-Levy-9 collided with Jupiter. When a comet comes within about three times the Earth's distance from the Sun, some of its icy surface vaporizes; the comet then develops a diffuse, luminous coma of dust and gas, and, normally, one or more tails. These tails can become longer than the distance between the Earth and the Sun. The curved dust tails are pushed away from the Sun by the pressure of solar radiation, and the straight ion tails are swept away by the solar wind. Most comets are invisible, stored far from the planetary system in two large reservoirs: the Kuiper belt beyond the orbit of Neptune, and the Oort comet cloud at near-interstellar distances. *See* Biela's Comet, *Deep Space 1*, Encke's Comet, *Giotto Mission*, Halley's Comet, Kuiper belt, long-period comet, Oort comet cloud, Shoemaker-Levy 9 Comet, short-period comet, solar wind, and *Stardust*.

Comet Beila: *See* Biela's Comet.

Comet Encke: *See* Encke's Comet.

Comet Halley: *See* Halley's Comet.

Comet Shoemaker-Levy 9: *See* Shoemaker-Levy 9 Comet.

Comet Swift-Tuttle: *See* Swift-Tuttle Comet.

Compton Gamma Ray Observatory: NASA launched the *Compton Gamma Ray Observatory,* abbreviated *CGRO,* on 5 April 1991, and it re-entered the Earth's atmosphere on 4 June 2000. It was the second of NASA's four great observatories is space, designed to study the Universe in an invisible, high-energy form of radiation known as gamma rays. Instruments aboard *CGRO* detected more than 2600 gamma-ray

TABLE 5   Comets - Great[a]

Name	Perihelion Date	Days Visible[b]	Perihelion Distance[c] (AU)	Brightest Apparent Magnitude[d]
Great Comet (1807 R1)	19 Sept. 1807	90	0.65	1 to 2
Great Comet (1811 F1)	12 Sept. 1811	260	1.04	0
Great March Comet (1843 D1)	27 Feb. 1843	48	0.006	1
Comet Donati (1858 L1)	30 Sept. 1858	80	0.58	0 to 1
Great Comet (1861 J1)	12 June 1861	90	0.82	0(−2?)
Great Southern Comet (1865 B1)	14 Jan. 1865	36	0.03	1
Comet Coggia (1874 H1)	09 July 1874	70	0.68	0 to 1
Great September Comet (1882 R1)	17 Sept. 1882	135	0.008	−2
Comet Comet (1901 G1)	24 Apr. 1901	38	0.24	1
Great January Comet (1910 A1)	17 Jan. 1910	17	0.13	1 to 2
1/P Comet Halley (1910)	20 Apr. 1910	80	0.59	0 to 1
Comet Skjellerup-Maristany (1927X1)	18 Dec. 1927	32	0.18	1
Comet Ikeya-Seki (1965 S1)	21 Oct. 1965	30	0.008	2
Comet Bennett (1969 Y1)	20 Mar. 1970	80	0.54	0 to 1
Comet West (1975 V1)	25 Feb. 1976	55	0.20	0
Comet Hyakutake (1996 B2)	01 May 1996	30	0.23	1 to 2
Comet Hale-Bopp (1995 O1)	01 Apr. 1997	215	0.91	−0.7

[a] Adapted from Donald K. Yeomans' *Great Comets in History*, at the Web site http://ssd.jpl.nasa.gov/great_comets.html.

[b] Days visible to the naked eye unaided by binoculars or a telescope.

[c] The perihelion distance is the distance from the Sun at the closest approach to the star, given in astronomical units, or AU, roughly the mean distance between the Earth and the Sun and about 150 billion, or 1.5×10^{11} meters.

[d] The apparent magnitude is a measure of the apparent brightness of a celestial object, in which brighter objects have smaller magnitudes. Sirius A, the brightest star other than the Sun, has an apparent visual magnitude of −1.5. The nearest star other than the Sun is about 0 on the magnitude scale, while Venus has an apparent magnitude of −4 when brightest and at its brightest Jupiter appears at magnitude −2.7

TABLE 6　Comets – short period[a]

Name	Orbit Period (years)	Perihelion Date[b] (year)	Perihelion Distance (AU)	Orbital Inclination (degrees)	Absolute Magnitude[c]
1P Halley	76.01	2061	0.59	162.2	5.5
2P Encke	3.30	2003	0.34	11.8	9.8
6P d'Arrest	6.51	2002	1.35	19.5	8.5
9P Tempel 1	5.51	20th century	1.50	10.5	12.0
19P Borrelly	6.88	2001	1.36	30.3	11.9
21P Giacobini-Zinner	6.61	2005	1.00	31.9	9.0
26P Grigg-Sjkellerup	5.11	2002	0.99	21.1	12.5
27P Crommelin	27.41	2011	0.74	29.1	12.0
46P Wirtanen	5.46	2002	1.06	11.7	9.0
55P Tempel-Tuttle	33.22	2031	0.98	162.5	9.0
73P Schwassmann-Wachmann 3	5.34	2001	0.94	11.4	11.7
81P Wild 2	6.39	2003	1.58	3.2	6.5
107P Wilson-Harrington[d]	4.30	2001	1.00	2.8	9.0

[a]Adapted from Lang, Kenneth R., *Astrophysical Data: Planets and Stars*. New York, Springer-Verlag 1992 and Gary M. Kronk's comet Web site http://cometography.com. Short-period comets that have either been visited by spacecraft in the past or have been considered for encounters in the future, listed in the order of their recognition, or periodic comet number.

[b]The given perihelion date is the first to occur in the 21st century, and the perihelion distance is in AU, roughly the mean distance between the Earth and the Sun and about 150 billion, or 1.5×10^{11}, meters.

[c]The absolute magnitude is a measure of the intrinsic brightness of a comet, and a smaller magnitude indicates a brighter comet.

[d]Also asteroid (4015).

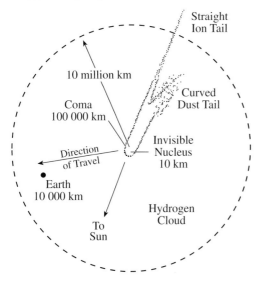

FIG. 23　Comet – anatomy The nucleus of a comet is usually invisible, unless a spacecraft is sent in to take a glimpse. A comet first becomes visible when it develops a coma of gas and dust. When the comet passes closer to the Sun, long ion and dust tails sometimes become visible, streaming out of the coma in the direction opposite to the Sun. When looking at a comet in ultraviolet light, the hydrogen atoms in its huge hydrogen cloud are detected.

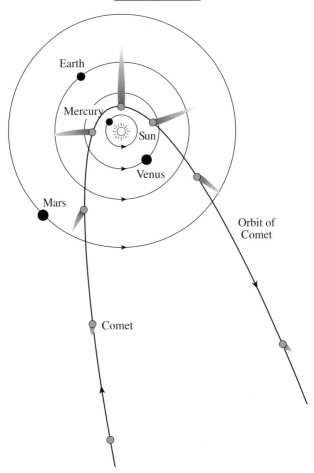

FIG. 24 Comet – trajectory and tails The path and changing shape of a typical comet as it enters the inner Solar System. Note that the tail of the comet is oriented away from the Sun, independent of the direction of travel of the comet.

TABLE 7 Comets - structure

Feature	Size[a]	Composition	Appearance
Nucleus	100 to 100,000 meters	Dust, ice and rock	Very dark
Coma	Up to 0.01 AU	Neutral (un-ionized) molecules and dust	Slightly yellow
Hydrogen cloud	Up to 0.1 AU	Hydrogen atoms	Ultraviolet radiation
Dust tail	Up to 0.1 AU	Dust particles	Yellow, curved
Ion tail	Up to 1 AU	Ionized molecules	Blue, straight

[a] One AU, roughly the average distance between the Earth and the Sun, is about 150 billion, or 1.5×10^{11}, meters.

bursts, as well as gamma rays from black holes, pulsars, quasars and supernovae. The gamma ray observatory was named after the American physicist **Arthur H. Compton (1892–1962)**, who is remembered for discovering the Compton effect in 1923, a phenomenon in which electromagnetic waves such as X-rays undergo an increase in wavelength after having been scattered by electrons. *See* gamma-ray burst.

Cone Nebula: A dark, tapering nebula in the constellation Monoceros surrounding the open star cluster NGC 2264.

Conjunction: The alignment of the Earth with two other bodies in the Solar System. Two other planets are said to be in conjunction when they are close together in the

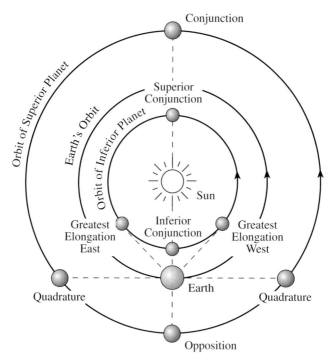

FIG. 25   Conjunction, elongation, and opposition When the Sun, the Earth and another planet are aligned, that planet is in conjunction. Venus and Mercury are inferior planets, with orbits between that of the Earth and the Sun, and they are in inferior conjunction when they lie between the Earth and the Sun. All of the other major planets are superior planets, with orbits outside that of the Earth, and they are in conjunction when on the other side of the Sun from the Earth. The superior planets are closest to the Earth when in opposition, aligned exactly opposite the Sun from the Earth. The elongation is the angular distance between the Sun and the planet, as viewed from the Earth, and the greatest elongations of Mercury and Venus are about 28 degrees and 47 degrees, respectively.

same part of the sky. Another planet is in conjunction with the Sun when the planet is on a line that passes through the Earth and the Sun, either on the opposite side of the Sun from the Earth, in superior conjunction, or on the same side of the Sun as the Earth, in inferior conjunction. *See* syzygy.

Constellation: A group of stars that forms a distinctive pattern or imaginary figure. traced on the sky, with a name linked to its shape that is often derived from Greek mythology. Twelve of the constellations are used to describe the zodiac. The list of 48 constellations given in Ptolemy's *Almagest* has grown to a modern list of 99 constellations, which vary in size. The stars in a constellation can be at very different distances, so the grouping lacks physical significance. *See* zodiac.

Continental drift: The idea that the continents are moving, or drifting, across the globe. The German meteorologist and geologist **Alfred Wegener (1880–1930)** developed the theory of continental drift, proposing that the continents were part of a single land mass, called *Pangea*, a Greek word meaning "all lands" that fragmented into today's continents, which started drifting apart 200 million years ago. This would explain the close jigsaw-fit between the coastlines on either side of the Atlantic Ocean and the geological and fossil similarities between Brazil and Africa. A mechanism for driving the continents across the globe was later realized with the discovery that continents are riding on the backs of large plates, which are pushed sideways by churning motions deep inside the Earth's hot interior and the spreading sea floor. *See* plate tectonics and sea-floor spreading.

Continuous spectrum: An unbroken distribution of radiation over a broad range of wavelengths. *See* continuum and spectrum.

Continuum: That part of a spectrum that has neither absorption nor emission lines, but only a smooth wavelength distribution of radiant intensity. *See* continuous spectrum and spectrum.

Convection: The transport of energy from a lower, hotter region to a higher, cooler region by the physical upwelling of hot matter. A bubble of gas that is hotter than its surroundings expands and rises vertically, resulting in transport and mixing. When it has cooled by passing on its extra heat to its surroundings, the bubble sinks again. Convection can occur when there is a substantial decrease in temperature with height such as the Earth's troposphere, the Sun's convective zone or a boiling pot of water. *See* convective zone and troposphere.

Convective zone: A layer in a star in which convection currents, or mass motions, are the main mechanism by which energy is transported outward. In the Sun, a convective zone extends from 0.713 of the solar radius to just below the photosphere. The opacity in the convective zone is so large that energy cannot be transported by radiation.

Copernicanism: The hypothesis that the Earth and the other planets orbit the Sun.

Core: The central region of a planet, star or other celestial object. In solar and stellar astronomy, the core is the central location where energy is generated by nuclear reactions. *See* fusion and proton-proton chain.

TABLE 8 Constellations[a]

Latin Name	R.A. h	Dec. °	Area °2	Genitive	Abbrev.	Translation
Andromeda	1	+40	722	Andromedae	And	Chained maiden
Antlia	10	−35	239	Antliae	Ants	Air Pump
Apus	16	−75	206	Apodis	Aps	Bird of Paradise
Aquarius	23	−15	980	Aquarii	Aqr	Water Bearer
Aquila	20	+5	652	Aquilae	Aql	Eagle
Ara	17	−55	237	Arae	Ara	Altar
Aries	3	+20	441	Arietis	Ari	Ram
Auriga	6	+40	657	Aurigae	Aur	Charioteer
Boötes	15	+30	907	Boötis	Boo	Herdsman
Caelum	5	−40	125	Caeli	Cae	Chisel
Camelopardalis	6	+70	757	Camelopardis	Cam	Giraffe
Cancer	9	+20	506	Cancri	Cnc	Crab
Canes Venatici	13	+40	465	Canum Venaticorum	CVn	Hunting Dogs
Canis Major	7	−20	380	Canis Majoris	CMa	Big Dog
Canis Minor	8	+5	183	Canis Minoris	CMi	Little Dog
Capricornus	21	−20	414	Capricorni	Cap	Sea Goat
Carina	9	−60	494	Carinae	Car	Ship's Keel[b]
Cassiopeia	1	+60	598	Cassiopeiae	Cas	Lady in Chair
Centaurus	13	−50	1060	Centauri	Cen	Centaur[c]
Cepheus	22	+70	588	Cephei	Cep	King of Ethiopia
Cetus	2	−10	1231	Ceti	Cet	Whale, Sea Monster
Chamaeleon	11	−80	132	Chamaeleontis	Cha	Chamaeleon
Circinus	15	−60	93	Circini	Cir	Compasses
Columba	6	−35	270	Columbae	Col	Dove
Coma Berenices	13	+20	386	Comae Berenices	Com	Berenice's Hair[c]
Corona Australis	19	−40	128	Coronae Australis	CrA	Southern Crown
Corona Borealis	16	+30	179	Coronae Borealis	CrB	Northern Crown
Corvus	12	−20	184	Corvi	Crv	Crow
Crater	11	−15	282	Crateris	Crt	Cup
Crux	12	−60	68	Crucis	Cru	Southern Cross
Cygnus	21	+40	804	Cygni	Cyg	Swan

TABLE 8 Constellations^a (continued)

Latin Name	R.A. h	Dec. °	Area °2	Genitive	Abbrev.	Translation
Delphinus	21	+10	189	Delphini	Del	Dolphin, Porpoise
Dorado	5	−65	179	Doradus	Dor	Swordfish, Dorado fish
Draco	17	+65	1083	Draconis	Dra	Dragon
Equuleus	21	+10	72	Equulei	Equ	Little Horse
Eridanus	3	−20	1138	Eridani	Eri	River Eridanus^c
Fornax	3	−30	398	Fornacis	For	Furnace
Gemini	7	+20	514	Geminorum	Gem	Heavenly Twins
Grus	22	−45	366	Gruis	Gru	Crane
Hercules	17	+30	1225	Herculis	Her	Kneeling Giant
Horologium	3	−60	249	Horologii	Hor	Clock
Hydra	10	−20	1303	Hydrae	Hya	Water Monster
Hydrus	2	−75	243	Hydri	Hyi	Sea Serpent
Indus	21	−55	294	Indi	Ind	Indian
Lacerta	22	+45	201	Lacertae	Lac	Lizard
Leo	11	+15	947	Leonis	Leo	Lion
Leo Minor	10	+35	232	Leonis Minoris	LMi	Little Lion
Lepus	6	−20	290	Leporis	Lep	Hare
Libra	15	−15	538	Librae	Lib	Balance, Scales
Lupus	15	−45	334	Lupi	Lup	Wolf
Lynx	8	+45	545	Lyncis	Lyn	Lynx
Lyra	19	+40	286	Lyrae	Lyr	Lyre, Harp
Mensa	5	−80	153	Mensae	Men	Table, Mountain
Microscopium	21	−35	210	Microscopii	Mic	Microscope
Monoceros	7	−5	482	Monocerotis	Mon	Unicorn
Musca	12	−70	138	Muscae	Mus	Fly
Norma	16	−50	165	Normae	Nor	Square, Level
Octans	22	−85	291	Octantis	Oct	Octant
Ophiuchus	17	0	948	Ophiuchi	Oph	Serpent bearer
Orion	5	+5	594	Orionis	Ori	Hunter
Pavo	20	−65	378	Pavonis	Pav	Peacock

TABLE 8   Constellations[a] (continued)

Latin Name	R.A. h	Dec. °	Area °2	Genitive	Abbrev.	Translation
Pegasus	22	+20	1121	Pegasi	Peg	Winged Horse
Perseus	3	+45	615	Persei	Per	Champion
Phoenix	1	−50	469	Phoenicis	Phe	Phoenix
Pictor	6	−55	247	Pictoris	Pic	Painter. Easel
Pisces	1	+15	889	Piscium	Psc	Fish
Pisces Austrinus	22	−30	245	Piscis Austrini	PsA	Southern Fish
Puppis	8	−40	673	Puppis	Pup	Ship's Stern[b]
Pyxis	9	−30	221	Pyxidis	Pyx	Ship's Compass[b]
Reticulum	4	−60	114	Reticuli	Ret	Net
Sagitta	20	+10	80	Sagittae	Sge	Arrow
Sagittarius	19	−25	867	Sagittarii	Sgr	Archer
Scorpius	17	−40	497	Scorpii	Sco	Scorpion
Sculptor	0	−30	475	Sculptoris	Sci	Sculptor
Scutum	19	−10	109	Scuti	Sct	Shield
Serpens	17	0	637	Serpentis	Ser	Serpent
Sextans	10	0	314	Sextantis	Sex	Sextant
Taurus	4	+15	797	Tauri	Tau	Bull
Telescopium	19	−50	252	Telescopii	Tel	Telescope
Triangulum	2	+30	132	Trianguli	Tri	Triangle
Triangulum Australe	16	−65	110	Trianguli Australis	TrA	Southern Triangle
Tucana	0	−65	295	Tucanae	Tuc	Toucan
Ursa Major	11	+50	1280	Ursae Majoris	UMa	Big Bear
Ursa Minor	15	+70	256	Ursae Minoris	UMi	Little Bear
Vela	9	−50	500	Velorum	Vel	Ship's Sail[b]
Virgo	13	0	1294	Virginis	Vir	Maiden, Virgin
Volans	8	−70	141	Volantis	Vol	Flying Fish
Vulpecula	20	+25	268	Vulpeculae	Vul	Little Fox

[a] Adapted from Lang, Kenneth R., *Astrophysical Data: Planets and Stars*, Springer-Verlag, 1992, pp. 158-160.

[b] Previously formed constellation Argo Navis, the Argonaut's ship.

[c] Proper names.

TABLE 9   Corona — emission lines[a]

Wavelength (nanometers)	Emitting Ion	Wavelength (nanometers)	Emitting Ion
332.8	Ca XII	530.3 (green line)	Fe XIV
338.8	Fe XIII	569.4 (yellow line)	Ca XV
360.1	Ni XVI	637.5 (red line)	Fe X
398.7	Fe XI	670.2	Ni XV
408.6	Ca XIII	706.0	Fe XV
423.1	Ni XII	789.1	Fe XI
511.6	Ni XIII	802.4	Ni XV

[a]Ca = calcium, Fe = iron, and Ni = nickel. Subtract one from the Roman numeral to get the number of missing electrons.

TABLE 10   Corona – holes and loops

Feature	Largest Extent (10^3 km)	Temperature (10^6 K)	Electron Density (m^{-3})
Coronal Holes	700 to 900	1.0 to 1.5	4×10^{14}
Active Region Loops	10	2 to 4	$(1 \text{ to } 7) \times 10^{15}$
X-ray bright point	5 to 20	2.5	1.4×10^{16}

Corona: A shimmering halo of pearl-white light seen momentarily at the limb, or apparent edge, of the Sun during a total solar eclipse, called the *corona,* from the Latin word for "crown." The outermost, high-temperature region of the solar atmosphere, above the chromosphere and transition region, consisting of almost fully ionized gas contained in closed magnetic loops, called coronal loops, or expanding out along open magnetic field lines to form the solar wind. In 1941 the Swedish astronomer **Bengt Edlén (1906–1993)** and the German astronomer **Walter Grotrian (1890–1954)** showed that emission lines from the solar corona were forbidden transitions of atoms that are highly ionized at an unexpectedly high, million-degree temperature. In 1958 the American physicist **Eugene N. Parker (1927–)** showed that the million-degree electrons and protons in the corona will overcome the Sun's gravity and accelerate to supersonic speeds, naming the resultant radial outflow the solar wind. Even near the Sun, the corona is a highly rarefied, low-density gas, with electron densities of less than 10^{16} electrons per cubic meter, heated to temperatures of millions of degrees. The corona is briefly visible to the unaided eye during a total eclipse of the Sun, for at most 7.5 minutes; at other times it can be observed in visible white light by using a special instrument called a coronagraph. The visible-light corona may be divided into the inner K or electron corona, with a continuous spectrum, and the outer F or dust corona that displays Fraunhofer absorption lines. The E corona is the emission-line corona. The corona can always be observed across the solar disk at X-ray and radio wavelengths. The shape of the corona is determined by the distribution of solar magnetic fields, which varies with the solar activity cycle.

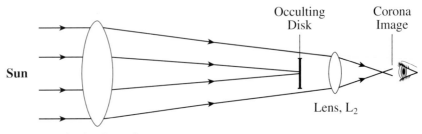

FIG. 26   Coronagraph Sunlight enters from the left, and is focused by an objective lens, L_1, on an occulting disk, blocking the intense glare of the photosphere. Light from the corona, which is outside the photosphere, bypasses this occulting disk and is focused by a second lens, L_2, forming an image of the corona. Other optical devices are placed along the light path to divert and remove excess light.

See coronagraph, coronal hole, coronal loop, coronal mass ejection, solar activity cycle, and solar wind.

Coronae: Large, elevated circular structures on Venus that are pushed up from below by rising molten rock trying to get out, from *corona,* the Latin word for "crown."

Coronagraph: An instrument, used in conjunction with a telescope, that makes it possible to mask the light from the Sun or other star, in order to observe gas, dust or larger objects very close to the star. A coronagraph is used to observe the faint solar corona in white light, or in all the colors combined, at times other than during a solar eclipse. The bright light of the Sun's photosphere is blocked out by an occulting disk, providing an artificial eclipse, with additional precautions for removing all traces of stray light. The French astronomer **Bernard Lyot (1897–1952)** invented the coronagraph in 1930. Even with a coronagraph located at a high site where the sky is very clear, scattering of light by the Earth's atmosphere is a problem. Coronagraphs therefore work best when placed on satellites above the Earth's atmosphere.

Coronal hole: An extended region in the solar corona where the density and temperature are lower than other places in the corona, and the coronal emission lines and the extreme ultraviolet and soft X-ray coronal emission are abnormally faint or absent. The weak, diverging and open magnetic field lines in coronal holes extend radially outward and do not immediately return back to the Sun. The high-speed part of the solar wind streams out from coronal holes. The low density of the gas makes these parts of the corona appear dark in extreme ultraviolet and soft X-ray images of the Sun, as if there were a hole in the corona. Coronal holes are nearly always present near the solar poles, and can also occur at lower solar latitudes.

Coronal loop: A magnetic loop that passes through the corona and joins regions, called footpoints, of opposite magnetic polarity in the underlying photosphere. Coronal loops can have exceptionally strong magnetic fields, and they often contain the dense, million-degree coronal gas that emits intense X-ray radiation. *See* footpoint.

Coronal mass ejection: Abbreviated CME, the transient ejection of plasma and magnetic fields from the Sun's corona into interplanetary space, detected by white-light coronagraphs. A large body of magnetically confined, coronal material being released from the corona. A coronal mass ejection contains 5 billion to 50 billion tons, or 5 million million to 50 million million kilograms, of gas, and can travel through interplanetary space at a high, supersonic speed of up to 1200 kilometers per second. It is often associated with an eruptive prominence, and sometimes with a strong solar flare. Coronal mass ejections produce intense shock waves, accelerate vast quantities of energetic particles, grow larger than the Sun in a few hours, and when directed at Earth can cause intense geomagnetic storms, disrupt communications, damage satellites and produce power surges on electrical transmission lines. *See* geomagnetic storm and prominence.

Coronal streamer: A magnetically confined, loop-like coronal structure in the low corona straddling a magnetic neutral line on the solar photosphere. These high-density, bright coronal structures have ray-like stalks that extend radially outward to large distances in the outer corona. Near the minimum in the 11-year solar activity cycle, coronal streamers are located mostly near the solar equator and appear to be the source of the slow-speed solar wind. *See* helmet streamer, solar activity cycle and solar wind.

Coronal transient: *See* coronal mass ejection.

Coronium: A supposedly unknown chemical element emitting unidentified emission lines in the spectrum of the solar corona. In 1941, the Swedish astronomer **Bengt Edlén (1906–1993)** and the German astronomer **Walter Grotrian (1890–1954)** showed that the mysterious lines are produced by highly ionized forms of known elements, such as iron, in the unexpectedly high, million-degree temperature of the solar corona.

Corpuscular radiation: Charged particles, mainly protons and electrons, emitted by the Sun. The stream of electrically charged particles was hypothesized in the 1950s and later renamed the solar wind. *See* solar wind.

COsmic Background Explorer: NASA launched the *COsmic Background Explorer,* abbreviated *COBE*, on 18 November 1989; instrument operations were terminated on 23 December 1993. These instruments measured the angular distribution and spectrum of the cosmic microwave background radiation, showing that it has a low-level anisotropy and a blackbody spectrum, with a temperature of 2.726 ± 0.010 kelvin.

Cosmic background radiation: *See* cosmic microwave background radiation.

Cosmic microwave background radiation: Isotropic microwave emission coming with almost equal intensity from all directions in space, discovered in 1965 by the German-born American radio engineer **Arno A. Penzias (1933–)** and the American radio astronomer **Robert W. Wilson (1936–)** while determining the sources of excess noise in a horn antenna at the Bell Telephone Laboratories. Measurements from the *COsmic Background Explorer* spacecraft, reported in 1990, demonstrated that the cosmic microwave background radiation has a blackbody spectrum corresponding to a temperature of 2.726 ± 0.010 kelvin. It is remnant radiation from the hot, early Universe

TABLE 11 Cosmic microwave background radiation

Radiation parameters [a, b]

Parameter	Name	Value
$T_0 = T_{cmb}$	Temperature	2.725 ± 0.002 K
n_γ	Photon density	410.4 ± 0.9 cm^{-3}
ρ	Mass-energy density	4.648×10^{-34} g cm^{-3}
T_1	Dipole anisotropy	$0.003\,346$ K ± 0.000017 K
T_1/T_0	Dipole anisotropy/Temperature	$0.001\,228$
V	Velocity of solar motion toward background	369.5 ± 3.0 km s^{-1}
RA (2000)	Direction of solar motion, right ascension	$11^h\ 11^m\ 57^S \pm 23^S$
Dec. (2000)	Direction of solar motion, declination	$-7.22° \pm 0.08°$
(l, b)	Direction of solar motion, galactic	$(263.85° \pm 0.1°,\ 48.25° \pm 0.04°)$
$\Delta T/T_0$	Anisotropy, dipole removed	$(1.1 \pm 0.1) \times 10^{-5}$
$T_2 = Q_{rms}$	Quadrupole moment	$(8 \pm 2) \times 10^{-6}$ K

Cosmological parameters [b]

Name	Meaning	Value
H_0	Hubble Constant	Current Expansion Rate	71 ± 4
t_0	Age of Universe	Time since Big Bang	$(13.7 \pm 0.2) \times 10^9$ years (billion years)
t_{dec}	Decoupling Time	Time Radiation Originated	$(3.79 \pm 0.08) \times 10^5$ years
		(After Big Bang, also Recombination Time and redshift 1089)	
t_r	Reionization Time	First Stars Originated (After Big Bang)	≈ 200 million years
Ω_{tot}	Omega	Matter and Energy Density	1.01 ± 0.02
Ω_Λ	Dark Energy	Amount of Dark Energy	0.73 ± 0.04
Ω_m	Total Matter	Amount of Total Matter	0.27 ± 0.04
Ω_b	Baryonic Matter	Amount of Baryons	0.044 ± 0.004
Ω_d	Dark Matter	Amount of Dark Matter	0.226 ± 0.044

[a] Adapted from Lang, Kenneth R., *Astrophysical Formulae II, Space, Time Matter and Cosmology,* Third Edition, New York, Springer-Verlag, 1999, pp. 269-273.

[b] From the **Wilkinson Microwave Anisotropy Probe** together with other astronomical observations, adapted from Charles L. Bennett et al., *Astrophysical Journal Supplement* **148**, 1 (2003). The parameter Ω is the ratio of the specified quantity to the critical amount required to keep the expansion of the Universe on the brink of closure. An $\Omega = 1.00$ is consistent with inflation and a Universe that is described by Euclidean geometry without space curvature.

released 379,000 years after the Big Bang, at the time of decoupling of radiation from cosmic matter, and subsequently thinned and cooled as the Universe expanded. In 2003 scientists reported measurements of a faint anisotropy, or temperature variations, in the radiation detected with the **Wilkinson Microwave Anisotropy Probe**, abbreviated **WMAP**, spacecraft, and combined them with other astronomical measurements to infer important cosmological parameters, including the age, energy and matter content, and expansion rate of the Universe. *See* blackbody radiation, **COsmic Background Explorer**, decoupling, isotropy, microwave, and **Wilkinson Microwave Anisotropy Probe**.

Cosmic rays: High-energy, charged, subatomic particles that enter the Solar System from interstellar space, moving at speeds approaching the speed of light and attaining energies greater than a million eV, typically a billion, or 10^9, eV and as large as 10^{20} eV. Cosmic rays were discovered in 1912 by the Austrian physicist **Victor F. Hess (1883–1964)**. In 1925 the American physicist **Robert Millikan (1868–1953)** conclusively showed that they come from outer space, giving them the name cosmic rays. In 1933 the American physicist **Arthur H. Compton (1892–1962)** demonstrated that the "rays" are charged particles. Cosmic rays can smash into atoms and split them apart, creating lighter elements, such as lithium, beryllium, and boron. Protons are the most abundant kind of cosmic rays, but they include lesser amounts of heavier atomic nuclei, such as alpha particles or helium nuclei, and electrons. Cosmic rays beyond the Earth's atmosphere are known collectively as primary cosmic rays. When they enter the Earth's atmosphere, cosmic rays collide with atmospheric atoms and produce various atomic and subatomic particles, called secondary cosmic rays, which can be detected at ground level. Sometimes the term galactic cosmic ray is used to distinguish those coming from interstellar space from high-energy, charged particles coming from the Sun. The latter have historically been called solar cosmic rays. *See* electron volt for eV.

Cosmogony: The study of the origin of the Universe and objects in it.

Cosmological constant: A controversial parameter introduced into the *General Theory of Relativity* by the German physicist **Albert Einstein (1879–1955)** in 1917 to provide an anti-gravity force of cosmic repulsion that opposes the attractive force of gravita-

TABLE 12   Cosmic rays – flux at the top of the atmosphere[a]

Type	Flux (particles m^{-2} s^{-1})
Hydrogen (protons)	640
Helium (alpha particles)	9
Electrons	<13
Carbon, Nitrogen, Oxygen	6

[a] The flux is in units of nuclei m^{-2} s^{-1} for particles with energies greater than 1.5 billion (1.5×10^9) eV per nucleon, arriving at the top of the atmosphere from directions within 30 degrees of the vertical.

tion, but abandoned by Einstein in 1931 after the discovery of the expanding Universe. The cosmological constant is designated by either λ or Λ, the lower or upper case Greek letters lambda. It was resurrected as a possible explanation for a difference between the expansion age of the Universe and the age of the oldest astronomical objects, and also as a possible explanation for dark energy. *See* critical mass density, dark energy, Hubble constant, Hubble time, and omega.

Cosmological principle: The assumption that the Universe is both homogeneous and isotropic, or the same in all directions, on a sufficiently large scale, introduced in 1934 by the English astrophysicist **Edward A. Milne (1896–1950)**. The high level of isotropy in the cosmic microwave background radiation supports the cosmological principle; but there are anisotropic clusters and voids in the distribution of galaxies as far as telescopes can see, stretching across billions of light-years. In 1948 the Austrian-born astrophysicists, then at Cambridge University, **Hermann Bondi (1919–2005)** and **Thomas Gold (1920–2004)** proposed the perfect cosmological principle, extending the principle to temporal as well as spatial homogeneity and using it as a basis for the Steady State cosmology. *See* homogeneous, isotropic, perfect cosmological principle, and Steady State.

Cosmological redshift: *See* redshift.

Cosmology: Conjectures about the beginning, composition, evolution, fate, and structure of the Universe as a whole, sometimes grounded in observational facts but often highly speculative. *See* Big Bang, inflation, and Steady State.

Coudé focus: A fixed focus along the polar axis of a telescope with an equatorial mount. A system of secondary mirrors is used to direct light reflected from the primary mirror around to the fixed coudé focus behind it, where large, massive immovable instruments, such as a spectrograph, are located. From the French word *coudé,* meaning "bent like an elbow".

Coulomb barrier: The electric field repulsion experienced during the approach of two charged subatomic particles, each of positive charge or both with negative charge. When the Coulomb barrier of two protons is overcome by the tunnel effect, nuclear reactions that produce the Sun's energy and radiation can occur. *See* tunnel effect.

Crab Nebula: The nebula M1, or NGC 1952, in the constellation Taurus, a remnant of a type II supernova explosion that occurred on 4 July 1054. It is about 6,500 light-years away, and contains a radio pulsar, or rotating neutron star, at its center, spinning 30 times a second. The English amateur astronomer **John Bevis (1695–1771)** discovered the nebula in 1731. The French astronomer **Charles Messier (1730–1817)** independently found it in 1758, listing it as the first entry in his famous catalogue. The Irish astronomer **William Parsons (1800–1867)**, the third **Earl of Rosse**, gave it its present name, the Crab nebula after sketching it to resemble a crab in 1845. In 1921 the American astronomer **John C. Duncan (1882–1967)** compared photographic plates taken 11.5 years apart, and found that the Crab Nebula was expanding at an average of about 0.2 seconds of arc per year, beginning roughly 900 years before. In 1928 the American

TABLE 13 Crab Nebula[a]

Expansion Center	R. A. (1950.0)	=	$05^h 31^m 32.2^s \pm 0.1^s$
	Dec.(1950.0)	=	$+21° 58' 50'' \pm 1''$
Explosion	Date	=	1054 A.D.
Progenitor Star Mass	M	≈	9 $M_\odot$ (main sequence)
Ejected Mass	M_{ejc}	=	2 to 3 $M_\odot$
Neutron Star Mass	M_N	=	1.4 $M_\odot$
Pulsar Position	R.A. (1950.0)	=	$05^h \quad 31^m \quad 31.405^s$
	Dec. (1950.0)	=	$+21° 58' 54.39''$
Pulsar Period	P	=	0.033 326 323 455 s
Pulsar Period Derivative	dP/dt	=	421.288×10^{-15} s s^{-1}
Angular Extent	θ	=	$4.5' \times 7.0'$
Distance	D	=	2.0 ± 0.1 kpc
Linear Radius	R	=	2.7 pc × 4.2 pc
Maximum Radial Velocity	V_{rmax}	=	1450 ± 40 km s^{-1}
Maximum Proper Motion	μmax	=	$0.222 \pm 0.002''$ yr^{-1}
Maximum Expansion Velocity	V_{exp}	=	1500 km s^{-1}
Radio Flux Density	S	=	1040 Jy at 1 GHz
Spectral Index	α	=	−0.30 (integrated radio)
X-ray Luminosity	L_x	=	$10^{37.38}$ erg s^{-1}
Total Luminosity	L	=	$10^{38.14}$ erg s^{-1}

[a]Adapted from Lang, Kenneth R., *Astrophysical Data: Planets and Stars*, New York, Springer-Verlag, 1992, p. 715. The mass values are in units of the Sun's mass $M_\odot = 1.989 \times 10^{33}$ grams.

astronomer **Edwin Hubble (1889–1953)** identified the Crab Nebula with a "guest star" recorded by the Chinese in the constellation Taurus in 1054 AD. In 1948 the Crab Nebula was identified as a strong source of radio radiation, named and listed as Taurus A and later as 3C 144. X-rays were detected from the Crab Nebula in 1963, and the Crab Radio Pulsar PSR 0531+21, with a period of 33 milliseconds, was discovered in 1968. The synchrotron radiation of the Crab Nebula extends from the radio domain into the visible. The *Chandra X-ray Observatory* has provided detailed X-ray observations of the Crab Nebula Pulsar. *See Chandra X-ray Observatory*, PSR 0531+02, supernova remnant, and synchrotron radiation.

Crab Nebula Pulsar: Designated PSR 0531+21, a radio pulsar with a period of 33 milliseconds that has also been observed at visible-light, radio, and X-ray wavelengths.

Crater: Round features excavated from the surface of an asteroid, satellite or planet by the explosive impact of a colliding meteorite arriving from interplanetary space. Craters have raised, circular rims and a floor depressed below the surrounding terrain. The force of a large meteorite creates a central peak in the floor of the biggest craters. Every solid planet or satellite contains impact craters, but in different amounts that depend on the ages of their surfaces; older surfaces have more craters. Impact craters on the Moon, Mercury and the icy satellites of the giant planets record an ancient, intense rain of meteorites about four billion years ago, and a continued, less intense cosmic bombardment since then. The largest lunar craters are called impact basins; they were filled with dark molten lava rising from the lunar interior after the impact. *See* Moon.

Critical mass density: The average mass density of matter in a homogeneous, isotropic Universe required to just barely continue its expansion forever, provided that there is no dark energy and the cosmological constant is zero. It amounts to about ten hydrogen atoms per cubic meter of space. If the Universe had a mass density greater than its critical value, then the Universe would stop expanding in the future, and then collapse back on itself. There is no observational evidence for such a dense Universe, and if it exists it would have to consist mainly of dark, invisible non-baryonic matter. The omega parameter is equal to the ratio of the actual mass density of the Universe to the critical density, so omega equals one when the actual mass density is just equal to the critical density. *See* closed Universe, dark matter, non-baryonic matter, omega, and open Universe.

Crux: A constellation in the southern hemisphere, popularly known as the Southern Cross, with four bright stars, Alpha, or Acrux, Beta, Gamma and Delta, arranged in a pattern that resembles the ends of a cross or the ends of the sticks in a kite. Its brightest star, Acrux, is located at distance of 320 light-years; it is a visible double star with a combined apparent visual magnitude of 0.77 and a combined absolute visual magnitude of −4.19. Crux is the smallest constellation in the sky, and it contains the dark nebula Coalstack and the open cluster NGC 4755, known as the Jewel Box from the varied colors of its stars. *See* Acrux, Coalstack, and Jewel Box.

Curvature of space: The distortion of space, described by non-Euclidean geometry, in the neighborhood of matter, one of the consequences of Einstein's *General Theory of Relativity*. This curvature makes rays of light follow curved paths when passing near a massive object, including the Sun and clusters of galaxies. *See General Theory of Relativity*.

CXO: Acronym for the ***Chandra X-ray Observatory***.

Cyclotron: *See* particle accelerator.

Cygnus A: The radio galaxy Cygnus A, also designated 3C 405, is the second brightest radio source in the sky, located in the constellation Cygnus. The American radio engineer **Grote Reber (1911–2002)** discovered the radio source in 1944. In 1954 two German-born American astronomers **Walter Baade (1893–1960)** and **Rudolph Minkowski (1895–1976)** used accurate positions supplied by English radio astronomers and the 200-inch (5.0-meter) **Hale telescope** on Mount Palomar to identify the discrete radio source with a distant galaxy about 750 million light-years away, discovering the prototype of radio galaxies whose radio luminosities are comparable to their optical, or visible-light ones. Like many other radio galaxies, Cygnus A consists of two radio lobes that extend about 200,000 light-years to either side of the optical galaxy. Jets emanating from the core of the visible-light counterpart, perhaps from a supermassive black hole, carry energetic electrons moving at the velocity of light, which feed the two lobes that shine by synchrotron radiation. *See* radio galaxy and synchrotron radiation.

Cygnus Loop: A vast nebulous region, about 3 degrees across. Its brightest part is known as the Veil Nebula, designated NGC 6692, and other parts of the Cygnus Loop that are also seen in visible light are known as NGC 6960, 6979, and 6995.

Cygnus X-1: An intense X-ray source in the constellation Cygnus, with a blue stellar companion seen in visible light. The millisecond time variations of the X-ray radiation from Cygnus X-1 were discovered in 1971 from observations with the **Uhuru** satellite. Cygnus X-1 is thought to be a stellar black hole, first identified in 1971 by the English astronomers **B. Louise Webster (1941–1990)** and **Paul Murdin (1942–)** and confirmed in 1972 by the Canadian astronomer **Charles Bolton (1943–)**. The unseen black hole is accreting material from its visible companion, producing X-rays as the material spirals into the black hole. The companion visible star is a blue O9 supergiant located at a distance of 8,000 light-years with an apparent visual magnitude of 8.8 and an absolute visual magnitude of –9.3 It has an estimated mass of 30 solar masses and a luminosity of about 400,000 times that of the Sun. The orbital parameters and mass of the blue supergiant indicate that the unseen black hole companion has a mass larger than 16 solar masses, or more massive than either a neutron star or white dwarf. *See* black hole, neutron star, stellar black hole, **Uhuru Mission,** and white dwarf.

Dark energy: In 1998 members of the Supernova Cosmology Project and the High-Z Supernova Search independently reported that observations of type Ia supernovae indicate that dark energy is counteracting the pull of gravity and apparently causing the expansion of the Universe to accelerate and increase in speed. Dark energy could account for as much as 73 percent of the matter-energy content of the Universe; only 23 percent is now supposed to reside in unseen dark matter and just 4 percent in atomic matter. *See* cosmic microwave background radiation, cosmological constant, dark matter, quintessence and supernovae.

Dark halo: The invisible, massive, outer region of the Milky Way that surrounds the luminous disk. The dark halo consists mostly of dark matter and outweighs the rest of our Galaxy by a factor of almost ten. *See* dark matter, galactic halo, Galaxy, and Milky Way.

Dark matter: Invisible, non-luminous matter that is inferred by measuring its gravitational influence on the motions of visible gas, stars or galaxies. Unseen dark matter surrounds bright spiral galaxies in haloes and exits in the space between galaxies within clusters of galaxies. In 1937 the Swiss astronomer **Fritz Zwicky (1898–1974)** showed that the Coma cluster of galaxies must contain ten times more invisible dark matter than the mass in visible galaxies to keep the cluster's galaxies from flying apart. In 1970 the Australian radio astronomer **Kenneth C. Freeman (1940–)** noticed that the hydrogen gas in the outer parts of spiral galaxies, observed by their twenty-one centimeter line radiation, is moving unexpectedly fast, suggesting that each spiral galaxy is enveloped by a halo of dark matter at least ten times more massive than the visible components. The American astronomer **Vera C. Rubin (1928–)** and her colleagues made similar conclusions in the early 1970s from observations of the fast-moving, visible-light material in the outer parts of spiral galaxies. Non-baryonic dark matter could account for about 23 percent of the matter content of the Universe, remaining unseen at any wavelength of electromagnetic radiation, but its composition remains unknown. *See* baryonic matter, cold dark matter, dark energy, hot dark matter, non-baryonic matter, and twenty-one centimeter line.

Dark nebula: Dense clouds of gas and dust in the Milky Way that are not illuminated. The American astronomer **Edward Barnard (1857–1923)** provided a photographic atlas of the dark markings, as he called them, in 1919, with a second atlas published in 1927 after his death. In 1923 the German astronomer **Maximilian Wolf (1863–1932)** showed that the dark nebulae are obscuring clouds containing solid dust particles, which absorb the light of distant stars. In the 1960s and 1970s, American radio astronomers discovered numerous molecules in the cold, dark nebulae, such as ammonia, carbon monoxide, formaldehyde and water. Present-day star formation takes place in massive molecular clouds associated with dark nebulae. *See* Coalstack, giant molecular cloud, Horsehead Nebula, and Keyhole Nebula.

Day: The length of time that a planet takes to spin once on its axis, often measured in units of Earth days. One Earth day is equal to 24 hours and to 1,440 minutes. By definition, one Julian century has 36,525 days and one day is defined as 86,400 seconds of International Atomic Time. The rotation of the Earth with respect to the Sun is called Universal Time or Solar Time, and the rotation of the Earth with respect to the stars is called Sidereal Time, or Star Time. A mean solar day is equal to 24 hours 03 minutes 56.553678 seconds of mean sidereal time, and a mean sidereal day is equal to 23 hours 56 minutes 04.090524 seconds of mean solar time. Thus, mean sidereal time gains 03 minutes 56.5553678 seconds per day on mean solar time, which means that relative to the Sun the stars rise and set four minutes earlier each day. The length of the day increases at the rate of about 0.002 seconds per century as the result of tidal friction. *See* International Atomic Time, sidereal time, tides, and Universal Time.

Deceleration parameter: A measure of the rate at which the expansion of the Universe may be slowing down, owing to the braking effect of the gravitational pull between galaxies.

Declination: The celestial latitude of a cosmic object, denoted by δ, the lower case Greek letter delta, or by Dec. The declination is the angular distance along a great circle from the celestial equator in a north (positive) or south (negative) direction. *See* celestial coordinates.

Decoupling time: The time, about 379,000 years after the Big Bang, when the expanding Universe had cooled enough, and become sufficiently rarefied, for the first atoms to form, resulting from the decrease in cosmic mass density as the Universe expanded into a large volume. The observed cosmic microwave background radiation was released at this time, when the radiation photons decoupled from regular interaction, or constant collisions, with particles of matter. *See* cosmic microwave background radiation.

Deep Impact: A NASA mission launched on 12 January 2005, making a deep crater in Comet Tempel 1 on 4 July 2005, to learn more about comets and the formation of the Solar System.

Deep Space Network: NASA's international network of antennas, abbreviated DSN, which supports interplanetary spacecraft missions, selected Earth-orbiting missions, and radio and radar astronomy observations for the exploration of the Solar System and the Universe.

Deep Space 1: Launched on 24 October 1998 the **Deep Space 1** spacecraft flight tested an ion engine and other advanced technologies, and was redirected to an encounter with Comet Borrelly on 22 September 2001, taking images of the comet nucleus and showing that it is an irregular chunk of ice and rock covered by dark slag that reflects only about four percent of the incident sunlight. The spacecraft was retired on 18 December 2001,

Degenerate matter: The state of matter existing in the interiors of stars in the final stage of their evolution when they no longer produce energy by nuclear reactions in their cores. A white dwarf star is supported by degenerate electron pressure, while neutron degeneracy supports a neutron star. This pressure does not depend on temperature and is only a function of mass density. It is a consequence of the Exclusion Principle, proposed in 1925 by the Austrian physicist **Wolfgang Pauli (1900–1958)**. In 1926 the English physicist **Ralph H. Fowler (1889–1944)** first showed that a white dwarf star is supported by electron degeneracy pressure. *See* exclusion principle, neutron star, and white dwarf.

Degree: The unit of angular measurement, equal to 1/360 of a full circle. There are 90 degrees from the zenith to the horizon, and the angle subtended by the Sun and the Moon is about one half of a degree. A degree is also a unit of temperature, as in degrees Celsius, Fahrenheit or Kelvin. *See* arc degree, Celsius, Fahrenheit, and Kelvin.

Deimos: The smaller and outermost of the two tiny satellites of Mars, discovered by the American astronomer **Asaph Hall (1829–1907)** in 1877. It is named *Deimos,* Greek for "flight, panic or terror", after one of the attendants of the Greek god of war, Ares, in Homer's *Illiad.* The dimensions of Deimos are 7.6 × 6.2 × 5.4 kilometers, with a mean mass density of about 1700 kilograms per cubic meter.

Delta Cephei: A pulsating star in the constellation Cepheus, which is the prototype for a class of variable stars known as the Cepheids. The English astronomer **John Goodricke (1764–1786)** noted the variability of Delta Cephei in 1784. Located at a distance of 930 light-years, Delta Cephei varies between apparent magnitude 3.5 and 4.3, and an absolute visual magnitude of –3.1 to –3.9, with a period of 5 days 8 hours 47 minutes and 32 seconds. *See* Cepheid variable.

Deneb: The brightest known A star in our Galaxy, shining in the tail of the constellation Cygnus, the Swan – *Deneb* means "tail" in Arabic. Also known as Alpha Cygni, it is a white A2 supergiant star with an apparent visual magnitude of +1.25, a distance of 3230 light-years, and an absolute visual magnitude of –8.7. Deneb is the most luminous white class A star known in our Galaxy, shining nearly a quarter million times brighter than the Sun.

Density: An object's mass divided by its volume, also known as the mass density. *See* mass density.

Deuterium: The rare, heavy isotope of hydrogen containing both a proton and a neutron in its nucleus, also known as heavy hydrogen. An atom of light hydrogen, which is much more abundant than deuterium, contains one proton and no neutron in its nucleus. Deuterium was discovered in 1931 by the American chemist **Harold Clayton Urey (1893–1981)**. *See* hydrogen, and isotope.

Diameter: The length of an imaginary straight line passing through the center of a spherical object, equal to twice the radius of the sphere.

Differential rotation: Rotation of the different parts of a non-solid body or collection of bodies at different rates. The outer layers of the giant planets and the Sun exhibit differential rotation, with their middle equatorial regions moving faster than the polar ones. Stars and interstellar gas also rotate differentially, revolving about the galactic center in independent orbits at speeds that can differ. The particles in the rings of

Saturn revolve around the planet at speeds that decrease with increasing distance, just as more distant planets move about the massive Sun with slower speeds and longer orbital periods. Stars that are nearer the galactic center also revolve about it at faster speeds than more distant stars.

Differentiation: The process by which the denser materials of a planetary body sink toward the center while the less dense materials rise toward the surface. Differentiation of the Earth resulted in a dense iron-rich core and a mantle and crust composed of less dense, silicate rocks.

Diffuse nebula: A luminous cloud of gas in interstellar space that cannot be resolved into individual stars. Emission nebulae, or H II regions, planetary nebulae, and reflection nebulae are examples of diffuse nebulae. *See* emission nebula, H II region, planetary nebula, and reflection nebula.

Dimension: A geometrical axis or direction in space or time.

Dipole: Magnetic fields that include both a north and a south magnetic pole.

Disk: The visible part of the Sun or other cosmic object.

Distance modulus: The distance modulus, m – M, of a celestial object is the difference between the apparent magnitude, m, and the absolute magnitude, M. It is related to its distance, D, in parsecs by the equation m – M = 5 (log D) – 5.

dM star: A red dwarf star at the lower end of the main sequence of the Hertzsprung-Russell diagram. A dMe star is a red dwarf with hydrogen lines in emission. *See* dwarf star, flare star, Hertzsprung-Russell diagram, M star, and red dwarf star.

Doppler effect: The change in the observed wavelength or frequency of sound or electromagnetic radiation due to the relative motion between the observer and the emitter along the observer's line of sight. The change is to longer wavelengths when the source of waves and the observer are moving away from each other, and to shorter wavelengths when they are moving toward each other. The Doppler effect produces the change in pitch of a siren as an ambulance speeds past. In astronomy, the Doppler effect is used to detect and measure relative motion along the line of sight, determining the radial velocity from the Doppler shift in wavelength. The effect is named after the Austrian physicist **Christian Doppler (1803–1853)**, who first described it in 1842. *See* blueshift, Doppler shift, radial velocity, and redshift.

Doppler shift: A change in the apparent wavelength or frequency of the radiation emitted from a moving source, caused by its relative motion along the line of sight, either toward or away from the observer. An object moving away from the observer will appear to be emitting radiation at a longer wavelength, or lower frequency, than if at rest or non-moving. A Doppler shift in the spectrum of an astronomical object is commonly described as a redshift when it is towards longer wavelengths (object receding) and as a blueshift when it is towards shorter wavelengths (object approaching). A measurement of the Doppler shift makes it possible to determine the radial velocity, or velocity along the line of sight. The redshifts caused by the expanding Universe can be attributed to a stretching of space, known as the cosmological redshift, which is not a Doppler shift due to radial velocity. *See* blueshift, cosmological redshift, radial velocity, and redshift.

Double star: Two stars that appear close to each other in the sky. When they are close to each other in space, the two stars revolve around each other, and the double star is known as a binary star. *See* binary star.

DSN: Acronym for Deep Space Network. *See* Deep Space Network.

Dumbbell Nebula: A large planetary nebula, designated M27 and NGC 6853, in the constellation Vulpecula. The French astronomer **Charles Messier (1730–1817)** discovered it in 1764. The Dumbbell Nebula has an apparent visual magnitude of 8, angular dimensions of 8 by 4 minutes of arc, and an estimated distance of about 1000 light-years. It obtains its name from the bipolar expulsion of material from the 13th magnitude central star. *See* planetary nebula.

Dust: Small solid particles found in interplanetary and interstellar space, with sizes of between 10^{-6} and 5×10^{-7} meters that are comparable to the wavelength of visible light. Interstellar dust reflects and reddens visible starlight, and in large quantities absorbs energetic ultraviolet light that might otherwise destroy interstellar molecules. Interplanetary dust reflects sunlight, accounting for the F corona. *See* F corona and zodiacal light.

Dwarf galaxy: A relatively small galaxy of low luminosity when compared to other galaxies. Dwarf galaxies are usually elliptical or irregular, and there are several dwarf elliptical galaxies in the Local Groups of galaxies. *See* Local Group.

Dwarf star: The most common type of star in the Galaxy, constituting 90 percent of its stars and 60 percent of its mass. All dwarf stars are located on the main sequence of the Hertzsprung-Russell diagram, and are therefore fusing hydrogen nuclei into helium nuclei in their core. A main-sequence dwarf star should not be confused with the much smaller white dwarf star, which is not on the main sequence and does not shine by nuclear reactions. The Sun is a typical dwarf star. *See* carbon-nitrogen-oxygen cycle, Hertzsprung-Russell diagram, hydrogen burning, main sequence, proton-proton chain, red dwarf, and white dwarf.

Dynamics: Study of the motion and equilibrium of systems under the influence of force.

Dynamo: An electric generator that employs a spinning magnetic field to produce electricity. The dynamo mechanism generates magnetic fields through the interaction of the convection of conducting matter with rotation and rotational shear. A dynamo converts the kinetic energy of a moving electrical conductor to the energy of electric currents and a magnetic field. It uses the motion of a conducting, convecting fluid to generate or sustain a magnetic field. The terrestrial magnetic field is supposed to be generated by such a dynamo, located in the Earth's molten core. The Sun's magnetic field is also sustained by an internal dynamo, perhaps located at a region of rotational shear just below the convection zone.

Eagle Nebula: A bright emission nebula located at a distance of about 7,000 light-years in the constellation Serpens. It measures 30 minutes of arc across, and surrounds the open star cluster designated as M16 or NGC 6611. The Eagle Nebula is itself designated IC 4703. The Eagle Nebula is the site of star formation, especially in dark, twisting columns of gas and dust nicknamed Elephant Trunks or Pillars of Creation. *See* emission nebula.

Earth: The third planet from the Sun, which we live on. The Earth has a mean radius of 6371 kilometers, a mean mass density of 5515 kilograms per cubic meter, and an age of 4.6 billion years. The mean distance between the Earth and the Sun is one Astronomical Unit, denoted by AU. The Earth's atmosphere is mainly composed of molecular nitrogen, at 77 percent, and molecular oxygen, at 21 percent. The greenhouse effect, caused by relatively small amounts of atmospheric carbon dioxide and water vapor, warms the surface of the Earth to an average temperature of about 288 kelvin (15 degrees Celsius and 59 degrees Fahrenheit), and without this effect the surface temperature would be 255 kelvin, below the freezing point of water at 273 kelvin. Global warming produced by human activity, such as the burning of coal, gas and oil to produce carbon dioxide, will continue to increase the Earth's surface temperature by the greenhouse effect. Liquid oceans cover 71 percent of the Earth's surface, but new oceans form and old ones close as the continents disperse and then reassemble. The bottom of the oceans is continually renewed as new sea-floor spills out of mid-ocean volcanoes, producing a spreading ocean floor that drives the drifting continents. The continents move sideways on the back of plates that grind against each other to produce earthquakes, collide to build mountains, and dive into the Earth to produce volcanoes that make continents grow at their edges. *See* astronomical unit, continental drift, global warming, greenhouse effect, plate tectonics, and sea-floor spreading.

Eccentricity: Measure of the departure of an orbit from a perfect circle, or how elliptical an orbit is. A circular obit has an eccentricity of 0.00; an elliptical orbit has an eccentricity between 0.00 and 1.00, and the closer this value is to 1.00, the longer and thinner the ellipse is. A parabolic orbit has an eccentricity of 1.00, and a hyperbolic orbit has an eccentricity greater than 1.00.

Eclipse: The partial or total obscuration of the light from an astronomical object by another such object. In a solar eclipse, the Moon passes between the Sun and the Earth, blocking the Sun's light. In a lunar eclipse, the Earth passes between the Moon and the Sun, blocking the solar light that the Moon reflects. The eclipse of a star by the Moon or by a planet or other body in the Solar System is called an occultation. A total solar eclipse can occur only at the new Moon, and it has a maximum duration of 7.5 minutes. During the brief moments of a total solar eclipse, darkness falls, and the outer parts of the Sun, the chromosphere and the corona, are seen. At any given point on the Earth's surface, a total solar eclipse occurs, on the average, once every 360 years. *See* lunar eclipse and solar eclipse.

TABLE 14   Earth

Mass	59.736×10^{23} kilograms
Mean Radius	6371 kilometers
Equatorial Radius	6378 kilometers
Mean Mass Density	5515 kilograms per cubic meter
Rotation Period	23 hours 56 minutes 04 seconds = 8.6 164 $\times 10^4$ seconds
Orbital Period	1 year = 365.24 days = 3.1557×10^7 seconds
Mean Distance from Sun	$1.495\ 98 \times 10^{11}$ meters = 1.000 AU
Orbital Eccentricity	0.0167
Tilt of Rotational Axis, or Obliquity	23.27 degrees
Age	4.6×10^9 years
Atmosphere	77 percent nitrogen, 21 percent oxygen
Surface Pressure	1.013 bar at sea level
Surface Temperature	288 to 293 kelvin
Magnetic Field Strength	0.305×10^{-4} tesla at the equator
Magnetic Dipole Moment	7.91×10^{15} tesla meters cubed

TABLE 15   Earth – abundant elements

Element	Symbol	Average abundance (percent by mass)
Iron	Fe	34.6
Oxygen	O	29.5
Silicon	Si	15.2
Magnesium	Mg	12.7
Nickel	Ni	2.4

Eclipsing binary: A double or binary star in which at least one of the two stars passes in front of and/or behind the other, crossing the line of sight from Earth, so that the system's total observed light periodically fades. Algol is an eclipsing binary. *See* Algol, binary star, double star, and Epsilon Aurigae.

Ecliptic: The projection of the Earth's orbit around the Sun on the celestial sphere, marking the Sun's apparent yearly path against the background stars. The ecliptic plane is the plane of the Earth's orbit around the Sun. It is inclined to the plane of the celestial equator by about 23.5 degrees, and intersects the celestial equator at the two equinoxes. The planets, most asteroids, and most of the short-period comets are in orbits with small or moderate inclinations relative to the ecliptic plane. *See* autumnal equinox, celestial equator, equinox, Vernal Equinox, and zodiac.

Effective temperature: The temperature an object would have if it emits blackbody radiation at the same energy and at the same wavelengths as the object. The effective

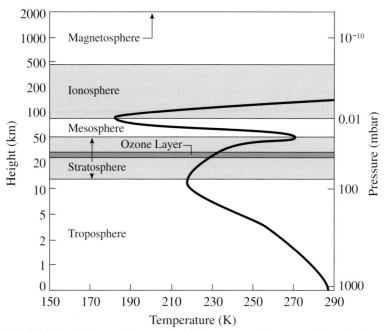

FIG. 27 Earth's layered atmosphere The pressure of our atmosphere (*right scale*) decreases with altitude (*left scale*). This is because fewer particles are able to overcome the Earth's gravitational pull and reach higher altitudes. The temperature (*bottom scale*) also decreases steadily with height in the ground-hugging troposphere, but the temperature increases in two higher regions that are heated by the Sun. They are the stratosphere, with its critical ozone layer, and the ionosphere. The stratosphere is mainly heated by ultraviolet radiation from the Sun, and the ionosphere is created and modulated by the Sun's X-ray and extreme ultraviolet radiation.

temperature is denoted by the symbol T_{eff}, and the effective temperature of the Sun is equal to 5780 kelvin. A star's effective temperature is a good approximation to the actual temperature of its visible disk. *See* blackbody radiation.

Effelsberg Radio Telescope: A radio telescope with a diameter of 100 meters (328 feet), completed in 1970, located near Bonn, Germany, and operated by the Max-Planck-Institut für Radioastonomie. The **Effelsberg Radio Telescope** remained the world's largest, fully steerable radio telescope until August 2000 when the United States National Radio Astronomy Observatory completed the construction of its 100 × 110 meter (328 × 360 inch) **Robert C. Byrd Green Bank Radio Telescope**, in Green Bank, West Virginia. The **Effelsberg Radio Telescope** is used either alone or as a component of a global radio interferometer. *See* National Radio Astronomy Observatory, radio telescope, and Very Long Baseline Interferometry.

Einstein Observatory: NASA launched the ***Einstein Observatory,*** the first fully imaging X-ray telescope put in space, on 12 November 1978, where it orbited Earth and obtained data until April 1981. The telescope had an angular resolution of a few seconds of arc, a field of view of tens of minutes of arc, and a sensitivity several hundred times greater than any X-ray mission before it. The **Einstein Observatory** provided the first high-resolution X-ray studies of supernova remnants, discovered unexpectedly intense coronal emission from normal stars, resolved numerous X-ray sources in nearby galaxies, and detected X-ray jets from Centaurus A and M87 aligned with radio jets. The observatory is named after the German-born United States physicist **Albert Einstein (1879–1955)** who revolutionized our understanding of space, matter and time.

Electromagnetic radiation: Radiation that carries energy and moves through vacuous space in periodic waves at the speed of light, propagating by the interplay of oscillating electrical and magnetic fields. The velocity of light is usually designated by the letter c and has a value of 299 792.458 kilometers per second. Electromagnetic radiation includes visible light, radio waves, infrared radiation, ultraviolet radiation, X-rays and gamma rays. Electromagnetic radiation, in common with any wave, has a wavelength, denoted by λ, and a frequency, denoted by ν; their product is equal to the velocity of light, or $\lambda\nu = c$, so the wavelength decreases when the frequency increases and vice versa. The energy associated with the radiation increases in direct proportion to frequency, and this energy, known as the photon energy and denoted by E, is given by $E = h\nu = hc/\lambda$, where Planck's constant $h = 6.6261 \times 10^{-34}$ Joule second. There is a continuum of electromagnetic radiation – from long-wavelength radio waves of low frequency and low energy, through visible-light waves, to short-wavelength X-rays and gamma rays of high frequency and high energy. *See* electromagnetic spectrum, gamma-ray radiation, Planck's constant, radio radiation, ultraviolet radiation, visible radiation, and X-ray radiation.

Electromagnetic spectrum: All types and wavelengths of electromagnetic radiation, from the most energetic and shortest waves, the gamma rays, to the least energetic and longest ones, the radio waves. From short wavelengths to long ones, the electromagnetic spectrum includes gamma rays, X-rays, ultraviolet radiation, visible light, infrared radiation and radio waves; visible light comprises just one small segment of this much broader spectrum. *See* gamma-ray radiation, optical spectrum, radio radiation, ultraviolet radiation, visible radiation, and X-ray radiation.

Electron: An elementary, negatively charged, subatomic particle that surrounds the positively charged nucleus of an atom, but can exist in isolation outside an atom. The English physicist **Joseph John "J. J." Thomson (1856–1940)** discovered the negatively charged "corpuscles" in 1897 while investigating cathode rays. The Dutch physicist **Hendrik Lorentz (1853–1928)** named them "electrons", a term previously introduced by the Irish physicist **George Johnstone Stoney (1826–1911)**. The interactions between electrons of neighboring atoms create the chemical bonds that link atoms together as molecules. A neutral, or uncharged, atom has as many electrons as positively charged protons, which reside in the nucleus. An ionized atom is usually positively charged and has fewer electrons than protons. Free electrons have broken away from their atomic bonds and are not bound to atoms. At the hot temperatures inside the Sun and other stars, and in the solar corona and solar wind, the electrons have been set free of their atomic bonds. The mass of an electron is 9.1094×10^{-31} kilograms, or 1/1836 of the mass of a proton. *See* ion and proton.

Electron neutrino: The type of neutrino that interacts with the electron. It is the only kind of neutrino produced by the nuclear reactions that make the Sun shine. *See* neutrino and solar neutrino problem.

Electron volt: A unit of energy, denoted by eV, often used for measuring the energies of particles and electromagnetic radiation. An electron volt is defined as the energy acquired by an electron when it is accelerated through a potential difference of 1 volt in a vacuum. $1\,\text{eV} = 1.602177 \times 10^{-19}$ Joule, where one Joule is equal to 10 million erg. Radiation with an energy of one electron volt has a wavelength of 1240 nanometers and a frequency of 2.42×10^{14} Hertz. The energies of X-rays are expressed in thousands of electron volts, abbreviated keV. Millions and billions of electron volts, respectively denoted by MeV and GeV, are used as units for very energetic charged particles, such as cosmic rays. An electron with a kinetic energy of a few MeV is traveling at almost the velocity of light.

Element: A chemically homogeneous substance that cannot be decomposed by chemical means. The atoms of a particular element all have the same number of protons in their nucleus, but the number of neutrons can vary, giving rise to different isotopes of the same element. There are 92 naturally occurring elements, with the number of protons, or atomic number, ranging from 1 for hydrogen to 92 for uranium. *See* atomic number and isotope.

Ellipse: A closed, plane curve in which the sum of the distances of each point along its periphery from two other points, its foci, is a constant. The orbits of planets, satellites and comets can be described by ellipses.

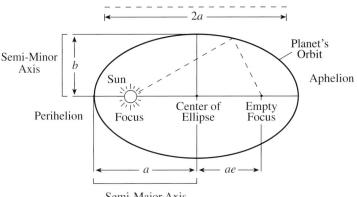

FIG. 28 Ellipse Each planet moves in an ellipse with the Sun at one focus. The length of a line drawn from the Sun, to a planet and then to the empty focus, denoted by the dashed line, is always *2a*, or twice the semi-major axis, *a*. The eccentricity, or elongation, of the planetary ellipse has been greatly overdone in this figure; planetary orbits look much more like a circle.

Elliptical galaxy: A galaxy of elliptical or round shape. Elliptical galaxies are comprised mostly of old stars and contain very little dust and low-temperature gas that can now form stars. An example of an elliptical galaxy is M87, in the constellation Virgo. *See* Virgo cluster.

Emission line: A bright spectral feature at a particular wavelength or narrow bands of wavelengths, emitted directly by a hot, luminous gaseous source, such as an emission nebula or planetary nebula, revealing by its wavelength a chemical constituent of its gas. *See* absorption line, emission nebula, and planetary nebula.

Emission nebula: A diffuse, luminous cloud of gas in interstellar space that displays emission lines in its visible-light spectrum. The English astronomer **William Huggins**

TABLE 16   **Emission nebulae**[a]

Name	M, NGC	RA (2000) h	RA (2000) m	Dec. (2000) °	Dec. (2000) ′	θ[b] ′	Distance (light-years)
Bubble Nebula	NGC 7635	23	20.7	+61	12	3	—
California Nebula	NGC 1499	04	00.7	+36	37	20 × 150	—
Cone Nebula	NGC 2264	06	40.8	+09	54	—	—
Eagle Nebula	IC 4703	18	18.8	−13	47	30	7000
(Surrounds open cluster M16, NGC 6611)							
Eta Carinae Nebula	NGC 3373	10	45.1	−59	41	120	8000
(Surrounds Homunculus Nebula, NGC 3372, and Keyhole Nebula)							
Lagoon Nebula	M8	18	03.8	−24	23	40 × 90	5200
(Also known as NGC 6523, central region Hourglass Nebula and star cluster NGC 6530)							
North America Nebula	NGC 7000	20	58.8	+44	20	100 × 120	3500
Omega Nebula	M17	18	20.8	−16	11	37 × 47	5000
(Also known as NGC 6618, Horseshoe Nebula, and Swan Nebula)							
Orion Nebula	M42	05	35.4	−05	27	60 × 66	1500
(Also known as NGC 1976, northwestern part is M43, NGC 1982)							
Pelican Nebula	IC 5070	20	50.8	+44	21	70 × 80	3500
Rosette Nebula	NGC 2244	06	30.3	+05	03		5000
(Also includes NGC 2237-39, surrounds star cluster NGC 2244)							
Tarantula Nebula[c]	NGC 2070	05	39	−69	07	30	170 000
(Also known as 30 Doradus and Great Looped Nebula)							
Trifid Nebula	M20	18	02.6	−23	02	27 × 29	5200
(Also known as NGC 6514)							

[a] Adapted from Lang, Kenneth R., *Astrophysical Data: Planets and Stars*, New York, Springer-Verlag 1992, pp. 407, 408.

[b] Angular diameter θ in minutes of arc, designated ′.

[c] The Tarantula Nebula is also known as 30 Doradus, and it is located in the Large Magellanic Cloud.

(1824–1910) discovered such emission lines in 1864, from the Orion Nebula. The Dutch astronomer **Herman Zanstra (1894–1972)** explained the hydrogen emission lines by supposing that the hydrogen atoms are ionized by the ultraviolet light of bright stars embedded in an emission nebula, and that visible light is emitted when the free electrons recombine with the protons to make hydrogen atoms. Emission nebulae are therefore often called ionized hydrogen regions, denoted as H II regions and pronounced H-two regions. As shown in 1928 by the American astronomer **Ira S. Bowen (1898–1973)**, other emission lines can arise from ionized atoms of oxygen and nitrogen undergoing "forbidden" transitions in the low-density nebulae; such transitions are very improbable in the higher-density laboratory situation. The mass density of the tenuous emission nebulae is typically only about one gram in a million cubic kilometers, and the green color of some of them is due to the oxygen emission lines. *See* Bubble Nebula, California Nebula, Cone Nebula, Eagle Nebula, Eta Carinae Nebula, Gum Nebula, H II region, Lagoon Nebula, North American Nebula, Orion Nebula, Pelican Nebula, Rosette Nebula, Tarantula Nebula, Trifid Nebula, and ultraviolet radiation.

Emission spectrum: A series of bright emission lines. *See* emission line and forbidden line.

Encke's Comet: A periodic comet with the shortest known period, of 3.3 years. It is named after the German astronomer **Johann Encke (1791–1865)** who first computed its orbit. The French astronomer **Pierre Mechain (1744–1804)** discovered the comet in 1786, and French astronomer **Jean Louis Pons (1761–1831)** found it again in 1818. In 1950, the American astronomer **Fred L. Whipple (1906–2004)** explained the acceleration of Comet Encke in terms of his icy conglomerate, or dirty snowball, comet model. When the comet's orbit brings it close to the Sun, it sublimates its surface ice and produces jets of gas and dust; when the comet is rotating these jets will either push the comet ahead or delay its journey.

Energy: The ability or capacity to do work.

Epicycle: A circular orbit around a point that itself orbits another point, used in Ptolemaic cosmology. *See* Ptolemaic cosmology.

Epsilon Aurigae: An eclipsing binary star with the longest known period, 27 years, with an apparent magnitude of 2.9, but 3.8 during eclipse. The primary component is a yellow-white F0 supergiant, and no light is detected from the secondary component. *See* eclipsing binary.

Epsilon Eridani: An orange K2 dwarf star in the constellation Eridanus with an apparent visual magnitude of 3.73 that makes it visible to the unaided eye. At 10.5 light-years away, Epsilon Eridani is the ninth-closest star to the Earth, with an apparent absolute magnitude of 6.19. At least one Jupiter-sized planet orbits the Sun-like star, which also contains large starspots and exhibits a 5-year activity cycle.

Equator: An imaginary line around the center of a body where every point on the line is an equal distance from the poles. The equator defines the boundary between the northern and southern hemispheres. *See* hemisphere.

Equilibrium: The unchanging state of a system resulting from the balance between competing forces or processes.

Equinox: Either of the two points where the Sun crosses the celestial equator and lies in the Earth's equatorial plane, when the Sun rises due east and sets due west and the day and night are equally long. The Vernal, or spring, Equinox occurs when the Sun crosses the equator from south to north on or near 21 March each year. The Vernal Equinox is also called the First point of Aries. The other equinox is the autumnal equinox, which occurs on or near 23 September when the Sun crosses the equator from north to south. The locations of the spring, or Vernal, and fall, or autumnal, equinoxes are slowly moving, or precessing, westward around the stellar background at the rate of about 50.3 seconds of arc per year. *See* autumnal equinox, precession, season, and Vernal Equinox.

Erg: A unit of energy equal to the work done by a force of 1 dyne acting over a distance of 1 centimeter. An energy of one Joule is equal to 10^7, or 10 million, erg.

Eros: An asteroid discovered on 13 August 1898 by the German astronomer **Gustav Witt (1866–1946)** and independently photographed by the French astronomer **Auguste Charlois (1864–1910)** the same night. Numbered 433 in order of asteroids with established orbits, 433 Eros is the first asteroid found to travel mainly inside the orbit of Mars, approaching within 22 million kilometers of the Earth, and an example of a near-Earth asteroid. The *NEAR-Shoemaker* spacecraft orbited 433 Eros for a year, beginning on 14 February 2000, taking images of its battered shape, establishing its dimensions as $31 \times 13 \times 13$ kilometers, and showing that it is a homogeneous, solid rock with a mass density of 2670 kilograms per cubic meter. *See NEAR-Shoemaker.*

ESA: Acronym for European Space Agency. *See* European Space Agency.

Escape velocity: The minimum outward velocity that a small object must attain to overcome the gravitational attraction of a larger body and leave on a trajectory that does not bring it back again. The escape velocity, denoted by V_{esc}, depends on the mass, denoted by M, and the radius, specified as R, of the larger body, and is given by $V_{esc} = [2GM/R]^{1/2}$, where the gravitational constant $G = 6.6726 \times 10^{-11}$ N m^2 kg^{-2}. If a rocket leaves the Earth it must move faster than the Earth's escape velocity, which is 11.2 kilometers per second, or 25 000 miles per hour.

ESO: Acronym for European Southern Observatory. *See* European Southern Observatory.

Eta Carinae: One of the most massive stars in the Galaxy, located at a distance of about 8000 light-years, with an absolute visual magnitude of −10 and an intrinsic luminosity more than 5 million times brighter than the Sun. Eta Carinae is situated in two dust lobes, nearly a light-year across, which are together known as the "Homunculus" from their vague resemblance to a human figure. They were most likely formed in 1843 when the star produced the Great Eruption, reaching apparent visual magnitude of −0.8 and rivaling Sirius in intensity; Eta Carinae has now faded to a relatively faint apparent visual magnitude of +5. *See* Homunculus Nebula.

TABLE 17   Escape velocity – Sun and planets

	Temperature[a] (K)	Escape Velocity (km s^{-1})
Sun	5780	617.8
Terrestrial planets		
Mercury	440	4.43
Venus	735	10.36
Earth	288–293	11.19
Mars	140–300	5.03
Giant planets		
Jupiter	165	59.54
Saturn	134	35.49
Uranus	76	21.29
Neptune	73	23.71

[a]Temperature of the visible solar disk, the photosphere, surface temperature for the terrestrial planets, and atmospheric temperatures for the giant planets at the level where the atmospheric pressure is one bar.

Eta Carinae Nebula: A bright emission nebula, designated NGC 3373, surrounding the star Eta Carinae. The emission nebula is about 2 degrees across and divided by a dark, V-shaped, obscuring dust lane. It contains the bright Homunculus Nebula and the dark Keyhole Nebulae. *See* dark nebula, emission nebula, Eta Carinae, Homunculus Nebula, and Keyhole Nebula.

Ether: The medium that light and gravitation were once supposed to propagate through. *See* aether and luminiferous ether.

Euclidean geometry: The geometry of non-curved space in which parallel lines always remain equidistant from each other, named for the Greeek mathematician **Euclid of Alexandria (lived c. 300 BC)**, who described the geometery in his book *The Elements*. Euclidean geometry is the three-dimensional analog of a flat surface, or a plane. In non-Euclidean geometry, space is curved, and parallel lines can meet or diverge in the three-dimensional analog of a sphere or a hyperbola. *See* geometry and non-Euclidean geometry.

Euclidean Universe: A universe without noticeable spatial curvature on the largest scales, and hence with a non-curved shape described by Euclidean geometry.

Europa: The smallest of Jupiter's Galilean satellites, and the second closet to the planet, discovered in 1610 by the Italian astronomer **Galileo Galilei (1564–1642)**. It has a radius of 1561 kilometers and a mass density of 3014 kilograms per cubic meter. The bright, smooth surface of Europa is covered with water ice, and has few impact craters, indicating a relatively young age. Dirty liquid water or soft water ice rise up from a global internal sea within Europa, filling long, deep fractures in the satellite's icy covering and lubricating large raft-like blocks of water ice that float across the surface. Europa also has a variable magnetic field, probably induced in its internal ocean of liquid water when Jupiter's magnetic field sweeps by. *See* Galilean satellites.

European Southern Observatory: An intergovernmental, European organization for astronomical research in the southern hemisphere, with headquarters in Garching, near Munich, Germany. It has eleven member countries: Belgium, Denmark, Finland, France, Germany, Italy, Portugal, the Netherlands, Sweden, Switzerland and the United Kingdom. The European Southern Observatory operates astronomical observatories at La Silla and Cerro Paranal, Chile. The major telescopes at the La Silla Observatory are the 3.6-meter (142-inch) and 2.2-meter (88-inch) reflectors, and the **New Technology Telescope**, which pioneered the use of active optics to continuously shape its 3.6-meter (142-inch) reflector. The **Very Large Telescope**, abbreviated **VLT**, at Cerro Paranal includes four unit telescopes, each 8.2 meters (328 inches) in diameter, operating at visible and infrared wavelengths. They can be operated individually, or together as an interferometer. Four auxiliary telescopes, each of 1.8 meter (72 inches) in diameter, will also be connected to this **Very Large Telescope Interferometer**. Two telescopes intended for imaging surveys from Paranal are the 2.6-meter (104-inch) **VLT Survey Telescope**, abbreviated **VST**, and the 4.0-meter (160-inch) **Visible and Infrared Telescope for Astronomy**, or **VISTA** for short. *See* New Technology Telescope and Very Large Telescope.

European Space Agency: An organization of 14 European nations, founded in 1975 to promote space research and technology for peaceful purposes. It is headquartered in Paris, and the member countries are Austria, Belgium, Denmark, Finland, France, Germany, Ireland, Italy, the Netherlands, Norway, Spain, Sweden, Switzerland, and the United Kingdom.

EUV: Acronym for Extreme Ultra-Violet, a portion of the electromagnetic spectrum from approximately 10^{-8} to 10^{-7} meters, or 100 to 1,000 Ångströms, in wavelength. *See* extreme ultraviolet radiation.

eV: Abbreviation of electron volt. *See* electron volt.

Event horizon: The horizon in the geometry outside a black hole, within which you can see no events, just as the Earth's horizon is the boundary for our vision. An observer outside a black hole cannot communicate with anything inside the event horizon. The radius of the event horizon is known as the Schwarzschild radius. *See* black hole and Schwarzschild radius.

Evershed effect: The radial flow of gases within the penumbra of a sunspot, discovered in 1908 by the English astronomer **John Evershed (1864–1956)** from the Doppler shift in the spectrum of a sunspot. The flow is outward at the level of the photosphere, but inward and downward at higher levels in the solar atmosphere. *See* photosphere and sunspot.

Excitation: An atomic process in which an atom or ion is raised to a higher-energy state, when one of its electrons goes from one orbit to another farther out from the nucleus of the atom.

Exclusion Principle: No two electrons can occupy the same quantum state, and no two neutrons can occupy the same quantum state. This principle was first announced in 1925 by the Austrian physicist **Wolfgang Pauli (1900–1958)**, and is hence also known as the Pauli exclusion principle. It gives rise to the degenerate electron pressure that can support a white dwarf star and the degenerate neutron pressure that can support a neutron star. *See* degenerate matter.

Exobiology: Investigations into the possibility of life outside the Earth.

Exoplanet: *See* extrasolar planet.

Expanding Universe: The continual increase, with time, in the amount of space separating galaxies from us and from each other, at a rate proportional to their distances, discovered in 1929 by the American astronomer **Edwin Hubble (1889–1953)**. The galaxies are receding from us at velocities that are proportional to their distance, a relation displayed in the Hubble diagram and known as the Hubble law. The expanding Universe is attributed to the Big Bang, occurring about 14 billion years ago. *See* Big Bang, Hubble constant, Hubble diagram, and Hubble law.

Extrasolar planet: Any planet located outside the Solar System and orbiting a star other than the Sun; also called an exoplanet. One pulsar, PSR 1257 + 12, has two Earth-sized planets orbiting it, and more than 100 Sun-like stars have at least one Jupiter-sized planet in tight orbit around them. *See* Arecibo Observatory, 51 Pegasi, 47 Ursae Majoris, millisecond pulsar and PSR 1257 + 12.

Extreme UltraViolet Explorer: Abbreviated *EUVE*, a NASA mission launched on 7 June 1992, which observed cosmic sources at extreme ultraviolet, or EUV, wavelengths, between 70 and 760 Ångströms or 7 and 76 nanometers, for 8.5 years, until 31 January 2001. It completed an all-sky survey, resulting in catalogues of more than a thousand cosmic EUV sources and the discovery of more than 500 new EUV sources, and carried out detailed EUV spectroscopy of specific sources.

Extreme ultraviolet radiation: Abbreviated EUV, the short-wavelength part of the ultraviolet portion of the electromagnetic spectrum which extends from approximately 10^{-8} to 10^{-7} meters, or 10 to 100 nanometers and 100 to 1,000 Ångströms, in wavelength.

Faber-Jackson relation: A correlation between the intrinsic luminosity of an elliptical galaxy and its velocity dispersion, discovered in 1976 by the American astronomers **Sandra M. Faber (1944–)** and **Robert E. Jackson (–)**. The Faber-Jackson relation is used to determine the distances of elliptical galaxies, through the inferred intrinsic luminosity and the observed apparent luminosity. *See* inverse square law.

Faculae: Bright regions of the photosphere, associated with sunspots, seen in white light and visible only near the limb of the Sun. They are brighter than the surrounding medium due to their higher temperatures and greater densities. Faculae appear some hours before the associated sunspots, in the same place, but can remain for months after the sunspots have gone. The word *facula* is Latin for "little torch." A plage appears in the chromosphere just above a facula. *See* plage.

Fahrenheit: A unit of temperature denoted by F or by °F for degrees Fahrenheit. It is named after the Polish-born Dutch physicist **Gabriel Daniel Fahrenheit (1686–1736)**, who invented the first accurate thermometer and devised the Fahrenheit scale of temperature. The freezing temperature of water is 32°F and the boiling temperature of water is 212°F. The equivalent temperature in degrees kelvin is °K = (5°F/9) + 255.22, and the equivalent temperature in degrees Celsius is °C = 5(°F −32)/9. *See* Celsius and Kelvin.

Feynman diagram: A schematic representation of an interaction between particles, invented by the American physicist **Richard Feynman (1918–1988)**.

Fibrils: Dark elongated features seen in hydrogen-alpha spectroheliograms of the chromosphere, forming a linear pattern thought to delineate the magnetic field. *See* chromosphere, hydrogen-alpha line, and spectroheliograph.

Field: Mathematical description of a force in each point of space around its source.

Field of view: The angular area of the celestial sphere that is detected by a telescope. The telescope can only detect objects in its field of view.

51 Pegasi: A yellow G2.5 dwarf star located at a distance of 59 light-years from Earth in the constellation Pegasus. It has an apparent visual magnitude of 5.49 and an absolute visual magnitude of 4.56. The main-sequence star 51 Pegasi is the first Sun-like star known to have a planet; a Jupiter-sized planet in a very close orbit and discovered in 1995 by the Swiss astronomers **Michel Mayor (1942–)** and **Didier Queloz (1966–)**. *See* extrasolar planet.

Filament: A mass of relatively cool and dense material suspended above the photosphere in the low corona by magnetic fields, generally along a magnetic inversion, or neutral, line separating regions of opposite magnetic polarity in the underlying photosphere. Filaments appear as dark, elongated features when observed on the disk in the light of hydrogen alpha or ionized calcium. A dark filament is seen in projection against the bright solar disk. When detected above the limb of the Sun, a filament is seen in emission against the dark sky, and it is then called a prominence. *See* chromosphere, hydrogen-alpha line, and prominence.

Fireball: An exceptionally bright meteor. *See* meteor, meteorite, and meteoroid.

First point of Aries: *See* Vernal Equinox.

Fission: The splitting, or breaking apart, of a heavy atomic nucleus into two or more lighter atomic nuclei, releasing energy. Fission is spontaneous during the decay of a radioactive element, and it can be induced by particle bombardment. Fission powers atomic bombs, and occurs in man-made nuclear power reactors on Earth.

Five-minute oscillations: Vertical oscillations of the solar photosphere with a period of five minutes, interpreted in terms of trapped sound waves. The five-minute oscillations are detected in the Doppler effect of solar absorption lines formed in the photosphere and in light variations of the photosphere. The American physicist **Robert Leighton (1919–1997)** and his colleagues discovered localized five-minute oscillations while investigating motions of the photosphere in 1960–61. Observations of the oscillations have been used to infer the composition, motions, structure and temperature of the solar interior through the technique of helioseismology. *See* acoustic waves, Doppler effect, Global Oscillation Network Group, helioseismology, photosphere, *SOlar and Heliospheric Observatory*, and supergranulation.

Flare: A sudden and violent release of matter and energy from the Sun or another star in the form of electromagnetic radiation, energetic particles, wave motions and shock waves, lasting minutes to hours. The frequency and intensity of flares on the Sun increase near the maximum of the solar activity cycle, and they occur in solar active regions. Many red dwarf stars also emit flares, and they are called flare stars. Solar flares accelerate charged particles into interplanetary space. The impulsive or flash phase of flares usually lasts for a few minutes, during which matter can reach temperatures of hundreds of millions of degrees. The flare subsequently fades during the gradual or decay phase lasting about an hour. Most of the radiation is emitted as X-rays, but flares are also observed at visible hydrogen-alpha wavelengths and radio wavelengths. They are probably caused by the sudden release of large amounts of magnetic energy, up to 10^{25} Joule, in a relatively small volume in the solar corona. *See* active region, burst, flare star, gradual flare, impulsive flare, solar flare, type I, type II, type III, and type IV radio bursts, and X-ray flare class.

Flare star: A variable star, usually a red dwarf star, which suddenly and unpredictably emits flares lasting a few minutes or hours, sometimes outshining the entire star at visible light or radio wavelengths. *See* dM star, dwarf star, flare, M star, red dwarf star, and UV Ceti.

Flash spectrum: The emission line spectrum of the solar chromosphere that is seen for a few seconds just before and after totality during an eclipse of the Sun. *See* chromosphere, emission line, and emission spectrum.

Flat Universe: *See* Euclidean Universe.

Flatness problem: The fact that the Universe may be neither dramatically open nor noticeably closed, but appears to be almost perfectly balanced between these two geometrical shapes and described by Euclidean geometry. *See* closed Universe, Euclidean Universe, and open Universe.

Flux: The rate of flow through a reference surface; the flux density is the flux measured per unit area.

Focal length: The distance between a lens or curved mirror and its focus.

Focal ratio: The ratio of the focal length of a lens or curved mirror to its diameter. A focal ratio of 10, is written as *f*/10.

Fomalhaut: A bright, white A3 dwarf, or main sequence, star that has a circumstellar ring of dust first detected at infrared wavelengths. The star Fomalhaut has an apparent visual magnitude of 1.17, a distance of about 25.1 light-years from the Earth, an absolute visual magnitude of +1.7, and an intrinsic luminosity of about 17 times that of the Sun. It is the brightest star in the constellation Piscis Austrinus, and hence has the alternative name of Alpha Piscis Austrinus.

Forbidden lines: Emission lines not normally observed under laboratory conditions because they have a low probability of occurrence, often resulting from a transition between a metastable excited state and the ground state. Under typical conditions on Earth, an atom in a metastable state will lose energy through a collision before it is able to decay to the ground state and emit line radiation. Under astrophysical conditions, including the solar corona and highly rarefied nebulae, the metastable state can last long enough for the "forbidden" lines to be emitted. *See* emission nebula and emission line.

Force: Agency responsible for a change in a system.

40 Eridani: A double star system located at a distance of 16.5 light-years, whose faint component, known as 40 Eridani B is one of the first white dwarf stars to be discovered. The brighter component, 40 Eridani A, is a K1 dwarf, or main sequence, star about half as luminous as the Sun, with a apparent visual magnitude of 4.43 and an absolute visual magnitude of 5.92. The white dwarf 40 Eridani B has a high disk temperature of 16 000 kelvin and a low absolute luminosity of just 1.2 percent of the Sun's, implying a radius of 0.0136 that of the Sun and comparable to the radius of the Earth. The American astronomer **Walter Adams (1876–1956)** reported the A0 spectral classification and high temperature of 40 Eridani B in 1914. The white dwarf has its own companion, 40 Eridani C, a dimmer 11th magnitude class M ordinary dwarf. *See* Stefan-Boltzmann law and white dwarf.

47 Ursae Majoris: A yellow G-type main-sequence star with a Jupiter-sized planet orbiting it every three years; the star is located at a distance of 46 light-years from the Earth. *See* extra solar planet.

Fossil: Geological remains of a previously living thing.

TABLE 18   Fraunhofer lines

Wavelength[a] (nanometers)	Fraunhofer letter (Color)[b]	Element name, symbol[c]
393.368	K	Ionized calcium, Ca II
396.849	H (extreme violet)	Ionized calcium, Ca II
410.175	h	Hydrogen, Hδ, delta transition
422.674	g	Neutral calcium, Ca I
431.0 ± 1.0	G (violet)	CH molecule
434.048		Hydrogen, Hγ, gamma transition
438.356	d	Neutral iron, Fe I
486.134	F (blue)	Hydrogen, Hβ, beta transition
516.733	b_4	Neutral magnesium, Mg I
517.270	b_2	Neutral magnesium, Mg I
518.362	b_1	Neutral magnesium, Mg I
526.955	E (green)	Neutral iron, Fe I
588.997[d]	D2 (yellow)[d]	Neutral sodium, Na I
589.594	D1 (yellow)	Neutral sodium, Na I
656.281	C (red)	Hydrogen, Hα, alpha transition
686.719	B (red)	Molecular oxygen, O_2, in our air
759.370	A (extreme red)	Molecular oxygen, O_2, in our air

[a] The wavelengths are in nanometer units, where 1 nanometer = 10^{-9} meters. Astronomers have often used the Ångström unit of wavelength, where 1 Ångström = 1 Å = 0.1 nanometers.

[b] The letters were used by **Joseph von Fraunhofer (1727-1826)** around 1815 to designate the spectral lines before they were chemically identified. Fraunhofer did not resolve the numbered components of the D and b lines. The lines A and B are produced by molecular oxygen in the terrestrial atmosphere.

[c] A Roman numeral I after an element symbol denotes a neutral, or un-ionized atom, with no electrons missing, whereas the Roman numeral II denotes a singly ionized atom with one electron missing.

[d] Un-ionized helium, He I, was discovered in emission during a solar eclipse in 1868 at a wavelength of 587.56 nanometers, near the two sodium lines. The helium line is sometimes designated as D3.

Fraunhofer lines: The dark absorption features in the solar spectrum, caused by absorption at specific wavelengths in the cooler layers of the Sun's atmosphere, including the photosphere and chromosphere. Although they were first observed in 1802 by the English chemist and physicist **William Hyde Wollaston (1766–1828)**, they were first carefully studied from 1814 by the German physicist and optician **Joseph von Fraunhofer (1787–1826)**, who catalogued the wavelengths of more than 300 of them, assigning Roman letters to the most prominent, such as the H and K lines of ionized calcium, the D line of neutral sodium, the E line of iron, and the C and F lines of hydrogen. *See* absorption line, H and K lines, hydrogen alpha line, and spectrum.

Fred L. Whipple Observatory: An observatory located near Amado, Arizona on Mount Hopkins, named for the American astronomer **Fred L. Whipple (1906–2004)**. It operates the 6.5-meter (256-inch) **Multiple Mirror Telescope** (now a single mirror), jointly with the University of Arizona, and has smaller telescopes of about 1 meter in diameter, operating at visible light or infrared wavelengths.

Free electron: An electron that has broken free of its atomic bond and is therefore not bound to an atom.

Frequency: The number of crests of a wave passing a fixed point each second usually measured in units of Hertz, where one Hertz is equal to one oscillation per second. *See* Hertz, Hz, and MHz.

F star: A yellow-white star of spectral type F, which is slightly hotter than the Sun. The brightest F-type stars in the night sky are Canopus and Procyon. *See* Canopus, Procyon, and stellar classification.

Fusion: The joining together of two or more light atomic nuclei to produce a heavier atomic nucleus, releasing energy. Fusion powers hydrogen bombs and makes stars shine, including the Sun. *See* carbon-nitrogen-oxygen cycle, proton-proton chain, and thermonuclear fusion.

G

Galactic astronomy: The study of our Milky Way Galaxy.

Galactic center: The central hub or nucleus of our Galaxy, located about 27 700 light-years or 8500 parsecs from Earth.

Galactic cluster: *See* open cluster.

Galactic cosmic rays: Cosmic ray particles that enter the Earth's atmosphere from interplanetary space. Galactic cosmic rays originate from outside our Solar System, but from within our Galaxy. *See* cosmic rays and solar cosmic rays.

Galactic disk: A flattened, plate-shaped, rotating disk of gas and young stars in a spiral galaxy, in which the spiral arms are found. Most of a spiral galaxy's stars and interstellar matter reside in this disk. *See* galactic plane and Milky Way.

Galactic halo: A spherical aggregation of metal-poor stars and globular star clusters, centered on the nucleus of the galaxy and extending beyond the galactic disk. There is also a more extensive dark halo that consists of invisible dark matter. The stellar halo of our Galaxy, containing the globular clusters, is mainly closer to the galactic center than the Sun, while the radius of the dark halo surrounding our Galaxy, the Milky Way, extends some 50 000 light-years from the galactic center. *See* dark halo, dark matter, galactic disk, and globular cluster.

Galactic plane: The plane that contains the disk of the Milky Way or the disk of any other spiral galaxy. *See* galactic disk and Milky Way.

Galactic nucleus: The central concentration of matter in a galaxy, typically no more than a few light-years across. *See* active galactic nucleus and supermassive black hole.

Galaxy: A collection of millions, billions, or thousands of billions of stars and varying amounts of interstellar gas and dust, which are held together in space by the force of gravity. In 1925 the American astronomer **Edwin Hubble (1889–1953)** showed that certain nebulae lie beyond the confines of our Milky Way, which he called extra-galactic nebulae and that are now known as galaxies. There are three major classifications of galaxies – spiral, elliptical and irregular, first described in 1926 by Hubble. Elliptical galaxies are round or elliptical systems, while spiral galaxies are flattened disk-shaped systems in which young stars and interstellar gas and dust are concentrated in spiral arms coiling out from a central bulge or nucleus. The galaxy that we reside in, known as the Milky Way Galaxy, or just Galaxy for short, is a spiral galaxy. It is designated with a capital G to distinguish it from all of the other galaxies. Our Galaxy is about 130 000 light-years in diameter. The Sun is located at the edge of one of the spiral arms, about 27 700 light-years from the center of our Galaxy. All of the galaxies are participating in the expansion of the Universe, discovered by Hubble in 1929, in which there is a continuous increase in the distances separating galaxies from us and from one another.

TABLE 19   Galaxies – bright[a]

Galaxy	RA (2000)	Dec. (2000)	Diam. '	Axial Ratio	B_T	$(B-V)_0$	Type	V_r[b]	Dist. (Mpc)	M_B	V_{rot}	Log M_{tot}
NGC 55	00^h 15^m	−39°13'	32.4	0.17	8.42	0.54	SBc/m III	+94	1.3	−18.13	83	9.87
NGC 205 = M110	00^h 40^m	+41°41'	21.9	0.50	8.92	0.82	Sph-E5p	−60	0.725	−15.60	—	—
NGC 224 = M31	00^h 43^m	+41°16'	190.5	0.32	4.36	0.68	Sb I-II	−121	0.725	−20.67	254	11.37
NGC 221 = M32	00^h 43^m	+40°52'	8.7	0.74	—	0.88	E2	−28	0.725	−15.34	—	—
NGC 247	00^h 47^m	−20°46'	21.4	0.32	9.67	0.54	SABc/d III-IV	+176	2.1	−17.98	92	10.02
NGC 253	00^h 48^m	−25°17'	27.5	0.25	8.04	—	SABc II	+251	3.0	−20.02	197	10.87
SMC	00^h 53^m	−72°49'	316.2	0.60	2.70	0.36	Im IV-V	+34	58.0	−16.10	50	—
NGC 300	00^h 54^m	−37°41'	21.9	0.71	8.72	0.58	Sc/d II-IV	+98	1.2	−16.88	100	9.89
NGC 598 = M33	01^h 34^m	+30°34'	70.8	0.60	6.27	0.47	Sc, cd II-III	−46	0.795	−18.31	101	1010
NGC 628 = M74	01^h 37^m	+15°47'	10.5	0.91	9.95	0.51	SAc II	+753	9.7	−20.32	face on	—
NGC 1068 = M77	02^h 43^m	−00°01'	7.1	0.85	9.61	0.70	SAb II	+1144	14.4	−21.39	face on	—
NGC 1291	03^h 17^m	−41°06'	9.8	0.83	9.39	0.91	SB0/a	+712	8.6	−20.26	face on	—
NGC 1313	03^h 18^m	−66°30'	9.1	0.76	9.20	0.48	SBc/d III-IV	+292	3.7	−18.60	136	10.31
NGC 1316 (Fornax A)	03^h 23^m	−37°12'	12.0	0.71	9.42	0.87	SAB0/a pec	+1674	16.9	−21.47	—	—
LMC	05^h 24^m	−69°45'	645.7	0.85	0.91	0.43	SBm-Ir III-IV	+119	0.055	−18.19	100	—
NGC 2403	07^h 37^m	+65°36'	21.9	0.56	8.93	0.39	Scd III	+226	4.2	−19.68	124	10.67
NGC 2903	09^h 32^m	+21°30'	12.6	0.48	9.68	0.55	SABbc I-II	+476	6.3	−19.85	196	10.91

Galaxy	RA (2000)	Dec. (2000)	Diam. ′	Axial Ratio	B_T	$(B-V)_0$	Type	V_r^b	Dist. (Mpc)	M_B	V_{rot}	Log M_{tot}
NGC 3031 = M81	$09^h 56^m$	$+69°04'$	26.9	0.52	7.89	0.82	SAab I-II	+69	1.4	−18.29	236	10.73
NGC 3034 = M82	$09^h 56^m$	$+69°41'$	11.2	0.38	9.30	0.79	I0/amorphous	+323	5.2	−19.42	—	—
NGC 3115	$10^h 05^m$	$-07°43'$	7.2	0.84	9.87	0.94	S0	+492	6.7	−19.18	—	—
NGC 3521	$11^h 06^m$	$-00°02'$	11.0	0.47	9.83	0.68	SABbc II	+673	7.2	−19.88	295	10.99
NGC 3627 = M66	$11^h 20^m$	$+13°00'$	9.1	0.46	9.65	0.60	SABb II	+643	6.6	−19.66	188	10.74
NGC 4258 = M106	$12^h 19^m$	$+47°18'$	14.8	0.39	9.10	0.55	SABbc II-III	+510	6.8	−20.59	213	11.17
NGC 4449	$12^h 28^m$	$+46°06'$	6.2	0.71	9.99	0.41	IB/Sm IV	+255	3.0	−17.66	—	—
NGC 4472 = M49	$12^h 30^m$	$+08°00'$	10.2	0.81	9.37	0.95	E1-2/S0	+846	16.8	−21.82	—	—
NGC 4486 = M87	$12^h 31^m$	$+12°23'$	8.3	0.79	9.59	0.93	cD, E0p	+1229	16.8	−21.64	—	—
NGC 4594 = M104 (Sombrero)	$12^h 40^m$	$-11°37'$	8.7	0.41	8.98	0.45	Sa/ab	+969	20.0	−22.98	369	11.38
NGC 4631	$12^h 42^m$	$+32°32'$	15.5	0.17	9.75	0.55	SBc/d III	+629	6.9	−20.12	142	10.70
NGC 4649 = M60	$12^h 44^m$	$+11°33'$	7.4	0.81	9.81	0.95	S0/E2	+970	16.8	−21.36	—	—
NGC 4736 = M94	$12^h 51^m$	$+41°07'$	11.2	0.81	8.99	0.72	SAab II	+360	4.3	−19.37	186	10.78
NGC 4826 = M64	$12^h 57^m$	$+21°41'$	10.0	0.54	9.36	0.71	Sab II	+403	4.1	−19.15	—	—
NGC 4945	$13^h 05^m$	$-49°28'$	20.0	0.19	9.30	—	SBcd IV	+383	5.2	−20.65	—	—
NGC 5055 = M63	$13^h 16^m$	$+42°02'$	12.6	0.58	9.31	0.64	SAbc II-III	+571	7.2	−20.42	224	11.15
NGC 5128 (Cen. A)	$13^h 25^m$	$-43°01'$	25.7	0.79	7.84	0.88	S0p	+398	4.9	−20.97	—	—

TABLE 19 Galaxies – bright[a] *(continued)*

Galaxy	RA (2000)	Dec. (2000)	Diam. '	Axial Ratio	B_T	$(B -)_0$	Type	V_r^b	Dist. (Mpc)	M_B	V_{rot}	Log M_{tot}
NGC 5194 = M55	13h 30m	+47°12'	11.2	0.62	8.96	0.53	SAbc I-IIp	+551	7.7	−20.75	—	—
NGC 5236 = M83	13h 37m	−29°52'	12.9	0.89	8.20	0.61	SBc II	+384	4.7	−20.31	face on	—
NGC 5457 = M101	14h 03m	+54°21'	28.8	0.93	8.31	0.44	SABcd I	+360	5.4	−20.45	face on	—
NGC 6744	19h 10m	−63°51'	20.0	0.65	9.14	—	Sbc II	+746	10.4	−21.39	185	11.37
NGC 6946	20h 35m	+60°09'	11.5	0.85	9.61	0.40	Scd II	+277	5.5	−20.78	153	10.81
NGC 7793	23h 58m	−32°35'	9.3	0.68	9.63	—	SAd IV	+228	2.8	−17.69	104	9.95

[a] Adapted from Cox, Arthur N. (ed). *Allen's Astrophysics Quantities: Fourth Edition*, New York, Springer, 2000, pg. 579. The radial velocity, V_r, is with respect to the center of our Galaxy and it is in units of kilometers per second. The distances, Dist., are in units of Megaparsec, abbreviated Mpc, where 1 Mpc = 3.26 million light-years, the absolute blue magnitude is denoted by M_B, the rotation velocity, V_{rot}, is in kilometers per second and the logarithm of the total mass, Log M_{tot}, is in units of the sun's mass $M_\odot = 1.989 \times 10^{33}$ grams.

The distances between galaxies are usually measured in units of Megaparsecs, abbreviated Mpc and equal to 3.26 million light-years. The mean space density of galaxies is about 0.055 galaxies per cubic Megaparsec. Galaxies can exist singly, in small groups, or in rich clusters of thousands of galaxies. *See* cluster of galaxies, elliptical galaxy, expanding Universe, irregular galaxy, Megaparsec, Milky Way Galaxy, spiral galaxy, and supercluster.

Galaxy cluster: *See* cluster of galaxies.

GALaxy Evolution EXplorer: NASA launched the *GALaxy Evolution EXplorer*, abbreviated **GALEX**, on 28 April 2003. It is designed to observe galaxies in ultraviolet light across 10 billion years of cosmic history, providing information on star formation and the evolution of galaxies.

Galaxy supercluster: *See* supercluster.

GALEX: Acronym for *GALaxy Evolution EXplorer*. *See* **GALaxy Evolution EXplorer**.

Galilean satellites: The Italian astronomer **Galileo Galilei (1564–1642)** discovered Jupiter's four largest moons in January 1610, using the newly invented telescope. These objects are now collectively called the Galilean satellites, and they retain the individual names given to them by the German astronomer **Simon Marius (1573–1624)**, also in 1610. In order of increasing distance from the giant planet, they are Io, Europa, Ganymede, and Callisto, all the names of mythological consorts of Zeus, the Greek equivalent of the Roman god Jupiter. All four moons were subject to detailed scrutiny with instruments aboard the *Galileo* spacecraft, between 1995 and 2004, including the volcanoes on Io, the internal ocean of liquid water on Europa, and the magnetic field generated within Ganymede. *See* Callisto, Europa, Ganymede, and Io.

Galileo Mission: The *Galileo* spacecraft was launched by NASA from the *Space Shuttle Atlantis* on 18 October 1989, and arrived at Jupiter on 7 December 1995, orbiting the

TABLE 20   Galilean satellites[a]

Satellite	Distance from Jupiter center (Jovian radii)	Orbital Period[b] (days)	Mean Radius (km)	Mass (10^{22} kg)	Mean Mass Density (kg m^{-3})
Io	5.90 R$_J$	1.769	1821.6	8.932	3528
Europa	9.48 R$_J$	3.551	1560.8	4.80	3014
Ganymede	15.0 R$_J$	7.155	2631.2	14.82	1942
Callisto	26.3 R$_J$	16.69	2410.3	10.76	1834

[a] The mean distances from the center of Jupiter are in units of Jupiter's equatorial radius, R$_J$ = 71 492 kilometers. The radii are given in units of kilometers, abbreviated km, the mass is given in kilograms, abbreviated kg, and the mass density is in units of kilograms per cubic meter, denoted kg m^{-3}. By way of comparison, the radius of our Moon is 1738 kilometers and the Moon's mean mass density is 3344 kg m^{-3}. The planet Mercury has a radius of 2440 kilometers and a mean density of 5427 kg m^{-3}.

[b] The orbital period of Europa is about twice that of Io, and the orbital period of Ganymede is nearly twice that of Europa.

giant planet and vastly increasing our understanding of Jupiter and the four Galilean satellites, Io, Europa, Ganymede and Callisto. The mission is named after the Italian astronomer **Galileo Galilei (1564–1642)** who discovered the four large moons of Jupiter, now known as the Galilean satellites, in 1610. The mission was terminated on 21 September 2003.

Gamma: A unit of magnetic field strength equal to 10^{-5} Gauss, 10^{-9} Tesla, and one nanoTesla.

Gamma ray: *See* gamma-ray radiation.

Gamma-ray burst: Intense radiation of cosmic gamma rays lasting for a few seconds or minutes, never coming from the same place twice, whose discovery using the *Vela* satellites of the United States Department of Defense was announced in 1973 by **Ray W. Klebesadel (1932–)**, **Ian B. Strong (1930–)**, and **Roy A. Olson (1924–)**. The gamma-ray bursts are emitted from distant galaxies, sometimes producing about as much energy in a few seconds as the visible-light energy emitted by all the stars in all the galaxies that we can observe.

Gamma-ray radiation: The most energetic form of electromagnetic radiation, with photon energies in excess of 100 keV or 0.1 MeV. The wavelength of gamma-ray radiation is less than 0.01 nanometer or 10^{-11} meters, the shortest possible. Because the Earth's atmosphere absorbs the radiation at that end of the electromagnetic spectrum, gamma ray studies of cosmic objects are conducted from satellites such as the *Compton Gamma Ray Observatory*. *See Compton Gamma Ray Observatory* and electromagnetic spectrum.

Ganymede: The largest of Jupiter's Galilean satellites, discovered in 1610 by the Italian astronomer **Galileo Galilei (1564–1642)**. Ganymede has a radius of 2631 kilometers and a mass density of 1942 kilograms per cubic meter. It is the largest satellite in the Solar System and bigger than the planet Mercury. Ganymede has an intrinsic magnetic field, discovered with instruments aboard the *Galileo* spacecraft in the late 1990s; and it is the only satellite that now generates its own magnetism. *See* Galilean satellites.

Gas pressure: The outward pressure caused by the motions of gas particles and increasing with their temperature. Gas pressure supports main-sequence stars like the Sun against the inward force of their immense gravity.

Gauss: The c.g.s., or centimeter-gram-second, unit of magnetic field strength, named after the German physicist and astronomer **Karl Friedrich Gauss (1777–1855)**, who showed in 1838 that the Earth's dipolar magnetic field must originate inside the planet. The SI, or Systeme International, unit of magnetic field strength is the Tesla, where 1 Tesla = 10,000 Gauss = 10^4 Gauss. The Tesla is named after the Croatian-born American electrical engineer **Nikola Tesla (1856–1943)**, who was a pioneer in the use of alternating-current electricity. *See* gamma and Tesla.

Gaussian distribution: *See* Maxwellian distribution.

Geminga: A bright gamma-ray source discovered in 1973 in the constellation Gemini with instruments aboard NASA's second *Small Astronomy Satellite*, abbreviated

SAS-2. The name Geminga was chosen because it is an abbreviation for the "Gemini gamma ray source" and because *Geminga* means "not there" in the Milanese dialect of Italy. For nearly 20 years the nature of Geminga was unknown. Then, in 1991, instruments aboard the **ROSAT** satellite detected variable X-ray emission with a period of 0.237 seconds, suggesting that Geminga is an X-ray pulsar. The pulse, or rotational period, and the rate at which Geminga is slowing down, indicate that the rotating neutron star formed about 340,000 years ago, presumably during a supernova explosion. The optical counterpart is very faint, of apparent visual magnitude 25.5, with a proper motion that is consistent with a distance of about 800 light-years. No radio pulsar has been detected.

Gemini Observatory: Twin 8.1-meter (320-inch) optical/infrared telescopes, one in the north and the other in the south, which together access the entire sky. The Gemini South Telescope is located at Cerro Pachón, Chile, and the Gemini North Telescope is located on Hawaii's Mauna Kea. They were built and are operated by a partnership of 7 countries, the United States, United Kingdom, Canada, Chile, Australia, Brazil, and Argentina. The international headquarters of the Gemini Observatory is located in Hilo, Hawaii at the University of Hawaii in Hilo's University Park, and it is managed by the Association for Universities for Research for Astronomy.

Gemini telescopes: *See* Gemini Observatory.

General Theory of Relativity: A theory in which gravity is interpreted in terms of the curvature in the geometry of space-time by mass-energy, postulated in 1915–17 by the German physicist **Albert Einstein (1879–1955)**. It includes the assumption that gravitational mass is equal to inertial mass. The *General Theory of Relativity* explained the anomalous precession of Mercury's perihelion, and predicted the bending of starlight by the Sun, the existence of gravitational lenses, and the existence of gravitational radiation. *See* binary pulsar, gravitational lens, gravitational radiation, mass, and non-Euclidean geometry.

Genesis mission: NASA launched the *Genesis* spacecraft on 21 October 2001, designed to collect samples of the solar wind and return them to the Earth in a capsule with a mid-air capture in 2004, which failed.

Geocentric: Centered on the Earth. In geocentric cosmology the Earth is situated at the center of the Universe.

Geomagnetic field: The magnetic field in and around the Earth. The strength of the magnetic field at the Earth's surface is approximately 3.2×10^{-5} Tesla, or 0.32 Gauss, at the equator and 6.2×10^{-5} Tesla, or 0.62 Gauss, at the North Pole. To a first approximation, the Earth's magnetic field is like that of a bar magnet, or a dipole, currently displaced about 500 kilometers from the center of the Earth towards the Pacific Ocean and tilted at 11 degrees to the rotation axis.

Geomagnetic storm: A rapid, worldwide disturbance in the Earth's magnetic field, typically lasting a few hours, caused by the arrival in the vicinity of the Earth of a coronal mass ejection. A substorm is a magnetic disturbance observed in Polar Regions only, and is associated with changes in direction of the interplanetary magnetic field as it encounters the Earth. Auroral activity and disruption of radio communications are common during intense geomagnetic storms. *See* coronal mass ejection and magnetic storm.

Geometry: The mathematics of lines drawn through space. *See* Euclidean geometry and non-Euclidean geometry.

Geosynchronous orbit: The orbit of a satellite that travels above the Earth's equator from west to east at an altitude where the satellite's orbital velocity is equal to the rotational velocity of the Earth. In this orbit, a satellite remains nearly stationary above a particular point on the planet. Such an orbit has an altitude of about 35 900 kilometers. *See* synchronous orbit.

Giant molecular cloud: An interstellar cloud of gas and dust, consisting mostly of molecular hydrogen, but containing other molecules like ammonia, carbon monoxide, formaldehyde and water, which are up to 150 light-years across and as massive as 200 000 times the mass of the Sun. Stars are now being born in giant molecular clouds. *See* dark nebulae, interstellar molecules, molecular cloud, and protostar.

Giant planet: A large planet with a size and mass noticeably bigger than those of the Earth. The four giant planets in our Solar System are Jupiter, Saturn, Uranus and Neptune. More than 100 Jupiter-sized planets have been discovered in tight orbit about relatively nearby, Sun-like stars. *See* exoplanet, 51 Pegasi, 47 Ursae Majoris, Jupiter, Neptune, Saturn, and Uranus.

Giant star: A large, massive, high-luminosity star that shines roughly a hundred times more brightly than the Sun, and designated by luminosity classes II, III and IV for bright, normal and sub giants. A giant star has evolved off and above the main sequence of the Hertzsprung-Russell diagram. A giant star is larger than a main-sequence, or dwarf, star of the same temperature, but giants of intermediate temperature are smaller than the hottest main-sequence stars. A main-sequence star will evolve into a giant star when it has depleted the hydrogen fuel in its core. Giant stars belong to a late and relatively short period of stellar evolution, and are therefore relatively rare stars. Because of their high luminosity, giant stars are nevertheless quite common among the brightest stars, such as Aldebaran, Arcturus and Capella. *See* Aldebaran, Arcturus, Capella, Hertzsprung-Russell diagram, main sequence, and stellar classification.

***Giotto* Mission**: ESA launched the *Giotto* spacecraft on 2 July 1985. It encountered Comet Halley on 13 March 1986, obtaining images of its dark, irregular jet-spewing nucleus and measuring the amount of water and dust emitted. The mission is named after the Florentine painter **Giotto di Bondone (c. 1267–c. 1337)**, whose frescoes in the Arena Chapel in Padua, Italy illustrate the lives of Jesus Christ and of the Virgin Mary, including the birth of Jesus with a comet in the sky.

Global Oscillation Network Group: Abbreviated GONG, a network of six identical solar telescopes located around the world and operated by the United States National Solar Observatory. The telescopes follow the Sun as the Earth rotates, providing almost continuous monitoring of the Sun's five-minute oscillations, or pulsations, and the associated sound waves that penetrate the Sun's interior. Beginning operations in 1995, the Global Oscillation Network Group has provided valuable information about the composition, motions, structure and temperature of the Sun's interior using the technique of helioseismology. *See* five-minute oscillations, helioseismology, and National Solar Observatory.

Global warming: A potentially dangerous increase in the temperature of the Earth due to an increase of heat-trapping gases, like carbon dioxide, in the atmosphere, resulting from human activity such as the burning of the fossil fuels coal, gas and oil. In 1896 the Swedish chemist **Svante Arrhenius (1859–1927)** showed that a doubling of the Earth's atmospheric carbon dioxide produces a global warming of 5 to 6 degrees Centigrade (9 to 11 degrees Fahrenheit), comparable to modern estimates. Measurements begun in 1958 by the American scientist **Charles Keeling (1928–2005)** have demonstrated an exponential rise in the Earth's atmospheric carbon dioxide over the subsequent two decades, and that by burning coal, gas and oil humans have increased the amount of carbon dioxide in the Earth's atmosphere by 30 percent since the industrial revolution. *See* greenhouse effect.

Globular cluster: A densely packed, gravitationally bound aggregation of 10 thousand to a ten million stars within a region that is about 100 light-years across. Globular clusters contain Population II stars that are at least 10 billion years old and have a low abundance of heavy elements. Globular clusters are usually spherically shaped, and can be found in the halos of spiral galaxies. As shown in 1918 by the American astronomer **Harlow Shapley (1885–1972)**, the globular clusters in our Milky Way Galaxy are distributed in a roughly spherical system that is centered far from the Sun and envelops the flattened stellar disk of the Milky Way. Our Galaxy has about 140 known globular clusters, distributed in a spherical halo at distances as far as 150 000 light-years from the galactic center, but most of them lie no further from the galactic center than the Sun. Their extreme age and distribution in the galactic halo indicate that globular clusters originated when our Galaxy was forming by the gravitational contraction of a very large concentration of gaseous material. Bright globular clusters in our Galaxy include Omega Centauri, 47 Tucanae, and M13. The largest elliptical galaxies may have thousands of globular clusters. *See* galactic halo, Population II star, and stellar populations.

GMT: Abbreviation for Greenwich Mean Time. *See* Greenwich Mean Time, solar time, and Universal Time.

Goddard Space Flight Center: NASA's Goddard Space Flight Center, abbreviated GSFC, is located in Greenbelt, Maryland, just a few miles northeast of Washington, D.C. The center manages many of NASA's Earth observation, astronomy and space physics missions. *See* National Aeronautics and Space Administration.

GONG: Acronym for Global Oscillation Network Group. *See* Global Oscillation Network Group.

Granulation: A mottled, cellular pattern visible at high spatial resolution in the white light of the photosphere. The solar granulation exhibits a non-stationary, overturning motion, a visible manifestation of hot gases rising from the Sun's interior by convection. *See* granule and supergranulation.

Granule: One of about a million bright regions, or cells, that cover the visible solar disk at any instant, and that comprise the granulation detected in high-resolution, white-light observations of the photosphere. The bright center of each granule, or convection cell, is the highest point of a rising column of hot gas. The dark edges of each granule are the cooled gas, which is descending because it is denser than the hotter gas. The mean angular distance between the bright centers of adjacent granules is about 2.0 seconds of arc, corresponding to about 1500 kilometers at the photosphere. Individual granules are often polygonal-shaped regions, and they appear and disappear on time scales of about ten minutes. *See* granulation.

Gravitational collapse: The collapse of a cosmic object, such as an interstellar cloud or a star, or even the entire Universe, when there is nothing to sufficiently support it against the combined gravitational attraction of its component parts.

Gravitational lens: In 1915 the German physicist **Albert Einstein (1879–1955)** predicted that the Sun curves space in its vicinity, bending a light ray passing the Sun's edge by 1.75 seconds of arc. The predicted gravitational bending of starlight passing near the Sun was first measured during the total solar eclipse of 29 May 1919, reported that year by the English astronomers **Frank W. Dyson (1868–1939)**, **Arthur Eddington (1882–1944)** and **Charles Davidson (1875–1970)**. It wasn't until 1936 that Einstein published earlier calculations of the complete, cosmic gravitational lens effect, in response to prodding by an amateur scientist, showing that the apparent luminosity of a distant star could be magnified by the gravitational lens effect of a nearby star. The discovery of double, triple and quadruple images of distant quasars in 1979 to 1985 put the concept of cosmic gravitational lenses on a firm observational footing. As suggested in 1937 by the Swiss astronomer **Fritz Zwicky (1898–1974)**, dark matter in a massive cluster of galaxies can act as a gravitational lens, diverting and focusing the light of more distant galaxies. The cluster of galaxies can produce multiple images or arcs and sometimes cause the background galaxy or quasar to look brighter than it would otherwise appear. In 1989 the American astronomers **Roger Lynds (1928–)** and **Vahé Petrosian (1938–)** reported observations indicating that a cluster of galaxies acts as a gravitational lens, spreading the light of remote background galaxies into an array of luminous arcs and providing information about dark matter in the cluster of galaxies. *See* microlensing.

Gravitational radiation: The *General Theory of Relativity,* introduced by the German physicist **Albert Einstein (1879–1955)** predicts that any non-spherical, dynamically changing, massive system must produce gravitational waves, which carry energy at the speed of light. As with electromagnetic waves, the amplitude of gravitational waves falls off as the inverse of the distance from their source. The expected effects of gravitational radiation were detected in the decreasing orbital period of the binary pulsar, discovered in 1974 by the American radio astronomers **Russell A. Hulse (1950–)** and **Joseph H. Taylor Jr. (1941–)**. The binary pulsar consists of a pair of compact neutron stars that are slowly moving toward each other, producing a decrease in the orbital period with a rate of energy loss that is consistent with gravitational radiation. However, no instrument has directly observed gravity waves as they ripple through the fabric of space-time. Short-lived bursts of detectable gravitational waves, lasting just a few minutes or less, might be produced by violent events in the distant Universe, such as the collision of two black holes, the coalescence and merging of two neutron stars, the gravitational collapse associated with the formation of a black hole or neutron star, or the shock waves generated during powerful stellar explosions called supernovae. They might be detected in the future with instruments like the Laser Interferometer Gravitational-wave Observatory, abbreviated LIGO, which began operation in 2001. *See* binary pulsar and Laser Interferometer Gravitational-wave Observatory.

Gravitational waves: *See* gravitational radiation.

Gravity: The universal force of attraction between all particles of matter. According to the English physicist **Isaac Newton (1643–1727)**, the force of gravitational attraction between two objects is directly proportional to the mass of each object and decreases by the square of the distance separating them. The force, denoted by F, between two masses, denoted M_1 and M_2, and separated by distance, D, is given by $F = GM_1M_2 / D^2$, where the gravitational constant $G = 6.6726 \times 10^{-11}$ Newton m^2 kg^{-2}. In his *General Theory of Relativity,* the German physicist **Albert Einstein (1879–1955)** interpreted gravity as a consequence of the curvature of space induced by the presence of a massive object. *See* curvature of space, and *General Theory of Relativity.*

Gravity Probe B: A gyroscope experiment launched by NASA on 20 April 2004, designed for precise tests of the *General Theory of Relativity.*

Great attractor: A distant, concentration of enormous mass, discovered in 1987 by the American astronomer **Alan Dressler (1948–)** and his colleagues, that is apparently pulling all the galaxies within 100 million light-years in the direction of the constellations Hydra and Centaurus. The Great Attractor lies about 150 million light-years away with an estimated mass equivalent to about 50 000 galaxies like our Milky Way Galaxy.

Great circle: The line traced out on the Earth by a plane passing through the center of the globe, dividing it into two equal hemispheres. The name comes from the fact that no greater circles can be drawn on a sphere. A great circle halfway between the North and South Poles is called the equator, because it is equally distant between both poles.

Great Observatories: NASA's four orbiting satellites carrying telescopes designed to study the Universe in both visible light and non-visible forms of radiation. The first in the series was the *Hubble Space Telescope,* launched on 24 April 1990 and operating mainly in visible light, the second was the *Compton Gamma Ray Observatory,* launched into Earth orbit on 5 April 1991, the third was the *Chandra X-ray Observatory,* launched 23 July 1999, and the fourth is the infrared *Spitzer Space Telescope* launched on 23 August 2003. *See Chandra X-ray Observatory, Compton Gamma Ray Observatory, Hubble Space Telescope,* and *Spitzer Space Telescope.*

Great Red Spot: A large oval area in the southern hemisphere of Jupiter, which is larger than two Earth diameters and has been observed with small telescopes for centuries. The Great Red Spot is a swirling, high-pressure storm cloud, known as an anti-cyclone, trapped between counter-flowing east-west and west-east jet winds that funnel smaller eddies toward the spot, helping to roll it around. The Great Red Spot can consume smaller eddies that pass in its vicinity, extracting their energy and varying from about 11 000 to 14 000 kilometers wide by 24 000 to 40 000 kilometers long. The earliest sightings of a large spot in the atmosphere of Jupiter have been credited to the English physicist **Robert Hooke (1635–1703)** in 1664 and to Italian-born French astronomer **Gian (Giovanni) Domenico (Jean Dominique) Cassini (1625–1712)** the following year. The most prominent storm cloud – the Great Red Spot – appears in records and drawings dating back to 1831, by the German astronomer **Samuel Heinrich Schwabe (1789–1875)**.

Green flash: A momentary green appearance of the uppermost part of the Sun's disk, lasting a few seconds just before sunset or just after sunrise, which results from atmospheric refraction and can be observed in the tropics at the ocean's horizon.

Greenhouse effect: The warming of a planet's surface by heat trapped by gases in its atmosphere, owing to their infrared opacity. The French mathematician **Joseph (Jean Baptiste) Fourier (1768–1830)** first proposed the mechanism in 1827, while wondering how the Sun's heat could be retained to keep the Earth hot. Incoming sunlight is absorbed at the surface and re-radiated at longer wavelengths, as infrared heat radiation. As the Irish physicist **John Tyndall (1820–1893)** showed in 1861, significant heat is absorbed in the Earth's atmosphere by carbon dioxide and water vapor. These greenhouse gases are not transparent to the infrared, and the heat is reflected back to the ground. The natural greenhouse effect warms the surface of the Earth by about 33 degrees Celsius and keeps the oceans from freezing. How much heating the greenhouse effect causes depends on how opaque the atmosphere is to infrared radiation. On Venus, the dense carbon dioxide atmosphere has raised the surface temperature to around 750 kelvin, hot enough to melt lead. Concern has mounted that global warming of the Earth will result from increased concentrations of carbon dioxide and other so-called "unnatural" greenhouse gases released by human activity, particularly the burning of fossil fuels such as coal and oil. Measurements begun in 1958 by the American scientist **Charles Keeling (1928–2005)** have demonstrated an exponential rise in the amount of carbon dioxide in the Earth's atmosphere over the subsequent two decades. *See* global warming.

Greenwich Mean Time: Abbreviated GMT, the mean solar time for the meridian of Greenwich, England, designated in 1884 as the prime meridian of zero longitude for timekeeping and navigation purposes. For scientific purposes, the Greenwich Mean Time is widely known as Universal Time, or UT for short. *See* Royal Greenwich Observatory, Solar Time, and Universal Time.

Gregorian telescope: A reflecting telescope using a parabolic primary mirror and a concave, elliptical secondary mirror that reflects light back through a hole in the primary mirror, not in common use today but of historical interest since it was described by the Scottish mathematician **James Gregory (1638–1675)** in 1663, five years before the English physicist **Isaac Newton (1643–1727)** constructed his first reflecting telescope consisting of a spherical primary mirror and a flat secondary mirror to reflect the light to the side, now in common use with a parabolic primary. *See* Newtonian telescope, reflector, reflecting telescope and telescope.

TABLE 21   Greenhouse gases – produced by human activity[a]

Greenhouse gas	Pre-industrial Concentration[b] (1860)	Recent Concentration[b] (2005)	GWPC	Lifetime (years)	Sources
Carbon Dioxide (abbreviated CO_2)	280 p.p.m.v.	374.9 p.p.m.v	1	5–200 (variable)	Burning coal, oil, and natural gas, deforestation, cement manufacture
Methane (abbreviated CH_4)	709 p.p.b.v.	1791 p.p.b.v.	23	12	Livestock, rice growing, natural gas and oil production, coal mining
Nitrous Oxide (abbreviated N_2O)	270 p.p.b.v.	318 p.p.b.v.	296	114	Nitrogen fertilizers, chemical manufacturing, waste treatment
Chlorofluorocarbons CFC-11 CFC-12	zero zero	254.5 p.p.t.v. 544.5 p.p.t.v.	4 600 10 600	45 100	Refrigeration, aerosol cans, foam insulation, industrial solvents
Sulfur Hexafluoride (abbreviated SF_6)	zero	5.21 p.p.t.v.	22 200	3200	Electrical equipment insulation, magnesium production, medical treatments

[a] From http://cdiac.esd.ornl.gov/pns/current_ghg.html. Water vapor produces sixty to seventy percent of the Earth's global warming, but it is not included here because water vapor is not directly produced by human activity.

[b] Averages of the measured amounts, all by volume, v; where p.p.m. = parts per million (10^6), p.p.b. = parts per billion (10^9), p.p.t. = parts per trillion (10^{12}).

[c] The GWP, or the Global Warming Potential, is used to contrast the radiative effects of different greenhouse gases relative to carbon dioxide. Values are the ratio of global warming from one unit mass of a gas to that of one unit mass of carbon dioxide over 100 years.

GSFC: Acronym for Goddard Space Flight Center. *See* Goddard Space Flight Center.

G star: A yellow star of spectral type G, such as Alpha Centauri, Capella, 47 Ursae Majoris, 51 Pegasi, and the Sun. Most G stars are dwarf, or main sequence, stars similar to the Sun, with disk temperatures of between 5000 and 6050 kelvin. *See* Alpha Centauri, Capella, 47 Ursae Majoris, Hertzsprung-Russell diagram, 51 Pegasi, Sun, and stellar classification.

Gum Nebula: An enormous, near-circular emission nebula in the constellations Puppis and Vela, about 36 degrees in angular diameter It is named after the Australian radio astronomer **Colin S. Gum (1924–1960)** who mapped the southern sky for radio sources and emission nebula, discovering the Gum Nebula in the process. The Gum Nebula may be an ancient supernova remnant, with an age exceeding a million years, a distance estimated at 1300 light-years, and a diameter of about 840 light-years. The much more recent Vela supernova remnant, perhaps 12 000 years old and containing the Vela pulsar, lies within the Gum Nebula. *See* emission nebula, supernova remnant, Vela pulsar, and Vela supernova remnant.

Gunn-Peterson effect: The absorption of the light of distant quasars by intergalactic matter. In 1965 the American astronomers **James E. Gunn (1938–)** and **Bruce Peterson (1941–)** showed that the absence of atomic hydrogen absorption in quasar radiation places stringent limits to the amount of intergalactic un-ionized hydrogen. In 1995, the American astronomer **Arthur F. Davidsen (1944–2001)** measured the absorption of ultraviolet quasar radiation by singly ionized helium in intergalactic space.

Gyr: Abbreviation for gigayear, equal to a billion years or to 10^9 years.

Hadar: Also known as Beta Centauri. The second brightest star in the constellation Centaurus, the star Hadar has an apparent magnitude of $+0.61$, a distance of 320 light-years, and an absolute visual magnitude of -4.3.

Hale's law: The leading, or westernmost spots of any sunspot group in one hemisphere of the Sun have the same magnetic polarity, while the following, or easternmost, spots have the opposite magnetic polarity. The direction of any bi-polar group in the southern hemisphere is the reverse of that in the northern hemisphere. All of the spots' magnetic polarities reverse each 11-year solar activity cycle.

Half-life: The time it takes for half of a given quantity of radioactive material to decay. The half-life for uranium, which decays into lead, is 704 million years.

Halley's Comet: The bright, periodic comet named after the English astronomer **Edmond Halley (1656–1742)**. Halley found that the orbit of the comet he observed in 1682 was similar to those of comets observed in 1607, by the German astronomer **Johannes Kepler (1571–1630)** and in 1531 by another German astronomer **Petrus Apianus (1495–1552)**. All three comets moved around the Sun in retrograde orbits with a similar orientation. Halley also knew that the Great Comet of 1456 had traveled in the retrograde direction, and in 1705 he concluded that all four comets were returns of the same comet in a closed elliptical orbit around the Sun with a period of about 76 Earth years. Halley confidently predicted its return in 1758, noting that he would not live to see it. We now know that this comet has returned to fascinate the world on at least thirty-two close approaches to the Sun, ever since 240 BC. In March 1986 the *Giotto* spacecraft flew close enough to image the nucleus of Halley's Comet, showing that it has an irregular shape, with dimensions of $16 \times 8 \times 7$ kilometers, a dark surface crust, and jets of dust and ice on its sunlit side. At the time, the comet was emitting about 20 tons, or 20 000 kilograms, of gas and 10 tons of dust every second. *See* comet and ***Giotto Mission***.

Halo: *See* galactic halo.

H-alpha: *See* hydrogen-alpha line.

Hard X-rays: Electromagnetic radiation with photon energies of between 10 keV and 100 keV and wavelengths between about 10^{-10} and 10^{-11} meters.

Hayashi track: The evolutionary path in the Hertzsprung-Russell diagram of a star that is forming as the result of gravitational contraction, lying above and to the right of the star's eventual location on the main sequence. It was first described in 1961 by the Japanese astrophysicist **Chushiro Hayashi (1920–)**, who realized that strong convection would enable a contracting star to release its energy much faster than by radiation, and that, consequently, pre-main-sequence stars would be considerably more luminous than their counterparts already on the main sequence. As Hayashi showed, the evolutionary track of a contracting star moves vertically downward until the convection zone retreats

TABLE 22   **Halley's Comet**[a]

240 BC	25 May
164	13 November
87	6 August
12 BC	11 October
66 AD	26 January
141	22 March
218	18 May
295	20 April
374	16 February
451	28 June
530	27 September
607	15 March
684	3 October
760	21 May
837	28 February
912	19 July
989	6 September
1066	21 March
1145	19 April
1222	29 September
1301	26 October
1378	11 November
1456	10 June
1531	26 August
1607	28 October
1682	15 September
1759	13 March
1835	16 November
1910	20 April
1986	9 February
2061[a]	28 July
2134[a]	27 March

[a] Date of perihelion passage, where the perihelion of a comet is the point in its orbit that is closest to the Sun. The future two perihelion passages of Halley's comet are predicted dates; all of the others have been recorded.

toward the stellar surface, and becomes sufficiently thin that the star's evolutionary track turns leftward until it reaches the main sequence. His theoretical description accounted for the Hertzsprung-Russell diagrams of extremely young clusters obtained by the American astronomer **Merle F. Walker (1926–)** in 1956 to 1961. *See* Hertzsprung-Russell diagram.

Heavy water: A form of water in which the abundant, light, hydrogen atoms, H, are replaced by less abundant, deuterium atoms, a heavy form of hydrogen denoted by D, to form D_2O; here O denotes an atom of oxygen.

Heliocentric: Centered on the Sun. In heliocentric cosmology, the Sun is situated at the center of the Universe.

Heliometer: A refracting telescope with the lens cut exactly in half to give two images. When the two halves are moved apart, their separation can be used to precisely measure angles. A heliometer was first used to measure the angular diameter of the Sun, hence the name heliometer, and to measure the angles between celestial bodies. The British optician **John Dollond (1706–1761)** made the first heliometer in 1754 and one was used in 1838 by the German astronomer **Friedrich Wilhelm Bessel (1784–1846)** to measure the annual parallax of the star 61 Cygni, making the first determination of the distance of any star other than the Sun. *See* annual parallax and 61 Cygni.

Heliopause: The place where the solar wind pressure balances other pressures found in interstellar space. The heliopause is located about 100 astronomical units from the Sun or at about 100 times the distance between the Earth and the Sun.

Helioseismology: The study of the interior of the Sun by the analysis of sound waves that propagate through the solar interior and manifest themselves by oscillations at the photosphere. These oscillations have periods of around five minutes and are observed spectroscopically as Doppler shifts in the absorption line spectrum. The technique of helioseismology has been used to infer the composition, motions, structure and temperature of the solar interior. Helioseismology is a hybrid name combining the Greek words *Helios* for the "Sun" and *seismos* for "tremor or quake." *See* five-minute oscillations, Global Oscillation Network Group, and *SOlar and Heliospheric Observatory*.

Heliosphere: A vast region, cavity, or bubble carved out of interstellar space by the solar wind, and extending to about 100 astronomical units from the Sun or to about 100 times the mean distance between the Earth and the Sun. The heliosphere is the region of interstellar space surrounding the Sun where the Sun's magnetic field and the charged particles of the solar wind control plasma processes. The heliosphere is immersed in the local interstellar medium, and defines the extent of the Sun's influence. The heliosphere contains our Solar System, and extends well beyond the planets to the heliopause, an outer boundary that marks the place where the solar wind pressure balances other pressures found in interstellar space. *See* heliopause.

Heliostat: A moveable flat mirror used to reflect sunlight into a fixed solar telescope.

Helium: After hydrogen, the second most abundant element in the Sun and the Universe. The nucleus of a helium atom is called an alpha particle; it contains two protons and two neutrons. Helium was first observed as a spectral line during the solar eclipse on 18 August 1868, by the French astronomer **Jules Janssen (1824–1907)** and the British astronomer **Norman Lockyer (1836–1920)**, and named helium by Lockyer after the Greek Sun god *Helios*. Helium was not found on Earth until 1895, when the Scottish chemist **William Ramsay (1852–1916)** discovered it as gaseous emission from a heated uranium mineral. Most of the helium in the Universe was made in the immediate aftermath of the Big Bang that gave rise to the expanding Universe, but helium is also synthesized from hydrogen in the cores of main-sequence stars. Helium is one of the rare, inert noble gases. Helium has two isotopes, the more common helium-4, containing two protons and two neutrons in its nucleus, and the rare helium-3 with two protons and one neutron. *See* Big Bang, carbon-nitrogen-oxygen cycle, isotope, noble gases, nucleosynthesis, and proton-proton chain.

Helium burning: The release of energy by fusing helium into carbon within a star.

Helix Nebula: A large planetary nebula, designated NGC 7293, in the constellation Acquarius. The Helix Nebula has an apparent visual magnitude of 6.5 and is almost half a degree across, or about the angular size of the Moon and the Sun. It is illuminated by a hot 13th magnitude central star, and located at a distance of about 400 light-years, the nearest planetary nebula to Earth. *See* planetary nebulae.

Helmet streamer: Named after spiked helmets once common in Europe, helmet streamers form in the low corona over the magnetic inversion, or neutral, lines in large active regions, with a long-lived prominence commonly embedded in the base of the streamer. The footpoints of the streamer are in regions of opposite polarity so the streamer itself straddles the prominence. Higher up, the magnetic field is drawn out into interplanetary space within long, narrow stalks. Gas flowing out along these open magnetic fields might help to create the slow-speed solar wind. *See* coronal streamer.

Herbig-Haro object: Nebulous material ejected at high velocity from young stars, independently discovered by the American astronomer **George Herbig (1920–)** and the Mexican astronomer **Guillermo Haro (1913–1988)** in 1951 and 1952, respectively.

Hercules X-1: An intense X-ray source in the direction of the constellation Hercules, whose 1.2-second X-ray pulsations were discovered in 1971 by instruments aboard the *Uhuru* satellite. The X-ray pulsation period is decreasing as time goes on. It is a rotating neutron star that is speeding up, gaining rotational energy by accreting material from a nearby companion star. *See* pulsar, and ***Uhuru Mission***.

Herschel Space Observatory: An ESA mission that is in development, with an expected launch in February 2007. Its 3.5-meter mirror will collect far infrared and sub-millimeter radiation from distant, young galaxies. The observatory is named after the English astronomer **William Herschel (1738–1822)** who discovered the planet Uranus and made many pioneering studies of stars and nebulae.

Hertz: A unit of frequency equal to one cycle per second, abbreviated Hz. One kilohertz, abbreviated kHz, is one thousand Hz, one megahertz, abbreviated MHz, is a million Hz, and one gigahertz, abbreviated GHz, is one thousand million Hz.

Hertzsprung-Russell diagram: Abosolute luminosity of stars plotted against their spectral type, color index or effective temperature, abbreviated H-R diagram. Brightness increases from bottom to top, and temperature decreases from left to right, through the spectral sequence O, B, A, F, G, K, M. Most stars, including the Sun, lie on the main sequence, which is a diagonal band that extends from the bright blue stars in the upper left to the faint red ones in the lower right of the H-R diagram. Such stars are also called dwarf stars. The giant stars lie in a luminous band above the main sequence. The white dwarf stars occupy the bottom left of the H-R diagram; these tiny stars are comparable to the Earth in size. The American astronomer **Henry Norris Russell (1877–1957)** published an early version of this diagram in 1914; the Danish astronomer **Ejnar Hertzsprung (1873–1967)** previously plotted such diagrams for the Pleiades and Hyades star clusters in 1911. *See* dwarf star, giant star, Hayashi track, instability strip, main sequence, and white dwarf star.

HESSI: Acronym for the *High Energy Solar Spectroscopic Imager*. See *RHESSI*.

High-energy physics: *See* particle physics.

HIgh Precision PARallax COllecting Satellite: ESA launched its *HIgh Precision PARallax COllecting Satellite*, abbreviated *HIPPARCOS*, on 8 August 1989 and operated it until August 1993. It was the first space mission dedicated to astrometry, measuring the accurate positions, distances, motions, brightness and colors of stars. Instruments aboard *HIPPARCOS* measured the parallax and proper motions of 120 000 stars with a precision of 0.001 to 0.002 seconds of arc, up to 100 times better than possible from the ground. The acronym also alludes to the Greek astronomer **Hipparchus of Nicea (lived c. 146 BC)**, who recorded accurate star positions more than two millennia previously. *See* parallax and proper motion.

HIPPARCOS: Acronym for the *HIgh Precision PARallax COllecting Satellite*. *See HIgh Precision PARallax COllecting Satellite*.

H I region: A cloud of neutral, or un-ionized, atomic hydrogen gas in interstellar space, pronounced H one region. Such clouds do not emit visible light, and hence are invisible at optical wavelengths, but they do emit radio waves that are 21 centimeters long. *See* neutral hydrogen and twenty-one centimeter line.

H II region: An interstellar cloud of ionized atomic hydrogen, or protons and electrons, pronounced H two region. An HII region is produced by the ultraviolet radiation of nearby hot stars. *See* emission nebula, ionized hydrogen, Orion nebula, and Strömgren radius.

High-velocity star: Rare stars with extremely high radial velocities, over 100 kilometers per second. As first explained in 1925 by the Swedish astronomer **Bertil Lindblad (1895–1965),** the high-velocity stars are in the galactic halo and revolving slowly around the galactic center, in comparison with the Sun and nearby stars that are in the plane of our Galaxy and revolving rapidly about the galactic center. Consequently, high-velocity stars are moving relatively slowly, and only appear to have a high velocity because the Sun and Earth are moving fast. Another type of runaway star has an excep-

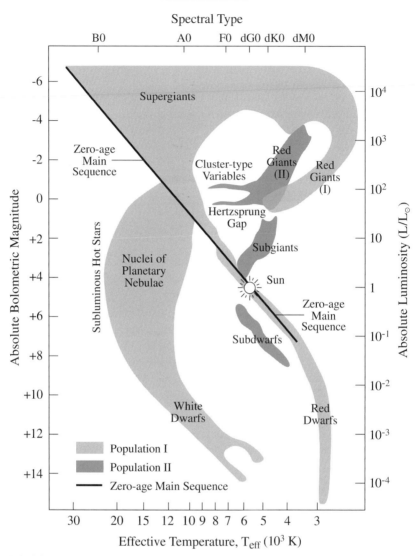

FIG. 29 Hertzsprung-Russell diagram – composite Stars of different absolute luminosity, L (*right axis in units of the Sun's absolute luminosity* $L_\odot$) and absolute magnitude, M, (*left axis*) are plotted as a function of the effective temperature, T_{eff}, of their visible disk (*bottom axis*). The vast majority, roughly 90 percent, of stars occupy the main sequence of the Hertzsprung-Russell diagram, running diagonally from the hot, luminous upper left to the cooler, less-luminous lower right. The hot, massive supergiants (*top left*) are the most luminous stars. The white dwarf stars (*lower middle*) are low-luminosity stars about the size of the Earth.

tionally high velocity due to its own motion through space, rather than the motion of the observer on Earth. *See* runaway star.

Hirayama families: Groups of asteroids that share very similar orbits, with common orbital inclinations and distances from the Sun, suggesting that they are the broken fragments of larger objects. They were discovered in 1928 by the Japanese astronomer **Kiyotsugu Hirayama (1874–1943)**, who called each group a family because he believed the members shared a common origin as the children of a bigger parent body. He also named a number of families after their largest member asteroids, such as the Eos, Flora, Koronis and Themis families. *See* asteroids.

Hobby-Eberly telescope: An optical or visible-light telescope consisting of 91 hexagonal mirror segments joined together to create an effective aperature with a 9.2-meter (360-inch) diameter. The **Hobby-Eberly telescope** is located on Mt. Fowlkes, Texas, and it began operation in October 1999. The telescope is named after **William P. Hobby (1932–)** and **Robert E. Eberly (1918–)**, American supporters of public education, and it is a joint project of five universities.

Homestake: The South Dakota gold mine where the chlorine neutrino detector is located. *See* chlorine experiment.

Homogeneity: Having the same physical properties at all points of space, in a system or in the Universe.

Homunculus Nebula: A cloud of dust surrounding the star Eta Carinae, and shed by the star in outbursts dating back to 1843. The Homunculus Nebula has an angular extent of about 12×17 seconds of arc. The name homunculus comes from its shape, which resembles a human figure. *See* emission nebula, Eta Carinae, and Eta Carinae Nebula.

Horizon: The maximum distance that an observer can see.

Horizon problem: Few of the particles of the early Universe would have had time to be in causal contact with one another at the outset of cosmic expansion. *See* inflation.

Horsehead Nebula: A dark nebula in the constellation Orion, with the shape of the head and mane of a horse, projected in silhouette against the emission nebula IC 434. The angular dimensions of the Horsehead Nebula are 4×6 minutes of arc, corresponding to about 3 light-years across. *See* dark nebula.

Hot dark matter: Subatomic particles that move almost as fast as light, such as the neutrino. *See* cold dark matter, dark matter, and neutrino.

Hourglass Nebula: A planetary nebula in the constellation Musca, designated MyCn 18. It is only 4 seconds of arc across and has an hourglass shape when imaged from the *Hubble Space Telescope*. *See* planetary nebula.

H-R diagram: Acronym for Hertzsprung-Russell diagram. *See* Hertzsprung-Russell diagram.

HST: Acronym for the *Hubble Space Telescope. See Hubble Space Telescope.*

Hubble constant: The rate at which the Universe expands and the recession velocities of galaxies increase with distance, denoted by the symbol H_0 where the zero subscript denotes the value at the present time. The Hubble constant is about 72 kilometers per second per Megaparsec. *See* Hubble diagram, Hubble law, and Hubble time.

Hubble diagram: Plot of the radial velocities, or redshifts, of galaxies against their distances, first made in 1929 by the American astronomer **Edwin Hubble (1889–1953)**. This was the first evidence of the expansion of the Universe. The Hubble constant is equal to the slope of the Hubble diagram. *See* expanding Universe and Hubble constant.

Hubble law: A relationship that states that the farther away from us a galaxy is, the faster it is receding from us. The linear increase of galaxy recession velocity, denoted by V_r, with galaxy distance, specified by D, according to the relation $V_r = H_0 D$, where H_0 is the Hubble constant. Although the American astronomer **Edwin Hubble (1889–1953)** did not specifically state it, this relation is now known as the Hubble law. *See* Hubble constant and Hubble diagram.

FIG. 30   Hubble diagram This plot of galaxy distance versus recession velocity is analogous to that obtained by **Edwin Hubble (1889–1953)**, in his 1929 discovery of the expansion of the Universe (Fig. 15). The slope of the linear fit (*solid line*) to the data (*dots*) measures the expansion rate of the Universe, a quantity called the Hubble constant, designated H_0. The data shown here, obtained by **Wendy Freedman (1957–)** and her colleagues, summarize eleven years of efforts to specify this constant by using the **Hubble Space Telescope** to measure the distances and velocities of Cepheid variable stars in galaxies. The distance, D, is in units of a million parsecs, or Mpc, where 1 Mpc is equivalent to 3.26 million light-years, and the radial velocity, V_r, is given in units of kilometers per second, denoted as km s^{-1}. The velocity and distance are related by Hubble's law $V_r = H_0 \times D$, where H_0 is Hubble's constant. The fit to these data indicate that $H_0 = 75 \pm 10$ km s^{-1} Mpc^{-1}, and that this constant lies well within the limits of 50 and 100 in the same units (*dashed lines*).

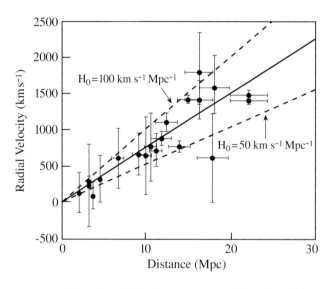

Hubble Space Telescope: NASA launched the ***Hubble Space Telescope***, abbreviated ***HST***, on 24 April 1990 from the ***Space Shuttle Discovery***, with servicing missions by shuttle astronauts in December 1993, February 1997, December 1999, and February 2002. It is the largest visible-light telescope ever put into space to look at the Cosmos, with a 2.4-meter (94.5-inch) primary mirror. The ***HST*** has observed newly formed galaxies when the Universe was half its present age, contributing to our understanding of the age and evolution of the Cosmos, watched super-massive black holes consuming the material around them, and helped astronomers determine how a mysterious "dark energy" has taken over the expansion of the Universe. The telescope was named for the American astronomer **Edwin Hubble (1889–1953)** who demonstrated that spiral nebulae are galaxies like our own and found the first evidence for the expansion of the Universe. The Space Telescope Science Institute, abbreviated STScI, is the astronomical research center responsible for operating the ***Hubble Space Telescope*** as an international observatory.

Hubble time: The reciprocal of the Hubble constant, denoted by $1/H_0$, which gives an estimate of the age of the expanding Universe, or the time since the Big Bang, on the assumption that there has been no slowing of the expansion of the Universe and that there is no cosmological constant. If there is no cosmological constant, the expanding Universe is slightly younger then the Hubble time, but the observable Universe can be much older if there is a non-zero value of the cosmological constant. *See* Big Bang, cosmological constant, and Hubble constant.

Hubble's variable nebula: A fan-shaped reflection nebula, designated NGC 2261, whose extreme variability was discovered in 1916 by the American astronomer **Edwin Hubble (1889–1953)**. The nebula extends northward from the Ae/Be star R Monocerotis, which is located at a distance of about 2500 light-years and exhibits emission features in its spectra. The youthful, high-mass star, estimated to be about ten times the mass of the Sun, is accreting mass from a disk of molecular hydrogen, revealed through radio radiation from carbon monoxide, that extends perpendicular to the fan-shaped nebula over 2.5 light-years from the central star.

Huygens Probe: ESA's element of the ***Cassini-Huygens Mission*** launched by NASA on 15 October 1997, with an initial encounter with Saturn in July 2004. The ***Huygens Probe*** descended into the hazy atmosphere of Titan, Saturn's largest moon, and landed on the satellite's surface on 14 January 2005, finding evidence for flowing liquid methane there. The ***Huygens Probe*** was named after **Christiaan Huygens (1629–1695)**, a Dutch astronomer who discovered Titan in 1655. *See **Cassini-Huygens Mission***.

Hyades: The nearest open star cluster, located in the constellation Taurus. The Hyades is 150 million light-years away from the Earth, and is about 600 million years old. *See* open cluster.

Hydrogen: The lightest, simplest and most abundant element in the Sun and in the Universe, and the first element in the periodic table, consisting of one proton and one electron. All of the hydrogen in the Universe was produced during the earliest moments of the Big Bang. Hydrogen is detected in the visible-light spectra of the Sun and other stars at wavelengths specified by the Balmer series, and at ultraviolet wavelengths by the Lyman series. Interstellar regions of neutral, un-ionized hydrogen atoms are called H I regions, and they emit radio radiation at 21-centimeters wavelength; bright stars ionize nearby interstellar hydrogen, producing H II regions. A hydrogen molecule consists of two hydrogen atoms, and molecular hydrogen is the most abundant molecule in interstellar space. Most of the mass of a hydrogen atom is in its nucleus, the proton, whose mass is 1.6726×10^{-27} kilograms or 1836 times the mass of an electron. A rare, heavy isotope of atomic hydrogen, known as deuterium, contains one proton and one neutron in its nucleus. Hydrogen is so light that it is found in only small amounts in Earth's atmosphere, but it is the main constituent of the massive giant planets Jupiter and Saturn. *See* alpha particle, Balmer series, Big Bang, deuterium, electron, Fraunhofer lines, H I region, H II region, hydrogen molecule, Lyman series, and twenty-one centimeter line.

Hydrogen-alpha line: The spectral line of neutral hydrogen in the red part of the visible spectrum, denoted by Hα. Light emitted or absorbed at a wavelength of 656.3 nanometers during an atomic transition in hydrogen, the lowest energy transition in its Balmer series. It is the dominant emission from the solar chromosphere. The wavelength also corresponds with a dark line produced by hydrogen absorption in the photosphere, designated with the letter C in 1814 by the German physicist and optician **Joseph von Fraunhofer (1787–1826)**. *See* Balmer series, chromosphere, filament, Fraunhofer lines, plage, prominence, spectral line, spectroheliograph, and spicule.

Hydrogen burning: The thermonuclear fusion of hydrogen nuclei into helium nuclei, and the process by which all main-sequence stars generate energy. Stars spend most of their lifetime burning hydrogen, and such stars are the most common kind of star. A star with the mass of the Sun spends around ten billion years in this stage; more massive stars spend less time. Stars more massive than 1.5 solar masses burn hydrogen by the CNO cycle; hydrogen burning by less massive stars proceeds by the proton-proton

TABLE 23   Hydrogen – lines[a]

Series m	Lyman ($n = 1$)	Balmer ($n = 2$)	Paschen ($n = 3$)	Brackett ($n = 4$)	Pfund ($n = 5$)
2	121.567				
3	102.572	656.280			
4	97.2537	486.132	1875.10		
5	94.9743	434.046	1281.81	4051.20	
6	93.7803	410.173	1093.81	2625.20	7457.8
7	93.0748	397.007	1004.94	2165.50	4652.5
8	92.6226	388.905	954.598	1944.56	3739.5
9	92.3150	383.538	922.902	1817.41	3296.1
10	92.0963	379.790	901.491	1736.21	3038.4

[a] Wavelengths of the m to n transitions of hydrogen for low quantum numbers m and n. The wavelengths are given in nanometers where 1 nanometer = 10^{-9} meters.

chain. *See* carbon-nitrogen-oxygen cycle, dwarf star, main sequence, and proton-proton chain.

Hydrogen molecule: A molecule, designated by the symbol H_2, composed of two hydrogen atoms, each designated by the capital letter H. Hydrogen molecules are the most abundant molecules in interstellar space. Although they do not emit radio waves, their distribution can be determined by observing the radio emission from carbon monoxide. *See* carbon monoxide and molecular cloud.

Hydrostatic equilibrium: The condition of stability in an atmosphere or stellar interior that exists when the inward gravitational force of the overlying material is exactly balanced by the outward force of the gas and radiation pressure. The pressure at each level in the gas supports the weight of all the overlying material.

Hypothesis: A possible explanation of something, which is less comprehensive and less well established than a theory.

Hz: Abbreviation for Hertz, a unit of frequency equivalent to cycles per second. *See* Hertz.

IAS: Acronym for the Instituto di Astrofisica Spaziale, Italy.

IAU: Acronym for the International Astronomical Union. *See* International Astronomical Union.

IC: Abbreviation for the two *Index Catalogues* produced in 1890 and 1908 by the Danish-born Irish astronomer **John Louis Emil "J. L. E." Dreyer (1852–1926)** as supplements to his *New General Catalogue,* abbreviated NGC, of nebulae and star clusters, published in 1888. Modern astronomers still designate galaxies and other cosmic objects by their numbers in the NGC and IC catalogues. *See* NGC.

Ice Age: A period of cool, dry climate on Earth causing a long-term buildup of extensive ice sheets far from the poles. The major ice ages last for about 100 000 years; they are separated by warmer interglacial periods that last roughly 10 000 years. The major ice ages seem to be triggered by rhythmic variations in the global distribution of sunlight with periods of 23, 41, and 100 thousand years. *See* little ice age and Milankovitch cycles.

Impact crater: *See* crater.

Indeterminacy principle. *See* **uncertainty principle**.

Index Catalogue: *See* IC.

Inertia: The tendency of any mass to remain either at rest or in constant motion unless acted upon by a force.

Inflation: A hypothetical, extremely rapid expansion of the very early Universe immediately after the Big Bang and before any matter was formed, in which the expansion occurred much more rapidly than it does today, at an exponential rather than a linear rate. An exponential expansion, based on quantum gravity, was proposed in 1979 by the Russian astrophysicist **Alexei A. Starobinsky (1948–)**. In 1981 the American physicist **Alan H. Guth (1947–)** noticed that such an expansion, which he named inflation, may be implied from elementary-particle theory, and that the rapid, exponential expansion would solve the flatness and horizon problems. Some difficulties with the Guth model were overcome in 1983 by the Russian astrophysicist **Andrei D. Linde (1948–)**, who proposed a chaotic inflation theory. Any such theory requires that the mass density of the Universe is equal to the critical density if there is no cosmological constant, so the Universe must be described by Euclidean geometry. It also requires substantial quantities of non-baryonic matter. *See* cosmological constant, critical density, flatness problem, horizon problem, and non-baryonic matter.

Infrared: *See* infrared radiation.

InfraRed Astronomical Satellite: NASA's *InfraRed Astronomical Satellite*, abbreviated *IRAS*, was launched on 26 January 1983, scanning most of the sky and making observations of 10 000 selected objects at infrared wavelengths. It operated for nine months, cataloguing 250 000 objects and discovering starburst galaxies and proto-planetary disks that surround young stars. *See* protoplanetary disk, starburst galaxy, and Vega.

Infrared radiation: Electromagnetic radiation with wavelengths slightly longer than red light, in the range between the visible red spectrum and radio waves and roughly between 10^{-6} and 10^{-4} meters in wavelength. The German-born English astronomer **William Herschel (1738–1822)** discovered radiant heat, or infrared radiation, in 1800, when he examined the solar spectrum using a prism and thermometer, noting that the hottest radiation was found when the thermometer was placed beyond the red end of the visible spectrum of sunlight. Infrared radiation is invisible to the human eye and cosmic infrared radiation is absorbed almost completely in the lower layers of the Earth's atmosphere, primarily by water vapor. For this reason, infrared astronomy observations have to be conducted from the highest mountain sites, or from aircraft or satellites. Relatively cool objects, such as dust and planets, radiate a greater percentage of their energy in infrared radiation than relatively hot objects such as stars. Infrared radiation penetrates interstellar dust much more readily than visible light does.

Instability strip: The narrow region in the Hertzsprung-Russell diagram where pulsating stars are located. *See* Hertzsprung-Russell diagram.

INTEGRAL: Acronym for *INTErnational Gamma Ray Astrophysics Laboratory*. *See INTErnational Gamma Ray Astrophysics Laboratory.*

Intensity interferometer: A method of interferometry in which electromagnetic radiation from a cosmic object is recorded at the same time at two or more independent telescopes, and correlated after detection to create an interferometer. The technique was pioneered at radio wavelengths by the English radio astronomer **Robert Hanbury Brown (1916–2002)** in the early 1950s. Brown built a visible-light, mirror intensity interferometer near Narrabri, Australia, and used it in 1956 to measure the angular diameter of the star Sirius. The intensity interferometer was used to measure the angular diameters of several main-sequence stars between 1958 and 1976.

Interferometer: A telescope that combines the signals from two or more smaller telescopes, so the radiation from one interferes with the radiation from the others, hence the name interferometer for an "interference meter." An interferometer yields much sharper views of cosmic objects than either one of its component smaller telescopes, with an angular resolution comparable in size to the separation between its components. In 1921, the American physicist **Albert Michelson (1852–1931)** and the American astronomer **Francis G. Pease (1881–1938)** published their optical, or visible-light, interferometer measurement of the angular size of Betelgeuse, the first determination of the size of any star other than the Sun. Modern optical interferometers include the Keck Telescope Interferometer and the Very Large Telescope Interferometer. This technique is also used to obtain angular resolutions as fine as 0.001 seconds of arc at radio wavelengths using Very Long Baseline Interferometry. *See* Betelgeuse, intensity interferometer, Keck Telescopes, Very Large Array, Very Large Telescope, Very Long Baseline Array, and Very Long Baseline Interferometry.

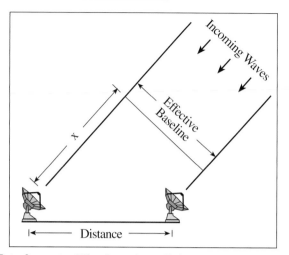

FIG. 31 Interferometer When incoming radiation approaches the Earth at an angle, the crests of the radiation will arrive at two separated telescopes at slightly different times. This delay in arrival time is the distance X divided by the velocity of light. If X is an exact multiple of the wavelength, then the waves detected at the two telescopes will be in phase and add up when combined. If not, they will be out of phase and interfere. The angular resolution of such an interferometer, or interference meter, is equal to the wavelength divided by the effective baseline. When the object being observed is directly overhead, the effective baseline is equal to the distance between the two telescopes.

Interferometry: The use of an interferometer to combine the radiation received from a cosmic object by two or more telescopes. *See* interferometer.

Intergalactic matter: Matter in the space between galaxies, creating absorption of the light from distant galaxies or quasars and producing X-ray emission from clusters of galaxies. *See* Gunn-Peterson effect and *Uhuru Mission*.

International Astronomical Union: Abbreviated IAU, the International Astronomical Union was founded in 1919. Its mission is to promote and safeguard the science of astronomy in all its aspects through international cooperation. Its individual members are professional astronomers all over the world, at the Ph.D. level or beyond and active in professional research and education in astronomy. The scientific and educational activities of the IAU are organized into 12 Scientific Divisions, 37 more specialized Commissions and 86 Working and Program Groups.

International atomic time: The fundamental unit of time is the atomic second, which is defined from worldwide observations of atomic clocks, combined by the Bureau International de l'Heure, Paris. By definition, a day has 86 400 seconds of International Atomic Time, or in French, Temps Atomique International and abbreviated TAI. *See* Universal Time.

INTErnational Gamma Ray Astrophysics Laboratory: ESA launched its **INTErnational Gamma Ray Astrophysics Laboratory**, abbreviated **INTEGRAL**, on 17 October 2002, dedicated to the spectroscopy and imaging of celestial gamma-ray sources.

International Ultraviolet Explorer: NASA's **International Ultraviolet Explorer**, abbreviated **IUE**, satellite obtained over 104 000 ultraviolet spectra of cosmic objects between 26 January 1978 and 30 September 1996.

Interplanetary magnetic field: The Sun's magnetic field carried into interplanetary space by the expanding solar wind. It has a magnetic field strength of about 6×10^{-9} Tesla, or 6 nanoTesla and 6×10^{-5} Gauss, at the Earth's orbital distance of about 1 astronomical unit from the Sun. The interplanetary magnetic field is carried radially outward by the solar wind, but since the magnetic field originates at the Sun, where the magnetic footpoints are located, solar rotation twists the interplanetary magnetic field into a spiral structure with the shape of an Archimedean spiral.

Interplanetary matter: *See* interplanetary medium.

Interplanetary medium: The medium or material in the space between the planets in the Solar System. It is composed of electrically charged particles, mainly electrons and protons, ejected from the Sun via the solar wind, dust particles from comets, and neutral gas from the interstellar medium.

Interplanetary scintillation: Fluctuations in the signal received from a distant radio source observed along a line of sight close to the Sun. The scintillation is caused by irregularities in the solar wind.

Interstellar cloud: A mass of gas and dust that lies in the space between the stars, containing great quantities of either atomic or molecular hydrogen, as well as other molecules, and sometimes the site of recently formed stars and planets.

Interstellar dust: Small, solid particles in the space between the stars in our Galaxy, with a size comparable to the wavelength of light and similar to smoke. Interstellar dust absorbs, reddens and reflects the light from distant stars. *See* dust, reddening, and reflection nebula.

Interstellar matter: Material in the space between the stars. *See* H I region, H II region, interstellar cloud, interstellar dust, interstellar medium, interstellar molecules, and molecular cloud.

Interstellar medium: The gas and dust in the space between the stars in our Galaxy. It is mainly composed of hydrogen atoms, molecules, and solid dust particles. Most of the hydrogen atoms are in a cool, un-ionized form at a temperature of 10 to 100 kelvin, but ultraviolet starlight ionizes the hydrogen atoms near hot, bright stars. Cold, dense molecular clouds consist mainly of hydrogen molecules, each with a mass of up to a million solar masses, but these clouds also contain hundreds of other molecules including ammonia, carbon monoxide, formaldehyde, and water. *See* H I region, H II region, interstellar dust, interstellar molecules, and molecular cloud.

TABLE 24   **Interstellar molecules**[a]

Chemical Symbol	Name of Molecule	Year of Discovery	Frequency[a]	Wavelength[a]
OH	Hydroxyl	1963	1665.4 MHz	18.0054 cm
CO	Carbon Monoxide	1970	115.27 GHz	2.5911 mm
H_2O	Water	1968	22.235 GHz	1.3483 cm
HCN	Hydrogen Cyanide	1970	88.632 GHz	3.3824 mm
NH_3	Ammonia	1968	23.694 GHz	1.2653 cm
H_2CO	Formaldehyde	1969	4829.7 MHz	6.2072 cm

[a]The transition frequencies are given in MegaHertz, abbreviated MHz, or GigaHertz, abbreviated GHz. One MHz = 10^6 Hz or a million Hertz and 1 GHz = 10^9 Hz or a billion Hertz. The transition wavelengths are given in units of centimeters, cm, or millimeters, mm = 0.1 cm.

Interstellar molecules: The most abundant atoms in the Cosmsos, such as hydrogen, H, carbon, C, nitrogen, N, and oxygen, O, combine to form molecules within cold, dark interstellar clouds. The most abundant cosmic molecule is molecular hydrogen, H_2, but ammonia, NH_3, carbon monoxide, CO, hydrogen cyanide, HCN, water, H_2O, formaldehyde, H_2CO, and hydroxyl, OH, are also found. *See* dark nebula, giant molecular cloud, and molecular cloud.

Interstellar space: The space located between the stars.

Intrinsic brightness: The amount of radiation an object emits, as opposed to how bright the object looks from Earth. *See* absolute luminosity.

Inverse square law: The apparent, or observed, intensity of radiation diminishes by the square of the distance from its source; also a reduction in gravitational attraction by the same factor. *See* absolute luminosity and apparent luminosity.

Invisible radiation: Those kinds of radiation to which the human eye is not sensitive; for example, radio waves, X-rays and gamma rays. The human eye is sensitive to visible light, also known as optical radiation since optics are used to detect it. *See* optical radiation and visible light.

Io: The innermost of Jupiter's four Galilean satellites, discovered in 1610 by the Italian astronomer **Galileo Galilei (1564–1642)**. Io has a radius of 1822 kilometers and a mass density of 3528 kilograms per cubic meter. Tidal forces from Jupiter squeeze Io's rocky interior in and out, making it molten inside and producing volcanoes that were discovered in 1979 when the *Voyager 1* spacecraft flew by Jupiter. Io is the most volcanically active body in the Solar System. Its volcanoes produce red and yellow flows of sulfurous lava and expel plumes of sulfur-dioxide gas. Jupiter's powerful magnetic field sweeps past Io, picking up a ton, or 1000 kilograms, of sulfur and oxygen ions every second, and directing them into a doughnut-shaped ring known as a plasma torus. Io is also connected to Jupiter by a flux tube that carries a vast current of 5 million amperes between the satellite and the planet's poles, producing aurora lights on both the satellite and the giant planet. *See* Galilean satellites and tides.

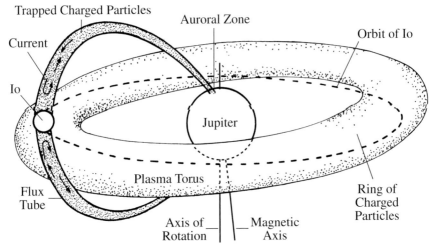

FIG. 32 Io's flux tube and plasma torus An electric current of five million amperes flows along Io's flux tube, connecting Io to the upper atmosphere of Jupiter. The plasma torus, centered near Io's orbit, is filled with energetic sulfur and oxygen ions. Because the planet's rotational axis is tilted with respect to the magnetic axis, the orbit of the satellite Io (*dashed line*) is inclined to the plasma torus.

Ion: An atom that has lost one or more electrons, or in less common instances gained an electron. An ion has a net electrical charge. By contrast, a neutral atom has an equal number of negatively charged electrons and positively charged protons, giving the atom a zero net electrical charge.

Ionization: The process in which a neutral atom or molecule is given a net electrical charge. The atomic process in which ions are produced by removing an electron from an atom, ion or molecule, typically by collisions with atoms or electrons, known as collisional ionization, or by interaction with electromagnetic radiation, called photoionization.

Ionization potential: The amount of energy required to remove the least tightly bound electron from a neutral atom or molecule is called the first ionization potential, denoted by I, and it is usually measured in electron volts, or eV. The additional energy needed to remove the next least tightly bound electron is the second ionization potential, and so on.

Ionized: State of an atom having fewer, or more, electrons than the number of protons, leaving it with an electrical charge.

Ionized hydrogen: A hydrogen atom that has lost its electron. Hydrogen in interstellar space is ionized by nearby hot O and B stars, forming H II regions such as the Orion Nebula. *See* H II region, Orion Nebula, and Strömgren radius.

Ionosphere: The upper region of a planet's atmosphere in which there are free electrons and ions produced when solar ultraviolet and X-ray radiation ionizes the constituents of the atmosphere. Most of the Earth's ionosphere lies between heights of about 50 and 300 kilometers above the ground, though the extent varies considerably with time, season and solar activity. The D region, between 50 and 90 kilometers, has low electron density. The E and F regions, at about 100 and 200 to 300 kilometers, form the main part of the ionosphere. The reflecting power of the ionosphere makes long-range broadcasting and telecommunication possible at radio frequencies up to about 30 MHz, or wavelengths longer than 10 meters.

IRAS: Acronym for the *InfraRed Astronomical Satellite*. *See* **InfraRed Astronomical Satellite**.

Iron: Element with atomic number 26, created mostly by type Ia supernovae. The Earth's core consists mostly of iron, and iron also makes your blood red.

Iron meteorite: About 5 percent of the meteorites that have been seen to fall and then recovered are chunks of metal called iron meteorites, composed of 100 percent nickel and iron and with a high mass density of between 7,600 and 7,900 kilograms per cubic meter. When polished and etched with acid, an iron meteorite displays a distinctive Widmanstätten pattern produced by regions of crystalline structure. Most of the other fallen meteorites, 94 percent, are stony meteorites. *See* meteorite, stony meteorite, and Widmanstätten pattern.

Irradiance: The solar irradiance is the amount of solar energy received at the Earth outside its atmosphere per unit area per unit time, with units of watts per square meter. *See* solar constant.

Irregular galaxy: A galaxy that appears disorganized and disordered. An irregular galaxy lacks symmetry and is without a distinct spiral or elliptical shape.

ISAS: Acronym for the Institute of Space and Astronautical Science, Japan.

Island Universe hypothesis: The assertion that spiral nebulae are huge stellar aggregates separated from both the Milky Way and each other by vast distances.

Isotope: Atom having the same number of protons in its nucleus as other atoms of an element, but a different number of neutrons, with the result that its mass differs though it may have the same number of electrons. One of two or more forms of the same chemical element, whose atoms all have the same atomic number, or number of protons in their nucleus, but a different number of neutrons and therefore a different mass or atomic mass number. The term, proposed in 1913 by the English chemist **Frederick Soddy (1877–1956)**, derives from the Greek *isos* for "equal" and *topos* for "place," and reflects the fact that such species occupy the same position in the periodic table of the elements. *See* atomic number and atomic mass number.

Isotropy: Quality of being the same in all directions, within a system or the Universe. Radiation coming with equal intensity from all directions is isotropic; a car light and a spotlight's beam are anisotropic, as is the distribution of galaxies, but the three-degree cosmic microwave background radiation is very isotropic, with only a low level of anisotropy. *See* anisotropy.

IUE: Acronym for *International Ultraviolet Explorer*. See *International Ultraviolet Explorer*.

J

James Clerk Maxwell Telescope: Abbreviated JCMT, a submillimeter-wave telescope, with a diameter of 15 meters (590 inches), located near the summit of Mauna Kea, Hawaii and capable at working at wavelengths between 0.002 and 0.0003 meters or between 2.0 and 0.3 millimeters. It is named for the Scottish physicist **James Clerk Maxwell (1831–1879)** who in 1864 laid the foundation for all subsequent studies of electromagnetic radiation, showing that it consists of electric and magnetic waves that oscillate at right angles to each other and to the direction of travel. The JCMT is a facility of the United Kingdom, Canada, and the Netherlands, operated on their behalf by the Joint Astronomy Center in Hilo Hawaii. *See* Mauna Kea Observatories.

James Webb Space Telescope: An orbiting infrared observatory, abbreviated *JWST,* with a 6.5-meter (255.6-inch) primary mirror, under study by NASA with a proposed launch date of August 2011. It is being planned as a collaborative project with ESA, and is foreseen as a successor to the *Hubble Space Telescope*. Initially named the *Next Generation Space Telescope*, or *NGST* for short, it has been renamed for **James E. Webb (1906–1992)**, the NASA administrator who guided the agency during the *Apollo* moon landings.

Jet Propulsion Laboratory: The Jet Propulsion Laboratory, abbreviated JPL, manages most NASA Solar System missions, such as the *Cassini-Huygens, Galileo, Mars Exploration Rovers Opportunity* and *Spirit, Mars Global Surveyor*, and *2001 Mars Odyssey Missions,* as well as NASA's Deep Space Network, and NASA's *Spitzer Space Telescope* and *GALaxy Evolution EXplorer*. The Jet Propulsion Laboratory is located in Pasadena, California, and it is managed for NASA by the California Institute of Technology. *See* National Aeronautics and Space Administration.

Jewel Box: The open star cluster NGC 4755 located 7600 light-years away in the direction of the constellation Crux, with stars of varied colors like jewels. *See* Crux.

Jodrell Bank Observatory: The **Jodrell Bank Observatory** is part of the University of Manchester, England, and was founded by English radio astronomer **Bernard Lovell (1913–)** in 1945. It includes a 76-meter (250-feet) radio telescope, which began operation in 1957, was upgraded in 1971 and 1987, and named the Lovell Telescope in 1987. It has played an important role in the early years of radio astronomy and helped pioneer long baseline interferometry. *See* **MERLIN**, radio interferometer, radio telescope, and Very Long Baseline Interferometry.

JPL: Acronym for the Jet Propulsion Laboratory. *See* Jet Propulsion Laboratory.

Julian century: The Julian century has 36525 days in 100 years.

Julian day number: The number of days that have elapsed since noon Greenwich Mean Time on 1 January 4713 BC, which is defined as day zero. The standard epoch

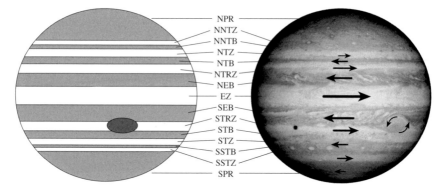

FIG. 33 Jupiter's wind-blown clouds The traditional nomenclature of Jupiter's light and dark bands of clouds (*left*) is given in abbreviated form (*center*). The dark bands are called belts, denoted by "B", the light bands are known as zones, or "Z", and the rest of each name is based on climatic regions at the corresponding latitudes on Earth. North, letter "N" is at the top, and south, denoted by "S", is at the bottom. The equatorial, or "E", bands are in the middle, the tropical, "TR", bands on each side of the equator, and the temperate, "T", ones at mid-latitudes. Far northern latitudes are denoted by "NN", far southern latitudes by "SS", and the polar regions by "P". The image of Jupiter (*right*) was taken from the *Cassini* spacecraft on 7 December 2000, when Jupiter's moon Europa cast a shadow on the planet. The arrows point in the direction of wind flow, and their length corresponds to the wind velocity, which can reach 180 meters per second near the equator.

that is now used in new star catalogues is 2000.0 with a Julian Date of 2 451 545.0. *See* Greenwich Mean Time and precession.

Jupiter: The fifth major planet from the Sun, and the largest planet in the Solar System, with a mass equal to 318 times the mass of the Earth and 0.001 solar masses. The planet Jupiter orbits the Sun at a mean distance of 5.2 astronomical units with an orbital period of 11.86 Earth years and an orbital speed of 12.5 meters per second. All we see on Jupiter are clouds, swept into parallel bands of bright zones and dark belts by the planet's rapid rotation, and counter-flowing, east-west winds, as well as anti-cyclonic storms such as the Great Red Spot, white ovals and smaller eddies. Jupiter is an incandescent globe that radiates 1.67 times the energy it receives from the Sun, probably as heat left over from when the giant planet formed. The most abundant element in Jupiter is hydrogen, which exists in liquid molecular and liquid metallic form within the planet. Rotationally driven, electrical currents within a shell of liquid metallic hydrogen generate Jupiter's powerful magnetic field, the largest permanent structure in the Solar System, even surpassing the Sun in size. The planet and its four largest moons were examined in close-up detail in 1979, when the ***Voyager 1*** and **2** spacecraft flew by the planet, and from 1995 to 2004 when the ***Galileo*** spacecraft orbited Jupiter. Instruments aboard ***Voyager 1*** conclusively demonstrated that Jupiter has a thin ring

TABLE 25   Jupiter[a]

Mass	$18\ 986 \times 10^{23}$ kilograms $= 317.70\ M_E$
Equatorial Radius at 1 bar	$71\ 492$ kilometers $= 11.19\ R_E$
Polar Radius at 1 bar	$66\ 854$ kilometers
Mean Mass Density	1326 kilograms per cubic meter
Rotation Period	9.9249 hours $=$ 9 hours 55 minutes 29.7 seconds
Orbital Period	11.86 Earth years
Mean Distance from Sun	7.7833×10^{11} meters $= 5.203$ AU
Age	4.6×10^9 years
Atmosphere	86.4 percent molecular hydrogen, 13.6 percent helium
Energy Balance	1.67 ± 0.08
Effective Temperature	124.4 kelvin
Temperature at 1-bar level	165 kelvin
Central Temperature	$17\ 000$ kelvin
Magnetic Dipole Moment	$20\ 000\ D_E$
Equatorial Magnetic Field Strength	4.28×10^{-9} tesla or $14.03\ B_E$

[a]The symbols M_E, R_E, D_E and B_E denote respectively the mass, radius, magnetic dipole moment, and magnetic field strength of the Earth. One bar is equal to the atmospheric pressure at sea level on Earth. The energy balance is the ratio of total radiated energy to the total energy absorbed from sunlight, and the effective temperature is the temperature of a blackbody that would radiate the same amount of energy per unit area.

composed of very small particles. In July 1994 Comet Shoemaker-Levy 9 collided with Jupiter. *See* Callisto, Europa, Galilean satellites, Ganymede, Great Red Spot, Io, and Shoemaker-Levy 9 Comet.

JWST: Acronym for ***James Webb Space Telescope***. *See **James Webb Space Telescope***.

K or °K: Abbreviation for degrees on the Kelvin scale of temperature. *See* Kelvin.

Kamiokande: A massive underground neutrino detector in Japan filled with water, replaced by the Super Kamiokande detector. *See* Super Kamiokande.

Kapteyn's Star: A faint red M1 dwarf star with an apparent visual magnitude of 8.86, located in the southern constellation of Pictor. As the name implies, the Dutch Astronomer **Jacobus Kapteyn (1851–1922)** discovered Kapteyn's Star in 1897. It has the second largest proper motion known, at 8.67 seconds of arc per year. It is one of the closest stars to the Sun, with a distance of 12.8 light-years, a parallax of 0.255 seconds of arc, and an absolute visual magnitude of 10.9. *See* parallax and proper motion.

Keck Observatory: *See* Keck telescopes.

Keck telescopes: Two telescopes located on the summit of Hawaii's dormant Mauna Kea volcano. Each telescope has a mirror that is 10 meters (394 inches) in diameter, consisting of 36 hexagonal segments, and they are the largest optical, or visible-light, telescopes in the world, with the greatest light-gathering power. The **Keck I telescope** began science observations in May 1993; **Keck II** began in October 1996. The Keck interferometer will combine the light of the two telescopes to obtain the resolving power of a single optical telescope with a 90-meter (3543-inch) mirror and an angular resolution as fine as 0.001 seconds of arc. The telescopes are named after the American philanthropist **William Myron Keck (1880–1964)**, founder of the Superior Oil Company, who in 1954 established the W. M. Keck Foundation that funded the construction and supports the operation of the telescopes. The California Association for Research in Astronomy operates the **W. M. Keck Observatory**. *See* Mauna Kea Observatories.

Kelvin: A unit for measuring absolute temperature abbreviated K or °K for degrees on the Kelvin scale, and named for the Irish physicist **William Thomson (1824–1907)**, **Lord Kelvin of Largs**. On the Kelvin scale, the coldest possible temperature is 0 degrees kelvin, where atomic or molecular motion stops. Water freezes at 273 kelvin, room temperature is about 295 kelvin, and the boiling point of water is 373 kelvin. To convert a temperature in kelvin to the equivalent temperature in degrees Celsius, subtract 273.15, or $°C = °K - 273.15$, and the equivalent temperature in degrees Fahrenheit is $°F = (9 °K/5) - 459.4$. *See* absolute temperature, Celsius, and Fahrenheit.

Kepler Mission: A mission, currently under study by NASA, designed to detect rocky, Earth-sized planets around other stars. It was named for **Johannes Kepler (1571–1630)**, the German astronomer who discovered the three laws of planetary motion that now have his name.

Kepler's laws: The three fundamental laws governing the motions of the planets around the Sun, first worked out by the German astronomer **Johannes Kepler (1571–1630)** between 1609 and 1619, and based on observations made by the Danish astronomer

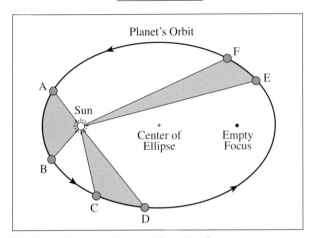

FIG. 34 Kepler's first and second laws The German astronomer **Johannes Kepler (1571–1630)** published his first two laws of planetary orbital motion in 1609. His first law states that the orbit of a planet about the Sun is an ellipse with the Sun at one focus. The other focus of the ellipse is empty. According to Kepler's second law, the line joining a planet to the Sun sweeps out equal areas in equal times. This is also known as the law of equal areas. It is represented by the equality of the three shaded areas ABS, CDS and EFS. It takes as long to travel from A to B as from C to D and from E to F. A planet moves most rapidly when it is nearest the Sun (*at perihelion*); a planet's slowest motion occurs when it is farthest from the Sun (*at aphelion*).

Tycho Brahe (1546–1601). They are: (1) The orbit of each planet is an ellipse with the Sun at one focus. (2) A line from the Sun to a planet sweeps out equal areas in equal time. (3) The square of a planet's orbital period is proportional to the cube of its average distance from the Sun. The second law means that a planet moves at a slightly faster speed when its orbit is closest to the Sun. This law is known as the law of areas. The third law indicates that more distant planets move around the Sun at slower speeds and take longer times to complete each orbit.

Kepler's star and supernova remnant: A brilliant star when discovered in 1604 by the German astronomer **Johannes Kepler (1571–1630)**, with a maximum visual apparent magnitude of –2.2, about the brightness of Jupiter. At an estimated distance of 18 000 light-years, Kepler's star had a maximum absolute visual magnitude of –18.9, but now it is almost invisible optically. It was the last stellar explosion, or supernova, seen in our Galaxy, and the expanding remnant of the explosion, called Kepler's supernova remnant, is a strong source of radio and X-ray radiation today.

keV: Abbreviation for kilo electron volts, or one thousand electron volts. A unit of energy with 1 keV = $1.602\,191\,7 \times 10^{-16}$ Joules. The wavelength of radiation with a photon energy in keV is 1.24×10^{-9} meters / energy in keV. *See* electron volt.

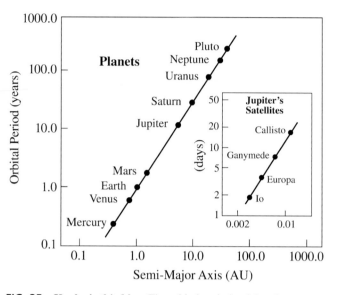

FIG. 35   Kepler's third law The orbital periods of the planets are plotted against their semi-major axes, using a logarithmic scale. The straight line that connects the points has a slope of 3/2, thereby verifying Kepler's third law that states that the square of the orbital periods increases with the cubes of the planetary distances. The German astronomer **Johannes Kepler (1571–1630)** published this third law in 1619. This type of relation applies to any set of bodies in elliptical orbits, including Jupiter's four largest satellites shown in the inset.

Keyhole Nebula: A dark nebula, designated NGC 3324, which is in the constellation Carina and is seen in silhouette against the bright Eta Carinae Nebula. *See* dark nebula and Eta Carinae Nebula.

Kilometer: A unit of distance, abbreviated km. It is equal to one thousand meters and to 0.6214 miles.

Kilometer per second: A unit of speed equal to 2237 miles per hour.

Kinetic energy: The energy that an object possesses as a result of its motion.

Kirchhoff's law: The ratio of the emission and absorption coefficients of a blackbody is equal to its brightness; a relation named for the German physicist **Gustav Kirchhoff (1824–1887)** who announced it in 1859.

Kirkwood gaps: Regions in the asteroid belt, between the orbits of Mars and Jupiter, which contain very few asteroids. The American astronomer **Daniel Kirkwood (1814–1895)** discovered these gaps in 1866. The Kirkwood gaps are located at places where an asteroid would have an orbital period that is a rational fraction, such as 1/4, 1/3 and 1/2, of Jupiter's orbital period, and repeated gravitational interactions with Jupiter seems to have tossed the asteroids out of these places. A complete explanation

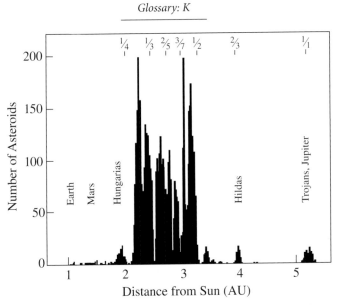

FIG. 36 Kirkwood gaps The number of asteroids at different distances from the Sun. Most of the asteroids are found in the asteroid belt that lies between 2.2 and 3.3 AU from the Sun. The gaps of missing asteroids are named for the American astronomer **Daniel Kirkwood (1814–1895)**, who discovered them in 1866. Repeated gravitational interactions with Jupiter seem to have tossed asteroids out of the Kirkwood gaps with orbital periods of 1/4, 2/7, 1/3, 2/5, 3/7 and 1/2 of Jupiter's orbital period. These fractions are placed above the relevant gap in the figure.

of the Kirkwood gaps was obtained in the 1980s when the American scientist **Jack Wisdom (1953–)** showed that Jupiter induces a chaotic zone in the vicinity of the Kirkwood gaps, eventually removing an asteroid from them.

Kitt Peak National Observatory: Founded in 1958, the Kitt Peak National Observatory, abbreviated KPNO, supports telescopes for nighttime optical and infrared astronomy and daytime study of the Sun. It is located near Tucson, Arizona, and is part of the United States National Optical Astronomy Observatory. The largest telescope at KPNO is the 4-meter (158-inch) Mayall telescope, opened in 1973. KPNO is also the site of the McMath Solar Telescope, the world's largest solar telescope opened in 1994; it has a 3.5-meter (138-inch) mirror and is part of the National Solar Observatory. *See* National Optical Astronomy Observatory and National Solar Observatory.

kpc: Abbreviation for kiloparsecs, or one thousand parsec, a unit of distance with 1 kpc = $3.085\,678 \times 10^{19}$ meters.

KPNO: Abbreviation of Kitt Peak National Observatory. *See* Kitt Peak National Observatory.

K star: An orange star of spectral type K, whose visible disk is slightly cooler than the Sun. Examples of K stars are the dwarf, or main sequence, star Epsilon Eridani and the giant stars Aldebaran, Arcturus and Pollux. *See* Aldebaran, Arcturus, Epsilon Eridani, Hertzsprung-Russell diagram, and stellar classification.

Kuiper airborne observatory: A jet aircraft, operated by NASA, used to make cosmic infrared observations for more than 20 years, ending its service in October 1995. It was named for the Dutch-born American astronomer **Gerard P. Kuiper (1905–1973)**.

Kuiper belt: A collection of 1 billion to 10 billion (10^9 to 10^{10}) or more icy bodies orbiting the Sun in low-eccentricity, low-inclination orbits beyond Neptune, extending out possibly to about 1000 astronomical units. The Kuiper belt provides the Solar System with some short-period comets, which revolve around the Sun with periods less than 200 Earth years. The Kuiper belt is named after the Dutch-born American astronomer **Gerard P. Kuiper (1905–1973)**, who was one of the first to propose its existence. The American astronomers **Jane X. Luu (1963–)** and **David C. Jewitt (1958–)** discovered the first trans-Neptunian, Kuiper-belt object in 1992; hundreds of such objects have now been found and scientists estimate that the Kuiper belt may contain as many as 35 000 objects with a radius of about 100 kilometers. Neptune's largest satellite, Triton, may have been captured from the Kuiper belt. The orbit of the small planet Pluto crosses inside and outside the orbit of Neptune, and might be regarded as another member of the Kuiper belt. *See* comet, Pluto, and Triton.

L

Lagoon Nebula: A bright emission nebula, designated M8 or NGC 6523, in the constellation Sagittarius, surrounding the star cluster NGC 6530. Both the nebula and star cluster are located at a distance of about 5200 light-years. The Lagoon Nebula measures 40 by 90 minutes of arc across. It receives its name from a dark lane that cuts across the bright nebula with the shape of a lagoon. The bright Hourglass Nebula is found in the central regions of the Lagoon Nebula. *See* emission nebula and Hourglass Nebula.

Lagrangian point: One of the five points at which an artificial satellite or a small cosmic object, such as an asteroid or natural satellite, can remain in a position of equilibrium with respect to two much more massive bodies orbiting each other. The Italian-born French mathematician **Joseph Lagrange (1736–1813)** first derived the five Lagrangian points in 1764 when investigating the mutual gravitation of the Earth, Sun and Moon. The Trojan asteroids are located along Jupiter's orbit at the fourth and fifth Lagrangian points, denoted L_4 and L_5, where the gravity of Jupiter and the Sun are equal. The *SOlar and Heliospheric Observatory* is located at the inner Lagrangian point, designated L_1, of the Earth-Sun system. It is located at a distance from the Earth of about 1.5×10^9, or 1.5 billion, meters towards the Sun, or about one one-hundredth of the way from the Earth to the Sun, where the gravitational pull of the Earth and Sun balance. A spacecraft at the inner Lagrangian point will orbit the Sun with the Earth without revolving about the Earth, and therefore never experiences night. *See SOlar and Heliospheric Observatory* and Trojan asteroid.

LANL: Acronym for the Los Alamos National Laboratory.

Large Magellanic Cloud: Abbreviated LMC, the nearest irregular galaxy that orbits the Milky Way, located 168 000 light-years away. Named for the Portuguese explorer **Ferdinand Magellan (1480–1521)** whose crew reported seeing the two Magellanic Clouds from the Southern Hemisphere during the first circumnavigation of the Earth, completed in 1522. The Large Magellanic Cloud contains the largest known emission nebula, the Tarantula Nebula, and the LMC is also the sight of Supernova 1987A, the first naked-eye supernova since 1604. *See* Magellanic Clouds, Supernova 1987A, and Tarantula Nebula.

Las Campanas Observatory: An optical, or visible-light, observatory near La Serena, Chile, run by the Carnegie Institution of Washington. The largest telescopes now in operation are the two 6.5-meter (256-inch) **Magellan Telescopes** built under collaboration between the Carnegie Observatories, the University of Arizona, Harvard University, the University of Michigan and the Massachusetts Institute of Technology. The first **Magellan Telescope** began operation in September 2000 and the second in September 2002. A **Giant Magellan Telescop**e, consisting of 7 mirrors, each 8.4 meters (336 inches) in diameter, is planned.

Laser Interferometer Gravitational-Wave Observatory: Abbreviated **LIGO**, a facility dedicated to the detection of cosmic gravitational waves, consisting of two

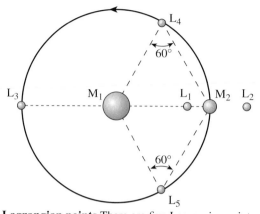

FIG. 37   **Lagrangian points** There are five Lagrangian points where
a celestial body can remain in position of equilibrium with respect to two
much more massive bodies, designated M_1, and M_2, orbiting each other.
These points are named for the Italian-born French mathematician **Joseph
Louis de Lagrange (1736–1813)**, who investigated them in 1772. At
these points the forces acting on the smaller body cancel out. Bodies at
points L_1, L_2, and L_3, on the line joining M_1 and M_2, are less stable than
bodies at L_4 and L_5, which form equilateral triangles with M_1 and M_2.

widely separated installations within the United States, operated in unison as a single
observatory. Each facility has laser beams that measure the distance between masses
with sufficient precision to possibly measure a displacement caused by a passing
gravitational wave.

Latitude: Distance north or south of the Earth's equator along a great circle connecting the poles. A great circle divides the sphere of the Earth in half, and the name comes
from the fact that no greater circle can be draw on the sphere.

Law: A widely tested theory that is thought to be always valid.

Leonids: A meteor shower that occurs from 15 to 20 November, with a maximum on
17 November, a radiant in the constellation of Leo and associated with Comet 55P/
Temple-Tuttle. *See* meteor shower.

Lick Observatory: An optical, or visible-light, observatory built on the summit of
Mount Hamilton near San Jose, California, between 1876 and 1887 from a bequest
by the California businessman **James Lick (1796–1876)**. Lick Observatory is owned
and operated by the University of California and currently managed by the University
of California, Santa Cruz. Its 3.0-meter (120-inch) telescope was completed in 1959,
and named the Shane telescope in 1977 in honor of the astronomer **Charles Donald
Shane (1895–1983)** who helped acquire funding for the telescope from the Califor-

nia Legislature. The 0.91-meter (36-inch) refractor has been in operation since the observatory opened in 1888. In 1898 the American astronomer **James Edward Keeler (1857–1900)** began systematic photography of nebulae using the 0.91-meter refractor, estimating that over one hundred thousand spiral nebulae could be detected this way. The telescope was subsequently used in a cooperative project with the 100-inch Mount Wilson telescope to determine the redshifts and magnitudes of hundreds of extra-galactic nebulae, published in 1956. Observations at Lick Observatory have most recently played a role in the discovery of extrasolar planets.

Light: The kind of radiation to which the human eye is sensitive. *See* optical radiation, velocity of light, and visible spectrum.

Light element: An element of very low atomic number, such as hydrogen, helium and lithium, which have atomic numbers of one, two and three. *See* atomic number.

Light-second: The distance that light travels through a vacuum in one second, equal to 299 792 kilometers and roughly 186 000 miles.

Light-time: The time it takes light to travel from a celestial object to the Earth. It takes 8.3 minutes for light to reach the Earth from the Sun, and 4.29 years for light to travel to the Earth from Proxima Centauri, the nearest star other than the Sun.

Light-year: The distance that light travels through a vacuum in one year. One light-year is equivalent to 9.4607×10^{15} meters or 9.4607 trillion kilometers, to 5.88×10^{12} miles, to 63,240 astronomical units, and to 0.3067 parsecs. Proxima Centauri is the nearest star, other than the Sun, with a distance of 4.29 light-years.

LIGO: Abbreviation for Laser Interferometer Gravitational-Wave Observatory. *See* Laser Interferometer Gravitational-Wave Observatory.

Line element: The squared distance between two nearby points of space-time.

Little Ice Age: A prolonged period of unusually cold weather in Europe, from about 1400 to 1800, overlapping two periods of unusually low sunspot numbers called the Maunder Minimum (1645 to 1715) and the Spörer Minimum (1420 to 1570). During the Little Ice Age, alpine glaciers expanded, the River Thames and the canals of Venice regularly froze over and Europe experienced unusually harsh winters. *See* Maunder Minimum.

LLNL: Acronym for the Lawrence Livermore National Laboratory.

Local Group: A small, gravitationally bound collection of at least 40 nearby galaxies, including our Milky Way Galaxy, the Andromeda galaxy M31, the Triangulum galaxy M33, the two Magellanic Clouds, LMC and SMC, and several dwarf galaxies, all within 5 million light-years of the Earth. Unlike more distant galaxies, the galaxies in the Local Group are not receding from us; Andromeda is approaching the Milky Way and several dwarf galaxies are orbiting the Milky Way. *See* dwarf galaxy and Andromeda galaxy.

Local supercluster: A supercluster of galaxies that includes our Milky Way Galaxy and the entire Local Group, centered on the Virgo cluster. *See* supercluster.

TABLE 26 Local Group[a]

Galaxy	Other Name	Year of Discovery	RA (2000)	Declination (2000)	Type	Subgroup	Distance (Mly)	V_r	Luminosity ($10^6 \, L_\odot$)	Mass ($10^6 \, M_\odot$)
M31	NGC 224	—	$00^h 42.7^m$	$+41° 16'$	SbI-II	M31	2.5	3.4	25000	700000
Milky Way, MW		—	$17^h 45.7^m$	$-29° 01'$	Sbc	MW	0.03	—	8300	350000
M33	NGC 598	—	$01^h 33.9^m$	$+30° 40'$	ScII-III	M31	2.7	5.9	3000	30000
LMC		—	$05^h 23.6^m$	$-69° 45'$	IrrIII-IV	MW	0.16	0.4	2100	20000
SMC	NGC 292	—	$00^h 52.7^m$	$-72° 50'$	IrrIV-V	MW	0.19	2.0	580	1000
WLM	DDO 221	1923	$00^h 02.0^m$	$-15° 28'$	IrrIV-V	LGC	3.0	10.4	500	150
M32	NGC 221	1749	$00^h 42.7^m$	$+40° 52'$	E2	M31	2.6	8.1	380	2120
NGC 205	M110	1864	$00^h 40.4^m$	$+41° 41'$	E5p/dSph-N	M31	2.6	8.1	370	740
NGC 3109	DDO 236	1864	$10^h 03.1^m$	$-26° 10'$	IrrIV-V	N3109	4.1	9.9	160	6550
IC 10	UGC 192	1895	$00^h 20.4^m$	$+59° 18'$	dIrr	M31	2.7	11.6	160	1580
NGC 185	UGC 396	1864	$00^h 39.0^m$	$+48° 20'$	dSph/dE3p	M31	2.0	9.1	130	130
NGC 147	DDO 3	1864	$00^h 33.2^m$	$+48° 31'$	dSph/dE5	M31	2.3	9.4	130	110
NGC 6822	DDO 209	1864	$19^h 44.9^m$	$-14° 48'$	IrrIV-V	LGC	1.6	9.1	94	1640
IC 5152	DDO 8	1895	$22^h 02.7^m$	$-51° 18'$	dIrr	LGC	5.2	11.2	70	400
IC 1613		1906	$01^h 04.9^m$	$+02° 08'$	IrrV	M31	2.3	9.6	64	795
Sextans A	DDO 75	1942	$10^h 11.1^m$	$-04° 43'$	dIrr	N3109	4.7	11.3	56	395
Sextans B	DDO 70	1955	$10^h 00.0^m$	$+05° 20'$	dIrr	N3109	4.4	11.4	41	885
Sagittarius		1994	$18^h 55.1^m$	$-30° 29'$	dSph-N	MW	0.08	4.0	18	—
Fornax		1938	$02^h 40.0^m$	$-34° 27'$	dSph	MW	0.45	7.6	16	68
Pegasus	DDO 216	1958	$23^h 28.6^m$	$+14° 45'$	dIrr/dSph	LGC	3.1	12.0	12	58
EGB 427+63	UGCA 92	1984	$04^h 32.0^m$	$+63° 36'$	dIrr	M31	4.2	13.9	9.1	
SagDIG	UKS1927-177	1977	$19^h 30.0^m$	$-17° 41'$	dIrr	LGC	3.4	13.5	6.9	9.6
And VII	Cassiopeia	1998	$23^h 26.5^m$	$+50° 42'$	dSph	M31	2.5	15.2	5.7	—

TABLE 26 Local Group[a] (continued)

Galaxy	Other Name	Year of Discovery	RA (2000)	Declination (2000)	Type	Subgroup	Distance (Mly)	V_r	Luminosity (10^6 L$_\odot$)	Mass (10^6 M$_\odot$)
UKS2323-326	UGCA 438	1978	$23^h 26.5^m$	$-32° 23'$	dIrr	LGC	4.3	13.8	5.3	—
Leo I		1955	$10^h 08.5^m$	$+12° 19'$	dSph	MW	0.81	10.1	4.8	22
And I		1972	$00^h 45.7^m$	$+38° 00'$	dSph	M31	2.6	12.8	4.7	—
GR 8	DDO 155	1956	$12^h 58.7^m$	$+14° 13'$	dIrr	GR8	4.9	14.4	3.4	7.6
Leo A	DDO 69	1942	$09^h 59.4^m$	$+30° 45'$	dIrr	MW	2.2	12.8	3.0	11
And II		1972	$01^h 16.5^m$	$+33° 26'$	dSph	M31	1.7	12.7	2.4	—
Sculptor		1938	$01^h 00.2^m$	$-33° 43'$	dSph	MW	0.26	8.5	2.2	6.4
Antlia		1985	$10^h 04.1^m$	$-27° 20'$	dIrr/dSph	N3109	4.1	14.8	1.7	12
And VI	Peg dSph	1998	$23^h 51.7^m$	$+24° 36'$	dSph	M31	2.7	14.1	1.4	—
LGS 3	Pisces	1978	$01^h 03.9^m$	$+21° 53'$	dIrr/dSph	M31	2.6	14.3	1.3	13
And III		1972	$00^h 35.3^m$	$+36° 31'$	dSph	M31	2.5	14.2	1.1	—
And V		1998	$01^h 10.3^m$	$+47° 38'$	dSph	M31	2.6	15.0	1.0	—
Phoenix		1976	$01^h 51.1^m$	$-44° 27'$	dIrr/dSph	MW	1.4	13.2	0.9	33
DDO 210	Aquarius	1959	$20^h 46.8^m$	$-12° 51'$	dIrr/dSph	LGC	2.6	14.7	0.8	5.4
Tucana		1985	$22^h 41.8^m$	$-64° 25'$	dSph	LGC	2.9	15.2	0.6	—
Leo II	DDO 93	1950	$11^h 13.5^m$	$+22° 09'$	dSph	MW	0.66	12.0	0.6	9.7
Sextans		1990	$10^h 13.1^m$	$-01° 37'$	dSph	MW	0.28	10.3	0.5	19
Carina		1977	$06^h 41.6^m$	$-50° 58'$	dSph	MW	0.33	10.9	0.4	13
Ursa Minor	DDO 199	1955	$15^h 09.2^m$	$+67° 13'$	dSph	MW	0.21	10.3	0.3	23
Draco	DDO 208	1955	$17^h 20.3^m$	$+57° 55'$	dSph	MW	0.27	10.9	0.3	22

[a] Adapted from Paul Murdin (ed.), *Encyclopedia of Astronomy and Astrophysics*, New York, Nature Publishing Group, 2001, p. 433. The members of the Local Group are listed in order of decreasing intrinsic brightness. Other than Andromeda or M31, Triangulum or M33, and the Large and Small Magellanic Clouds, LMC and SMC, most of the members of the Local Group are dwarf galaxies, designated by a "d" in the Type column, and the absolute, or intrinsic, luminosity and mass are in units of a million times the solar values of $L_\odot = 3.85 \times 10^{33}$ erg s^{-1} and $M_\odot = 1.989 \times 10^{33}$ grams.

Longitude: Distance along the Earth's equator, measured east or west from the circle of longitude intersecting Greenwich, England to the equatorial intersection of the circle of longitude that passes through the point. A circle of longitude passes around the Earth from pole to pole perpendicular to the equator. *See* Royal Greenwich Observatory.

Long-period comet: A comet whose orbital period around the Sun is greater than 200 years. Long-period comets originate in the Oort comet cloud. *See* comet and Oort comet cloud.

Lookback time: The time at which the radiation we see from an object was emitted by that object, equal to the object's distance divided by the velocity of light in a Universe described by Euclidean geometry. The radiation we see from more distant objects was emitted a longer time ago, and hence at an earlier epoch in the evolution of the Universe.

Lorentz contraction: Decrease in the length of an observed object along the direction of its motion, as perceived by an external observer who does not share its motion. The effect is only noticeable at speeds approaching the velocity of light. The theory was proposed in 1904 by the Dutch physicist **Hendrik Lorentz (1853–1928)** to account for the negative result of the Michelson-Morley experiment. *See* Michelson-Morley experiment and *Special Theory of Relativity*.

Lorentz factor: A parameter, often denoted by the symbol γ, equal to the amount of length contraction, mass increase and time dilation at high speeds. If v is the object's velocity and c denotes the velocity of light, then the Lorentz factor is equal to $1 / \sqrt{(1 - (v/c)^2)}$, so it is equal to 1.00 at zero velocity and becomes very large when the velocity approaches the speed of light.

Lowell Observatory: An optical, or visible-light, observatory near Flagstaff, Arizona, founded in 1894 by the wealthy Bostonian **Percival Lowell (1855–1916)**, primarily to observe Mars and to search for signs of life on the red planet. The American astronomer **Vesto Slipher (1875–1969)** used the 0.61-meter (24-inch) refractor at Lowell Observatory, installed in 1896, to discover the large outward velocities of spiral nebulae, published by him in 1917. The American astronomer **Clyde William Tombaugh (1906–1997)** discovered the small planet Pluto on 18 February 1930 as the result of a systematic search using a 0.33-meter (13-inch) photographic telescope at the Lowell Observatory. *See* Pluto.

Luminosity: The amount of energy radiated per unit time by a glowing object, often measured in units of watts. This intrinsic brightness of an object is also called the absolute luminosity, and it differs from how bright the object looks, which is known as the apparent luminosity. One Joule per second is equal to one watt of power, and to ten million ergs per second. The absolute luminosity, L, of the thermal emission from a blackbody is given by the Stefan-Boltzmann law in which $L = 4\pi\sigma R^2 T_{eff}^4$, where $\pi = 3.14159$, the Stefan-Boltzmann constant, $\sigma = 5.67051 \times 10^{-8}$ Joule m^{-2} °K^{-4} s^{-1}, the radius is R and the effective temperature is T_{eff}. The brightness of radiation falls off with the inverse square of the distance, D, so the apparent luminosity $l = L/D^2$, and an

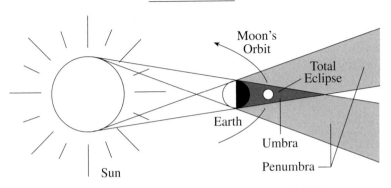

FIG. 38 **Lunar eclipse** During a lunar eclipse the initially full Moon passes through the Earth's shadow. A total lunar eclipse occurs when the entire Moon moves into the umbra. A partial lunar eclipse occurs when the Moon's orbit takes it only partially through the umbra or only through the penumbra.

object of a given absolute luminosity looks fainter when it is located farther away. *See* absolute luminosity, apparent luminosity, brightness, inverse-square law, and Stefan-Boltzmann law.

Luna Missions: A series of lunar probes launched by the Soviet Union. *Luna 2,* launched on 13 September 1959, was the first spacecraft to directly sample the solar wind, and *Luna 3*, launched on 4 October 1959, obtained the first photographs of the invisible, backside of the Moon. In 1966, *Luna 9* and *Luna 10* respectively became the first spacecraft to achieve a soft landing on the Moon and to enter into lunar orbit. In 1970, *Luna 16* made the first automatic return of a lunar soil sample to Earth.

Lunar eclipse: During a lunar eclipse, the Moon's orbital motion carries the initially full Moon through the Earth's shadow. This occurs once or twice in a typical year, and can be seen from anywhere on Earth where the Moon is above the horizon, or from about half the Earth. The Moon does not become totally dark during a total lunar eclipse, but instead turns red since it is partially illuminated by sunlight refracted through the Earth's atmosphere.

Lunar Prospector: NASA launched the *Lunar Prospector* spacecraft on 6 January 1998, which orbited the Moon from 15 January 1998 to 31 July 1999, determining the global distribution of its elements, magnetic field and gravity, detecting the lunar core, and obtaining some evidence for water ice at the Moon's poles. *See* Moon.

Lyman series: A series of ultraviolet spectral lines of the hydrogen atom, named after the American physicist **Theodore Lyman (1874–1954)** who first observed the ultraviolet spectrum of hydrogen in 1906 and announced in 1914 that the wavelengths of three of them are described by the same formula that specifies the Balmer series of hydrogen transitions, seen in visible light. The quantized electron orbits of the Bohr atom explain the Balmer and Lyman series. In the spectra of quasars, the Lyman series is redshifted into visible-light wavelengths. *See* Balmer series and Bohr atom.

M: Abbreviation for the *Messier catalogue* of just over 100 bright nebulous objects and star clusters, published in 1771 and 1784 by the French astronomer **Charles Messier (1730–1817)**.

MACHO: Acronym for MAssive Compact Halo Object, dark stars or planets that may make up the Milky Way's dark halo. *See* dark halo and dark matter.

Magellanic Clouds: Two nearby, irregular galaxies that orbit our Galaxy, the Milky Way. They are known as the Large Magellanic Cloud, abbreviated LMC, and the Small Magellanic Cloud, or SMC for short. The Large Magellanic Cloud is 168 000 light-years from Earth. Both galaxies can be observed with the unaided eye from the Earth's Southern Hemisphere, but cannot be seen from the United States. They are named for the Portuguese explorer **Ferdinand Magellan (1480–1521)** whose crew reported seeing them during the first circumnavigation of the globe, completed in 1522. The Large Magellanic Cloud contains the Tarantula Nebula, one of the largest emission nebulae known, and the LMC is also the sight of Supernova 1987A, the first naked-eye supernova since 1604. *See* irregular galaxy, Large Magellanic Cloud, Small Magellanic Cloud, Supernova 1987A, and Tarantula Nebula.

Magellan Mission: NASA launched its *Magellan* spacecraft to Venus on 4 May 1989. It arrived at the planet on 10 August 1990 and orbited the planet four years, using radar to map 98 percent of the planet's surface with an unprecedented detail of 100-meter vertical resolution and 50-kilometer horizontal resolution. The *Magellan* radar images revealed rugged highlands, plains smoothed out by global lava flows, ubiquitous volcanoes, and sparse, pristine impact craters.

Magellan Telescopes: *See* **Las Campanas Observatory**.

Magnetic dynamo: *See* dynamo.

Magnetic field: A magnetic force field around the Sun, a planet, and any other magnetized body, generated by electrical currents. The Sun's large-scale magnetic field, like that of Earth, exhibits a north and south pole linked by lines of magnetic force, but the Sun also contains numerous dipolar sunspots linked by magnetic loops.

Magnetic field lines: Imaginary lines that indicate the strength and direction of a magnetic field. The lines are drawn closer together where the field is stronger. Charged particles move freely along magnetic field lines, but cannot move across them.

Magnetic polarity: *See* polarity.

Magnetic pressure: A type of pressure inherent in magnetic fields, acting on a plasma and given by B^2/μ_0 where B is the magnetic field strength and the permeability of free space is $\mu_0 = 1.2566 \times 10^{-6}$ N A^{-2}.

TABLE 27 Magnetic fields – planets[a]

Planet	Magnetic Dipole Moment[b] (Earth = 1)	Magnetic Field at the Equator, B_0 (10^{-4} T)	Tilt[c] of Magnetic Axis (degrees)	Offset from Planet Center (R_P)	Bow-Shock Distance, R_{MP} (R_P)	Planet Equatorial Radius, R_P (km)
Mercury	0.0007	0.0033	$+14°$	$0.05\ R_M$	$1.5\ R_M$	$R_M = 2439$
Earth	1	0.305	$+11.7°$	$0.07\ R_E$	$10\ R_E$	$R_E = 6378$
Jupiter	20 000	4.28	$-9.6°$	$0.14\ R_J$	$42\ R_J$	$R_J = 71\,492$
Saturn	600	0.22	$<1.0°$	$0.04\ R_S$	$19\ R_S$	$R_S = 60\,268$
Uranus	50	0.23	$-58.6°$	$0.3\ R_U$	$25\ R_U$	$R_U = 25\,559$
Neptune	25	0.14	-40.8	$0.55\ R_N$	$24\ R_N$	$R_N = 24\,764$

[a]The magnetic field strengths are given at the surface of Mercury and the Earth and at the cloud tops for the giant planets. Venus and Mars have no detected global, dipolar magnetic field, with respective upper limits of 2×10^{-9} and 10^{-8} tesla.

[b]magnetic dipole moment, $(= B_0 R_P^3)$ is given in units of the Earth's magnetic dipole moment of 7.91×10^{15} T m^3. Here we use the SI unit for magnetic field strength, the tesla, abbreviated T. The c.g.s. unit of magnetic field strength, the gauss or G for short, can be computed from one tesla = 1 T = 10^4 gauss = 10^4 G. The nanotesla or nT, is also used, with 1 nT = 10^{-9} T, and the nanotesla has historically also been called the gamma. A dipole moment of 1 T m^3 = 10^{10} G cm^3, where m and cm respectively denote meter and centimeter. The equivalent unit of 1 G cm^3 = 10^{-3} A m^2 is also used, where the current is in units of amperes, abbreviated A.

[c]The tilt is the angle between the magnetic axis and the rotation axis.

Magnetic reconnection: A change in the topology of the magnetic field where the magnetic field lines reorient themselves via new connections. During magnetic reconnection, some magnetic field lines are broken and then rejoined in a new configuration. This can occur when two oppositely directed magnetic fields move toward each other and reconnect at the place that they touch. Magnetic reconnection is an important mechanism for releasing magnetic energy to heat the Sun's corona and to power explosive phenomena on the Sun, such as coronal mass ejections and solar flares.

Magnetic storm: A disturbance in the Earth's magnetic field observed all over the Earth due to collision by a coronal mass ejection. A substorm is a magnetic disturbance observed in Polar Regions only. *See* geomagnetic activity and geomagnetic storm.

Magnetism: One aspect of electromagnetism, a fundamental force of nature, whereby a magnetized object can force a charged particle into a new direction of motion and attract or repel other magnetized objects.

Magnetogram: A computer image, picture or map of the strength, direction and distribution of magnetic fields across the solar photosphere, based on Zeeman-effect measurements. *See* magnetograph and Zeeman effect.

Magnetograph: An instrument used to map the strength, direction and distribution of magnetic fields across the solar photosphere using the Zeeman effect. Normally, the longitudinal, or line-of-sight, magnetic field is measured, but the transverse component is also measured with a vector magnetograph. A magnetogram is a map of the measured magnetic field. *See* magnetogram and Zeeman effect.

Magnetosphere: The region of space surrounding a planet in which the planet's magnetic field predominates over the solar wind, and controls the motions of charged particles in it. The magnetosphere is shaped by interactions between a planet's magnetic field and the solar wind. The magnetosphere shields the planet from the solar wind, preventing or impeding the direct entry of the solar wind particles into the magnetic cavity. *See* bow shock, magnetopause, magnetosheath, magnetotail, and Van Allen belts.

Magnetotail: The portion of a planet's magnetosphere formed on the planet's dark night side by the pulling action of the solar wind. The magnetotail is an elongated extension of the planet's magnetic fields on the side of the planet opposite to the Sun. Magnetotails can extend hundreds of planetary radii.

Magnification: The increase in angular size of an object when viewed with a telescope. The magnification of a telescope is usually preceded by a multiplication $\times$ sign.

Magnitude: A measure of the brightness of a star or other celestial object. On the magnitude scale, the lowest magnitudes refer to objects of greatest brightness. In the 2nd century BC the Greek astronomer **Hipparchus (c. 190–c. 120 BC)** classified the stars into six magnitudes; the brightest stars are of the first magnitude while those of sixth magnitude are just visible to the unaided human eye. In the 19th century a scale of magnitudes was adopted in which a star of first magnitude is exactly 100 times as bright as one of sixth magnitude. The ratio of brightness between one magnitude and

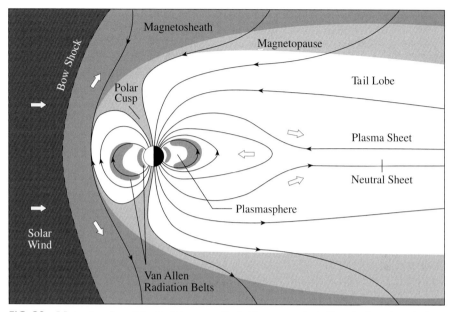

FIG. 39 Magnetosphere The Earth's magnetic field carves out a hollow in the solar wind, creating a protective cavity called the magnetosphere. The Earth, its atmosphere and ionosphere, and the two Van Allen radiation belts all lie within this magnetic cocoon. Similar magnetospheres are found around other magnetized planets. A bow shock forms at about ten Earth radii on the sunlit side of our planet. Its location is highly variable since it is pushed in and out by the gusty solar wind. The magnetopause marks the outer boundary of the magnetosphere, at the place where the solar wind takes control of the motions of charged particles. The solar wind is deflected around the Earth, pulling the terrestrial magnetic field into a long magnetotail on the night side. Electrons and protons in the solar wind are deflected at the bow shock (*left*), and flow along the magnetopause into the magnetic tail (*right*). Electrified particles can be injected back toward the Earth and Sun within the plasma sheet (*center*).

the next is thus the fifth root of 100, or 2.512. The apparent magnitude is the brightness of an object as seen from Earth. Absolute magnitude is the true brightness of an object, taking into account its distance from Earth; it is defined as the apparent magnitude an object would have at the arbitrary distance of ten parsecs. The brightness of a star as observed from the Earth, and hence its apparent magnitude, depends on both its absolute magnitude and its distance. The absolute visual magnitude of the Sun is + 4.82 magnitudes and its apparent visual magnitude is −26.72 magnitudes. *See* absolute magnitude, apparent magnitude, and visual magnitude.

Main sequence: The region in the Hertzsprung-Russell diagram that extends diagonally from the upper left to the lower right, or from the high-luminosity, high temperature stars to the low-luminosity, low-temperature stars. The stars on the main sequence are known as dwarf stars, but they are also called main-sequence stars. They are the most common types of star in the Galaxy, constituting 90 percent of its stars and 60 percent of its mass. The Sun is a main-sequence, or dwarf, star. Stars spend the majority of their lifetimes on the main sequence, during which they produce energy

TABLE 28   Main-sequence stars[a]

Spectral Type	Effective Temperature (K)	Mass $(M/M_\odot)$	Luminosity $(L/L_\odot)$	Radius $(R/R_\odot)$	Lifetime (years)
O5	44 500	60	7.9×10^5	12	3.7×10^6
B0	30 000	17.5	5.2×10^4	7.4	1.1×10^7
B5	15 400	5.9	8.3×10^2	3.9	6.5×10^7
A0	9 520	2.9	5.4×10	2.4	2.9×10^8
F0	7 200	1.6	6.5	1.5	1.5×10^9
G0	6 030	1.05	1.5	1.1	5.1×10^9
K0	5 250	0.79	0.42	0.85	1.4×10^{10}
M0	3 850	0.51	0.077	0.60	4.8×10^{10}
M5	3 240	0.21	0.011	0.27	1.4×10^{11}

[a] The mass, M, is in units of the Sun's mass $M_\odot = 1.989 \times 10^{33}$ grams, the absolute luminosity is in units of the Sun's absolute luminosity, $L_\odot = 3.85 \times 10^{33}$ erg s^{-1} = 3.85×10^{26} W, and the radius, R, in units of the Sun's radius, $R_\odot = 6.922 \times 10^8$ m. The lifetimes are the amount of time required to exhaust the nuclear hydrogen fuel that supplies the energy of stars on the main-sequence.

from the fusion of hydrogen into helium in their cores. The zero-age main sequence is where stars lie when they first start to burn hydrogen. The position of a star on the main sequence depends on its mass, the most massive stars being the brightest. The more massive a star, the sooner it evolves off the main sequence. *See* dwarf star and Hertzsprung-Russell diagram.

Main-sequence star: A star, like the Sun, that fuses hydrogen nuclei into helium nuclei in its core, and lies on the main sequence of the Hertzsprung-Russell diagram. The main-sequence stars can have any spectral type, and are designated by luminosity class V. They are also known as dwarfs, or dwarf stars, since they are smaller than the giant or supergiant stars. Ninety percent of all stars are main-sequence stars. *See* Alpha Centauri, Altair, Hertzsprung-Russell diagram, main sequence, Sirius, stellar classification, and Vega.

Major planet: Any one of the eight largest planets of the Solar System; in order of increasing distance from the Sun they are Mercury, Venus, Earth, Mars, Jupiter, Saturn, Uranus and Neptune. The asteroids are known as minor planets. Although Pluto was once regarded as a major planet, we now know that it is just one of many smaller bodies orbiting the Sun in the Kuiper belt outside the orbit of Neptune. *See* Kuiper belt.

Maria: Large dark, smooth features on the Moon, created between 3.9 and 3.2 billion years ago when magma from the Moon's interior filled large impact basins, formed between 4.3 and 3.9 billion years ago, and solidified into lava. The word *maria* is Latin for "seas", and the singular form of the word is mare. *See* Moon and volcanism.

Mariner Missions: A series of United States NASA space probes to the plants. The ***Mariner 2*** spacecraft, launched on 7 August 1962, measured the density and velocity of the solar wind on its way to Venus, and in December 1962 confirmed the emission of intense microwaves from the planet's exceptionally hot surface. The *Mariner 4* space-craft flew by Mars in July 1965, photographing craters on the red planet's surface. *Mariner 5* flew by Venus in October 1967, confirming the high temperature and pressure at the surface of the planet. NASA launched the *Mariner 9* spacecraft to Mars on 30 May 1971. It became the first spacecraft to orbit a planet, circling around the red planet for nearly a year, from 13 November 1971 to 27 October 1972, and obtaining close-up images of towering volcanoes, vast canyons, such as the Valles Marineris named for the spacecraft, and the planet's two small satellites, Phobos and Deimos. *Mariner 10* flew by Venus in February 1974, and continued to Mercury, which it encountered three times, in March and September 1974 and March 1975, revealing its heavily cratered, Moon-like surface. *See* Deimos, Mars, Phobos, solar wind, Valles Marineris, and Venus

Mariner Valley: *See* Valles Marineris.

Mars: The fourth major planet from the Sun. The red planet Mars is about half the size of the Earth and located 1.5 times as far from the Sun as the Earth is. Mars now has a thin, cold carbon-dioxide atmosphere, and a frozen surface, but there is abundant evidence that liquid water once flowed across the planet. It has powerful and perva-sive winds, polar caps, remnant magnetism, and towering volcanoes. Mars has two small moons, named Phobos and Deimos. Phobos is heading toward eventual collision with Mars. *See* Deimos, *Mariner 9*, Mariner Valley, *Mars Exploration Rover Missions, Mars Global Surveyor, Mars Odyssey, Mars Pathfinder*, Olympus Mons, Phobos, red planet, Syrtis Major, and *Viking Missions*.

Mars Exploration Rover Missions: NASA's twin rovers, *Spirit* and *Opportunity*, launched on 10 June and 7 July 2003, respectively. *Spirit* landed in Gusev Crater on 4 January 2004 and *Opportunity* landed in Meridiani Planum on 25 January 2004, obtaining evidence that liquid water once flowed across Mars.

Mars Express: ESA launched its *Mars Express* spacecraft to Mars on 2 June 2003, with an orbiter that has obtained images of the planet's surface and an unsuccessful *Beagle 2* lander.

Mars Global Surveyor: NASA launched its *Mars Global Surveyor* on 7 November 1996, placing it into orbit around the red planet on 12 September 1997. High-resolution images, beginning on 4 April 1999, provided abundant evidence for past and recent water flow, dust storms and volcanic activity. The global topography was determined with a laser altimeter, and a magnetometer determined the global distribution of remnant magnetic fields preserved in the ancient highland crust.

Mars Odyssey: NASA launched its *Mars Odyssey* spacecraft on 7 April 2001, placing it into orbit around the red planet on 24 October 2001. It has mapped the amount and distribution of chemical elements and minerals on the Martian surface, and provided evidence for substantial subsurface water ice.

Mars Pathfinder: The *Mars Pathfinder* lander and its roving vehicle, *Sojourner*, were launched by NASA on 4 December 1996, and landed on Chryse Planitia near the mouth of an outflow channel on 4 July 1997, finding evidence that liquid water once flowed

TABLE 29   Mars[a]

Mass	6.4185×10^{23} kilograms $= 0.1074$ M_E
Mean Radius	3389.93 kilometers $= 0.532$ R_E
Mean Mass Density	3933.5 kilograms per cubic meter
Surface Area	1.441×10^{14} square meters $= 0.2825$ A_E
Rotation Period or Length of Sidereal Day	24 hours 37 minutes 22.663 seconds $= 8.864\ 27 \times 10^4$ seconds
Orbital Period	686.98 Earth days $= 668.60$ Mars solar days
Mean Distance from Sun	2.2794×10^{11} meters $= 1.52366$ AU
Orbital Eccentricity	0.0934
Tilt of Rotational Axis, the Obliquity	25.19 degrees
Distance from Earth	5.6×10^{10} meters to 3.99×10^{11} meters
Angular Diameter at Closest Approach	14 to 24 seconds of arc
Age	4.6×10^9 years
Atmosphere	95.32 percent carbon dioxide, 2.7 percent nitrogen, 1.6 percent argon
Average Global Surface Pressure	0.0056 bars $= 560$ Pascals
Surface Pressure at *Viking 1* and *2* Sites	0.0067 to 0.0088 bars and 0.0074 to 0.010 bars
Surface Temperature Range	140 to 300 kelvin
Average Surface Temperature	210 kelvin
Magnetic Field Strength (Remnant)	$\pm 1.5 \times 10^{-6}$ tesla $= \pm 0.05$ B_E
Magnetic Dipole Moment	less than 10^{-4} that of Earth

[a] Adapted from H. H. Kieffer, B. M. Jakosky, C. W. Snyder and M. S. Matthews (eds.): *Mars*, University of Arizona Press, Tucson 1992. The symbols M_E, R_E, A_E, and B_E respectively denote the mass, radius, surface area, and magnetic field strength of the Earth.

across Mars and suggesting that the planet's atmosphere was once thicker, warmer and wetter than at present, but perhaps several billion years ago.

Mascon: An abbreviation for mass concentrations, on the Moon, discovered from precise radio tracking of spacecraft whose orbits were deflected when passing over the mascons. Mascons are buried under virtually all of the maria on the near side of the Moon, and they are also found beneath impact basins on both the near and far side of the Moon. The mascons are most likely the upward bulging of high-density mantle rocks that rose after impacts from large meteorites blasted away the lighter crust.

Maser: Acronym for microwave amplification by stimulated emission of radiation. Hydroxyl, OH, and water, H_2O, molecules in interstellar space can act as masers.

Mass: A measure of the total amount of matter contained within an object. Physicists distinguish between weight, which depends on the gravitational pull of a larger mass, and mass, which does not vary. The more mass an object has, the stronger is its gravitational pull on other objects. Inertial mass indicates an object's resistance to changes in its state of motion, while gravitational mass indicates its response to the gravitational

TABLE 30   Mars – satellites[a]

	Phobos	**Deimos**
Mass (kg)	1.08×10^{16}	1.8×10^{15}
Radii of triaxial ellipsoid (km)	$13.3 \times 11.1 \times 9.3$	$7.6 \times 6.2 \times 5.4$
Mean mass density (kg m^{-3})	$1,905 \pm 53$	$1,700 \pm 500$
Mean distance from Mars (10^3 km)[b]	$9.378 = 2.766\ R_M$	$23.479 = 6.926\ R_M$
Sidereal period (Martian days)[c]	0.3189	1.26244

[a] Adapted from H. H. Kieffer, B. M. Jakosky, C. W. Snyder and M. S. Matthews (eds): *Mars*, University of Arizona Press, Tucson 1992. The mass is in kilograms, abbreviated kg, the radius and distance are in kilometers, abbreviated km, and the mass density is in kilograms per cubic meter, abbreviated kg m^{-3}.

[b] The radius of Mars is $R_M = 3389.9$ kilometers.

[c] One Mars day = 24 hours 37 minutes 22.663 seconds, so the orbital periods of Phobos and Deimos are 7 hours 39 minutes 13.84 seconds and 30 hours 17 minutes 54.87 seconds, respectively. Both satellites are locked into synchronous rotation with a rotation period equal to their orbital periods.

TABLE 31   Mars – oppositions[a]

Opposition Date	**Right Ascension (hours minutes)**		**Declination (degrees minutes)**		**Diameter (seconds of arc)**	**Distance**[b] **(10^{10} meters)**
2001 June 13	17	28	-26	30	20.5	6.82
2003 August 28	22	38	-15	48	25.1	5.58
2005 November 7	02	51	$+15$	53	19.8	7.03
2007 December 28	06	12	$+26$	46	15.5	8.97
2010 January 29	08	54	$+22$	09	14.0	9.93
2012 March 3	11	52	$+10$	17	14.0	10.08
2014 April 8	13	14	-05	08	15.1	9.29
2016 May 22	15	58	-21	39	18.4	7.61
2018 July 27	20	33	-25	30	24.1	5.77
2020 October 13	01	22	$+05$	26	22.3	6.27
2022 December 8	04	59	$+25$	00	16.9	8.23
2025 January 16	07	56	$+25$	07	14.4	9.62
2027 February 19	10	18	$+15$	23	13.8	10.14
2029 March 25	12	23	$+01$	04	14.4	9.71
2031 May 4	14	46	-15	29	16.9	8.36
2033 June 27	18	30	-27	50	22.0	6.39
2035 September 15	23	43	-08	01	24.5	5.71

[a] An opposition occurs when the Earth moves between Mars and the Sun, and the two planets are closest. Adapted from William Sheehan: *The Planet Mars*, University of Arizona Press, Tucson 1996, Appendix 1, pp. 277-279.

[b] The distance between the Earth and Mars at opposition in units of ten billion (10^{10}) meters. Because the Martian orbit is more elliptical than Earth's, the distance between the two planets at different oppositions varies as much as 50 billion meters.

force. In the *General Theory of Relativity* gravitational and inertial mass are equal to each other. *See General Theory of Relativity* and solar mass.

Mass density: The amount of matter per unit volume, or an object's mass divided by its volume, sometimes just called the density. The mass density of water is 1000 kilograms per cubic meter. Air has a low density, while rocks have a high one. Giant stars have a low mass density, while compact white dwarfs or neutron stars have a very large mass density.

Mass-energy equivalence: Mass, denoted by m, can be transformed into energy, E, and every energy has a mass, according to the formula $E = mc^2$, where c is the velocity of light, first derived in its exact form by the German physicist **Albert Einstein (1879–1955)** in 1907.

Mass-luminosity relation: The relation between the absolute luminosity, denoted by L, and the mass, specified by M, for stars on the main sequence of the Hertzsprung-Russell diagram, first derived in 1924 by the English astronomer **Arthur Eddington (1882–1944)**. It is expressed as $L = \text{constant} \times M^x$, where the exponent x lies between 3 and 4.

Mass number: *See* atomic mass number.

Mass-to-light ratio: The amount of mass of an object divided by its absolute luminosity, usually measured in solar units.

Mauna Kea Observatories: At least thirteen working telescopes are located near the summit of the extinct Mauna Kea volcano at a height of about 4,200 meters above sea level, on the island of Hawaii. Nine of the telescopes operate at optical and infrared wavelengths, three of them are for submillimeter astronomy, and one is for radio astronomy. The unique, extremely dry, and cloud-free site is on land leased from the State of Hawaii by the University of Hawaii and developed for astronomy observations by its Institute of Astronomy. The telescopes include the twin **Keck** 10-meter (394-inch) telescopes, the largest visible-light telescopes in the world, and the 15-meter (590-inch) **James Clerk Maxwell Telescope** which operates in the submillimeter wavelength region of the electromagnetic spectrum. Other large optical/infrared telescopes on Mauna Kea include the 8.2-meter (323-inch) Japanese **Subaru Telescope** and one of the 8.1-meter (320-inch) **Gemini telescopes** – the other one is in Cerro Pachón, Chile. *See* Gemini Observatory, James Clerk Maxwell Telescope, Keck telescopes, and Very Long Baseline Array.

Maunder Minimum: The 70-year period between 1645 and 1715 when few sunspots were observed. It is named after the English astronomer **E. Walter Maunder (1851–1928)**, who in 1922 provided a full account of the missing sunspots, previously noticed by the German astronomer **Gustav F. W. Spörer (1822–1895)**. A prolonged period of unusually cold weather in Europe, known as the Little Ice Age, overlaps the Maunder Minimum and another period of unusually low sunspot numbers, the Spörer Minimum from 1420 to 1570, suggesting that a decrease in sunspots and associated solar activity may be associated with colder temperatures on Earth. *See* Little Ice Age.

Maxwellian distribution: Distribution of particle velocities for a gas in thermal equilibrium, characterized by a well-defined temperature, T, and root-mean-square velocity, $V_{rms} = (3kT/M)^{1/2}$ for a particle of mass M, where Boltzmann's constant $k = 1.38066 \times 10^{-23}$ Joule per degree kelvin. The distribution is named for the Scottish physicist **James Clerk Maxwell (1831–1879)**, who derived it in 1860.

McDonald Observatory: The astronomical observatory of the University of Texas, located in the Davis Mountains near Austin, Texas. The observatory includes the **Hobby-Eberly telescope**. *See* Hobby-Eberly telescope.

M dwarf: A main-sequence star of spectral type M. *See* M star.

Megahertz: A unit of frequency, abbreviated MHz. It is equal to one million Hertz. *See* Hertz.

Megaparsec: A unit of intergalactic distances. One Megaparsec, or 1 Mpc, is equivalent to a million parsecs, 3.0857×10^{22} meters and 3.26 million light-years.

Megaton: The energy released in the explosion of one million tons of TNT, equal to 4.2×10^{15} Joules, or 10^{22} ergs.

Mercury: The innermost planet of the Solar System, and the smallest of the four inner, terrestrial planets. Due to its small size and proximity to the Sun, Mercury is very difficult to observe from the Earth. Its true rotation period wasn't established until 1965, when American scientists **Gordon H. Pettengill (1926–)** and **Rolf B. Dyce (1929–)** sent radar pulses to Mercury from the Arecibo Observatory in Puerto Rico, using the return signal to show that Mercury's rotation period is 59.6 Earth days, or 2/3 of its orbital year of 88 Earth days. In 1974–75 the *Mariner 10* spacecraft flew by the planet three times, revealing its cratered, Moon-like surface. *See* radar astronomy.

TABLE 32   Mercury[a]

Mass	3.302×10^{23} kilograms $= 0.0553$ M_E
Mean Radius	2440 kilometers $= 0.382$ R_E
Mean Mass Density	5427 kilograms per cubic meter
Rotation Period	58.646 Earth days
Orbital Period	87.969 Earth days
Mean Distance from Sun	5.79×10^{10} meters $= 0.387$ AU
Age	4.6×10^9 years
Atmosphere	Very tenuous (helium, sodium, potassium)
Surface Pressure	2×10^{-13} bars
Surface Temperature	90 to 740 kelvin
Magnetic Field Strength	0.0033×10^{-4} tesla at the equator $= 0.01$ B_E
Magnetic Dipole Moment	5.54×10^{12} tesla meters cubed

[a] The symbols M_E, R_E and B_E respectively denote the mass, radius and magnetic field strength of the Earth.

MERLIN: Acronym for **Multi-Element Radio-Linked Interferometer Network**, which consists of six radio telescopes linked to the Jodrell Bank Observatory, with a maximum baseline of 217 kilometers. *See* interferometer, Jodrell Bank Observatory and Very Long Baseline Interferometry.

Messier object: Any of the emission nebulae, star clusters and galaxies that appear in the final catalogue published by the French astronomer **Charles Messier (1730–1817)** in 1781. Astronomers still designate these objects by the capital letter M followed by the number in Messier's list. Some of the best-known Messier objects include:

M1: The Crab Nebula, a supernova remnant. *See* Crab Nebula.

M31: The Andromeda Nebula, a spiral galaxy that lies 2.4 million light-years away from the Earth. *See* Andromeda galaxy and Local Group.

M32: An elliptical galaxy that is in orbit around Andromeda, M31. *See* Local Group.

M33: The Triangulum or Pinwheel galaxy, a spiral galaxy located 2.6 million light-years away from the Earth. It is the third largest member of the Local Group, after Andromeda and the Milky Way. *See* Local Group.

M42: The Orion Nebula, a region of star formation in the constellation Orion. *See* emission nebula and Orion Nebula.

M51: The Whirlpool Galaxy, a spiral galaxy in the constellation Canes Venatici.

M87: A giant elliptical galaxy in the Virgo cluster. *See* radio galaxy and supermassive black hole.

Metal: In astronomy, a metal is any element heavier than helium. In this definition, elements like oxygen and neon are metals, as is iron.

Metallicity: An object's abundance of metals, often the abundance of iron since it is easy to measure. *See* metal.

Meteor: The brilliant trail of light across the night sky produced when a small, solid object from space, a meteoroid, enters the Earth's upper atmosphere at high speed. Most meteors are fragile fragments of comets and never reach the ground. Those that do reach the ground are called meteorites, and they are rocky fragments of asteroids. At certain times of the year, meteors are more numerous than usual, and these meteor showers occur when the Earth passes through the orbit of a comet. *See* meteor shower and meteorite.

Meteor Crater: A crater near Flagstaff, Arizona, also known as the Arizona Meteor Crater and the Barringer Meteorite Crater, after the mining engineer **Daniel Moreau Barringer (1860–1929)** who unsuccessfully tried to mine the site for large quantities of iron. In the early 1960s, the American astronomer **Eugene M. Shoemaker (1928–1997)** demonstrated the similarity between the crater and those produced during nuclear test explosions in Nevada, demonstrating its origin by external impact. The circular crater is about 1200 meters in diameter and 180 meters deep. It is attributed to

TABLE 33 Messier Objects[a]

Messier M	NGC IC	Type[b]	Con	R. A, (2000) (h m)	Dec. (2000) (° ′)	m_V	Name
1	1952	Crab	Tau	05 34.5	+22 01	8.4	Crab Nebula
2	7089	Glob	Aqr	21 33.5	–00 49	6.5	
3	5272	Glob	CVn	13 42.2	+28 23	6.4	
4	6121	Glob	Sco	16 23.6	–26 32	5.9	
5	5904	Glob	Ser	15 18.6	+02 05	5.8	
6	6405	Op Cl	Sco	17 40.1	–32 13	4.2	Butterfly
7	6475	Op Cl	Sco	17 53.9	–34 49	3.3	
8	6523	Neb	Sgr	18 03.8	–24 23	5.8	Lagoon Nebula
9	6333	Glob	Oph	17 19.2	–18 31	7.9	
10	6254	Glob	Oph	16 57.1	–04 06	6.6	
11	6705	Op Cl	Sct	18 51.1	–06 16	5.8	Wild Duck Cluster
12	6218	Glob	Oph	16 47.2	–01 57	6.6	
13	6205	Glob	Her	16 41.7	+36 28	5.9	Hercules
14	6402	Glob	Oph	17 37.6	–03 15	7.6	
15	7078	Glob	Peg	21 30.0	+12 10	6.4	
16	6611	Op Cl	Ser	18 18.8	–13 47	6.0	Eagle Nebula
17	6618	Neb	Sgr	18 20.8	–16 11	7.5	Omega or Swan Nebula
18	6613	Op Cl	Sgr	18 19.9	–17 08	6.9	
19	6273	Glob	Oph	17 02.6	–26 16	7.2	
20	6514	Neb	Sgr	18 02.6	–23 02	8.5	Trifid Nebula
21	6531	Op Cl	Sgr	18 04.6	–22 30	5.9	
22	6656	Glob	Sgr	18 36.4	–23 54	5.1	
23	6494	Op Cl	Sgr	17 56.8	–19 01	5.5	
24	6603	Op Cl	Sgr	18 16.9	–18 29	4.5	
25	I 4725	Op Cl	Sgr	18 31.6	–19 15	4.6	
26	6694	Op Cl	Sct	18 45.2	–09 24	8.0	
27	6853	Plan	Vul	19 59.6	+22 43	8.1	Dumbbell Nebula
28	6626	Glob	Sgr	18 24.5	–24 52	6.9	
29	6913	Op Cl	Cyg	20 23.9	+38 32	6.6	
30	7099	Glob	Cap	21 40.4	–23 11	7.5	
31	224	Gal Sb	And	00 42.7	+41 16	3.4	Andromeda Nebula
32	221	Gal E	And	00 42.7	+40 52	8.2	
33	598	Gal Sc	Tri	01 33.9	+30 39	5.7	
34	1039	Op Cl	Per	02 42.0	+42 47	5.2	
35	2168	Op Cl	Gem	06 08.9	+24 20	5.1	
36	1960	Op Cl	Aur	05 36.1	+34 08	6.0	
37	2099	Op Cl	Aur	05 52.4	+32 33	5.6	
38	1912	Op Cl	Aur	05 28.7	+35 50	6.4	
39	7092	Op Cl	Cyg	21 32.2	+48 26	4.6	
40		2 stars	UMa	12 22.4	+58 05	8.0	Winnecke 4

TABLE 33 *continued*

Messier M	NGC IC	Type[b]	Con.	RA (2000) (h m)	Dec. (2000) (° ′)	m_V	Name
41	2287	Op Cl	CMa	06 46.1	−20 46	4.5	
42	1976	Neb	Ori	05 35.4	−05 27	4.0	Orion Nebula
43	1982	Neb	Ori	05 35.6	−05 16	9.0	Orion Nebula
44	2632	Op Cl	Cnc	08 40.2	+19 43	3.1	Praesepe (Beehive)
45		Op Cl	Tau	03 47.0	+24 07	1.2	Pleiades
46	2437	Op Cl	Pup	07 41.8	−14 49	6.1	
47	2422	Op Cl	Pup	07 36.6	−14 30	4.4	
48	2548	Op Cl	Hya	08 13.8	−05 48	5.8	
49	4472	Gal E	Vir	12 29.8	+08 00	8.4	
50	2323	Op Cl	Mon	07 02.8	−08 23	5.9	
51	5194	Gal Sc	CVn	13 29.9	+47 12	8.1	Whirlpool
52	7654	Op Cl	Cas	23 24.2	+61 35	6.9	
53	5024	Glob	Com	13 12.9	+18 10	7.7	
54	6715	Glob	Sgr	18 55.1	−30 29	7.7	
55	6809	Glob	Sgr	19 40.0	−30 58	7.0	
56	6779	Glob	Lyr	19 16.6	+30 11	8.2	
57	6720	Plan	Lyr	18 53.6	+33 02	9.0	Ring Nebula
58	4579	Gal SBb	Lyr	12 37.7	+11 49	9.8	
59	4621	Gal E	Vir	12 42.0	+11 39	9.8	
60	4649	Gal E	Vir	12 43.7	+11 33	8.8	
61	4303	Gal Sc	Vir	12 21.9	+04 28	9.7	
62	6266	Glob	Oph	17 01.2	−30 07	6.6	
63	5055	Gal Sb	CVn	13 15.8	+42 02	8.6	Sunflower
64	4826	Gal Sb	Com	12 56.7	+21 41	8.5	Black Eye
65	3623	Gal Sa	Leo	11 18.9	+13 05	9.3	
66	3627	Gal Sb	Leo	11 20.2	+12 59	9.0	
67	2682	Op Cl	Cnc	08 51.4	+11 49	6.9	
68	4590	Glob	Hya	12 39.5	−26 45	8.2	
69	6637	Glob	Sgr	18 31.4	−32 21	7.7	
70	6681	Glob	Sgr	18 43.2	−32 18	8.1	
71	6838	Glob	Sge	19 53.8	+18 47	8.3	
72	6981	Glob	Aqr	20 53.5	−12 32	9.4	
73	6994	Op Cl	Aqr	20 58.9	−12 38		4 stars
74	628	Gal Sc	Psc	01 36.7	+15 47	9.2	
75	6864	Glob	Sgr	20 06.1	−21 55	8.6	
76	650	Plan	Per	01 42.4	+51 34	11.5	Little Dumbbell
77	1068	Gal Sb	Cet	02 42.7	−00 01	8.8	
78	2068	Neb	Ori	05 46.7	+00 03	8.0	
79	1904	Glob	Lep	05 24.5	−24 33	8.0	
80	6093	Glob	Sco	16 17.0	−22 59	7.2	

TABLE 33 *continued*

Messier M	NGC IC	Type[b]	Con.	RA (2000) (h m)	Dec. (2000) (° ')	m_V	Name
81	3031	Gal Sb	UMa	09 55.6	+69 04	6.8	
82	3034	Gal Irr	UMa	09 55.8	+69 41	8.4	
83	5236	Gal Sc	Hya	13 37.0	−29 52	10.1	
84	4374	Gal E	Vir	12 25.1	+12 53	9.3	
85	4382	Gal So	Com	12 25.4	+18 11	9.3	
86	4406	Gal E	Vir	12 26.2	+12 57	9.2	
87	4486	Gal Ep	Vir	12 30.8	+12 24	8.6	Radio galaxy
88	4501	Gal Ep	Com	12 32.0	+14 25	9.5	
89	4552	Gal E	Vir	12 35.7	+12 33	9.8	
90	4569	Gal Sb	Vir	12 36.8	+13 10	9.5	
91	4548	Gal S	Com	12 35.4	+14 30	10.2	
92	6341	Glob	Her	17 17.1	+43 08	6.5	
93	2447	Op Cl	Pup	07 44.6	−23 52	6.2	
94	4736	Gal Sb	CVn	12 50.9	+41 07	8.1	
95	3351	Gal SBb	Leo	10 44.0	+11 42	9.7	
96	3368	Gal Sa	Leo	10 46.8	+11 49	9.2	
97	3587	Plan	UMa	11 14.8	+55 01	11.2	Owl Nebula
98	4192	Gal Sb	Com	12 13.8	+14 54	10.1	
99	4254	Gal Sc	Com	12 18.8	+14 25	9.8	
100	4321	Gal Sc	Com	12 22.9	+15 49	9.4	
101	5457	Gal Sc	UMa	14 03.2	+54 21	7.7	Pinwheel
102	5866	Gal Sa	Dra	15 06.5	+55 46	10.0	
103	581	Op Cl	Cas	01 33.2	+60 42	7.4	
104	4594	Gal Sa	Vir	12 40.0	−11 37	8.3	Sombrero
105	3379	Gal E	Leo	10 47.8	+12 35	9.3	
106	4258	Gal Sb	CVn	12 19.0	+47 18	8.3	
107	6171	Glob	Oph	16 32.5	−13 03	8.1	
108	3556	Gal Sb	UMa	11 11.5	+55 40	10.0	
109	3992	Gal SBc	UMa	11 57.6	+53 23	9.8	

[a] Adapted from *The Observer's Handbook,* Royal Astronomical Society of Canada, 1991.
[b] Op Cl = open cluster; Glob = globular cluster; Plan = planetary nebula; Neb. = emission nebula; Gal = galaxy (with classification).

the impact of a nickel-iron meteorite about 40 meters across, which collided with the Earth about 50 000 years ago.

Meteorite: An extraterrestrial chunk of rock and metal, which survives its fiery descent through the atmosphere to reach the ground. More than 20 thousand meteorites have been recovered from the ice at Antarctica, many of them spending about half a million years embedded in the ice; many fewer meteorites have been found at other locations on Earth. Radioactive dating indicates that meteorites formed about 4.6 billion years ago. Most meteorites are fragments of asteroids, but a small number are from Mars, called SNC meteorites, or from the Moon. There are three main classes of meteorites, the stony meteorites, mainly composed of silicates, the iron meteorites, consisting of iron and nickel, and stony-irons, composed of about half silicates and half iron and nickel. Almost all of these kinds of meteorites have a greater mass density than a typical rock on the Earth's surface. Most of the meteorites are stones, and about 90 percent of these are classified as chondrites, a name derived from the ancient Greek word, *chondros,* meaning "grain" or "seed". The chondrites contain small spherical masses of material known as chondrules. The other 10 percent of the stony meteorites are achondrites, which show signs of past igneous activity. The rare carbonaceous chondrite meteorites are composed of fragile, low-density material and contain appreciable amounts of carbon and water. *See* carbonaceous chondrite meteorite, iron meteorite, SNC meteorite, stony meteorite, and Widmanstätten pattern.

Meteoroid: A solid object in space, smaller than an asteroid. A meteoroid becomes a meteor when it lights up in the Earth's atmosphere and a meteorite if it reaches the ground.

Meteor shower: Comets strew particles along their orbital path as they loop around the Sun, and when the Earth passes through one of these meteor streams a meteor shower occurs, recurring at about the same time every year. The Italian astronomer **Giovanni Schiaparelli (1835–1910)** first proposed in 1866 that meteor showers are the result of the disintegration of comets, establishing a connection between the Perseid meteor shower and Comet Swift-Tuttle, discovered in 1862. The meteors in a shower appear to come from the same point in the sky, the radiant, and the meteor shower is named after the constellation in which its radiant lies. Meteor showers are sometimes called shooting stars or falling stars, but they have nothing to do with stars and are composed of fragile comet material that burns up in the atmosphere and never reaches the ground. *See* Leonids, meteor, Perseids, Quadrantids, and Swift-Tuttle Comet.

Meteor stream: Meteoroids circling the Sun in a common orbit, usually the debris of a comet with the same orbit. A meteor shower occurs when the Earth passes through a meteor stream. *See* meteoroid.

Methane: A molecule denoted by the symbol CH_4, where C is an atom of carbon and H is a hydrogen atom. We cook and heat some homes with gaseous methane, and it is found in abundance within Uranus, Neptune, and interstellar space.

TABLE 34   Meteor showers[a]

Shower	Maximum Date	Radiant Position[b]		Visibility Dates	Meteors Per Hour[c]
		RA	Dec.		
Quadrantids	3−4 January	15h 28m	+50°	1−6 January	110
Alpha Aurigids	6−9 February	04h 56m	+43°	Jan.−Feb.	10
Virginids	12 April	14h 04m	−09°	March−April	5
Lyrids	21−22 April	18h 08m	+32°	19−25 April	15
Eta Aquarids	5 May	22h 20m	−01°	24 April−20 May	35
Alpha Scorpiids	28 April	16h 32m	−24°	20 April−19 May	5
Ophiuchids	9 June	17h 56m	−23°	May−June	5
Alpha Cygnids	21 July	21h 00m	+48°	June−August	5
Capricornids	8−15 July	20h 44m	−15°	5 July−20 Aug.	5
Alpha Capricornids	2 August	20h 36m	−10°	15 July−20 Aug.	5
Delta Aquarids	29 July	22h 36m	−17°	15 July−20 Aug.	25
Iota Aquarids	6 August	22h 10m	−15°	July−August	10
Piscis Australids	31 July	22h 40m	−30°	July−August	5
Perseids	12 August	03h 04m	+58°	25 July−20 Aug.	80
Alpha Aurigids	28 August	04h 56m	+43°	August−October	10
Piscids	8−9 September	00h 36m	+07°	Sept.−Oct.	10
Orionids	21 October	06h 24m	+15°	15 Oct.−2 Nov.	30
Taurids	3 November	03h 44m	+14°	15 Oct.−25 Nov.	10
Leonids	17 November	10h 08m	+22°	15−20 November	45
Geminids	13−14 December	07h 28m	+32°	7−15 December	70
Ursids	22−23 December	14h 28m	+78°	19−24 December	10

[a] Adapted from Bone, N.: *Meteors*, Cambridge, Mass., Sky Publishing Co. 1993, p. 86, and Lang, Kenneth R., *Astrophysical Data: Planets and Stars*. New York, Springer Verlag 1992, p. 99.

[b] The celestial coordinates of the radiant are right ascension, abbreviated as RA, in hours, hr, and minutes, m, and declination, or Dec. for short, in degrees °.

[c] The maximum hourly frequency of meteors is under the assumption that the radiant is at the zenith, but this rate can vary from year to year because of the non-uniform distribution of the relevant meteoroid stream.

MeV: A unit of energy equal to one million electron volts and $1.602\ 177 \times 10^{-13}$ Joule. This unit is used to describe the total energy of high-velocity particles or energetic radiation photons. *See* cosmic rays, electron volt, and gamma-ray radiation.

MHz: Abbreviation for MegaHertz, or a million Hertz. *See* Hertz and Megahertz.

Michelson-Morley experiment: German-born American physicist **Albert Michelson (1852–1931)** and American chemist **Edward Morley (1838–1923)** use an interferometer in 1887 to try to detect any difference in the velocity of light in two directions at right angles to each other, failing to measure any difference and showing that the veloc-

TABLE 35   Meteor showers – comets[a]

Meteor Shower[b]	Comet
Lyrids	Thatcher C/1861 G1
Eta Aquarids	1P/Halley
Scorpiids-Sagittariids	2P/Encke
Bootids	7P/Pons-Winnecke
Perseids	109P/Swift-Tuttle
Aurigids	Kiess C/1911 N1
Draconids	22P Giacobini-Zinner
Orionids	1P/Halley
Epsilon Geminids	Ikeya C/1964 N1
Taurids	2P/Encke
Adromedids	3P/Biela
Leonids	55P/Temple-Tuttle
Geminids	3200 Phaethon[c]
Monocerotids	Mellish C/1917 F1
Ursids	8P/Tuttle

[a] Meteor showers that occur when the Earth intersects the orbit of the comet given here. Adapted from Lang, Kenneth R., *Astrophysical Data: Planets and Stars*. New York, Springer Verlag 1992, p. 98.

[b] The visibility dates of these showers were given in Table 34.

[c] 3200 Phaethon is cataloged as an asteroid, but it may be an inactive comet nucleus.

ity of light is not affected by the motion of the Earth through a hypothetical aether. *See* aether, ether, luminiferous ether, and *Special Theory of Relativity*.

Microlensing: The gravitational distortion and amplification of a star's light caused by an object, such as a planet, passing between the star and an observer. The object curves light rays from the star and concentrates them towards the observer. *See* gravitational lens.

Micron: A unit of length equal to a millionth, or 10^{-6}, of a meter. Visible light has wavelengths of between 0.4 and 0.7 microns.

Microwave: Electromagnetic radiation with wavelengths between 0.1 and 6 centimeters, or frequencies from 5 to 300 GHz, at the short-wavelength end of the radio spectrum.

Microwave background: *See* cosmic microwave background radiation.

Milankovitch cycles: Three rhythmic fluctuations in the wobble and tilt of the Earth's rotational axis and the shape of the Earth's orbit that set the major ice ages in motion by altering the seasonal distribution of the Sun's light and heat on Earth. They are the 23 000-year precessional wobble of the Earth's rotational axis, the 41 000-year variation in the Earth's axial tilt, and the 100 000-year change in the eccentricity or shape of the Earth's orbit. These cycles are named after the Yugoslavian astronomer **Milutin**

Milankovitch (1879–1958) who developed the relevant mathematical theory between 1920 and 1941. *See* ice age.

Mile: A unit of distance equal to 1.609 34 kilometers and 5280 feet.

Milky Way: The band of light that stretches across our night sky, produced by the light of numerous stars as we look out through the disk of our Galaxy. The starlight is obscured in places by clouds of interstellar gas and dust.

Milky Way Galaxy: The Galaxy that contains the Milky Way and the Sun, distinguished from other galaxies by a capital G. The Milky Way Galaxy is a spiral galaxy with a massive center located about 27 700 light-years from the Sun, which is located in one of the spiral arms. The Galaxy contains hundreds of billions of stars, mostly located in a disk that is about 130 000 light-years across. A dark halo that surrounds the disk of our Galaxy accounts for about ten times as much mass as the disk itself. The Milky Way Galaxy, or Galaxy for short, belongs to the Local Group of galaxies. *See* dark halo, galactic disk, Galaxy, and Local Group.

Million: One thousand thousand, written as 1 000 000 or 10^6.

Millisecond pulsar: A pulsar with a rapid period of about one millisecond or 0.001 second. In 1982 the American radio astronomer **Donald C. Backer (1943–)** discovered the first millisecond pulsar, designated PSR 1937+21, with a period of 1.56 milliseconds. Another millisecond pulsar, PSR 1257+12 with a period of 6.2 milliseconds, was discovered in 1990 by the American astronomer **Aleksander Wolszczan (1946–)**; accurate timing observations of this pulsar resulted in the 1992 discovery of two Earth-sized planets orbiting PSR 1257+12. The pulsar PSR 1913+16, with a somewhat longer period of 59 milliseconds, was discovered in 1974 by American radio astronomers **Russell A. Hulse (1950–)** and **Joseph H. Taylor Jr. (1941–)**; precise timing observations of its radio pulses indicated that the binary pulsar is losing energy at the rate expected by gravitational radiation. All three pulsars were discovered at the **Arecibo Observatory**. *See* Arecibo Observatory, extrasolar planet, gravitational radiation, PSR 1257+12, PSR 1913+16 and PSR 1937+12.

Mimosa: Also known as Beta Crucis, a blue-white giant star of spectral type B0.5, apparent magnitude +1.25, a distance of 490 light-years, and an absolute magnitude of −4.6. *See* stars.

Minkowski metric: The line element, ds, in space-time without gravity, derived in 1908 by the Lithuanian-born German mathematician **Hermann Minkowski (1864–1909)**, and given by $ds^2 = -c^2\,dt^2 + dx^2 + dy^2 + dz^2$, where c is the velocity of light, dt is the time, and dx, dy, and dz are the three space coordinates. *See Special Theory of Relativity*, and velocity of light.

Minor planet: *See* asteroid.

Minute of arc: *See* arc minute.

Mira: A red M7 giant star in the constellation Cetus, whose apparent visual magnitude ranges from 3 to 10 and back to 3 again every 330 days. Mira, the "Miracle Star", is also

TABLE 36   Milky Way Galaxy

Radius of galactic disk = 20 kpc = 65 000 light-years, diameter 130 000 light-years

Thickness of galactic disk = 3.7 kpc = 12 000 light-years

Sun's distance from galactic center R_0 = (8.5 ± 1.1) kpc,
    where 8.5 kpc = 27 700 light-years.

Sun's rotation velocity about galactic center V_{rot} = 220 km s^{-1}

Sun's rotation period about galactic center P = 2.4×10^8 years

Oort's constants A = (+ 14.4 ± 1.2) km s^{-1} kpc^{-1} and B = (−12.0 ± 2.8) km s^{-1} kpc^{-1}.

Galactic Center: RA (2000) = 17^h 45^m 37.1991^s Dec.(2000) = −28° 56' 10.221"

North Galactic Pole: RA(2000) 12^h 51^m 26.2755^s Dec.(2000) = +27° 07' 41.704"

Absolute Magnitude M = −20.5

Absolute Luminosity L_B = (2.3 ± 0.6) $\times 10^{10}$ $L_\odot$ or $L_B \approx 2 \times 10^{10}$ $L_\odot$,
    where the Sun's absolute luminosity $L_\odot$ = 3.85×10^{26} J s^{-1} = 3.85×10^{33} erg s^{-1}

Mass   M = 9.5×10^{10} $M_\odot$ within R_0 = 8.5 kpc = 27 700 light-years,
    or a mass equivalent to 100 billion stars like the Sun,
    where the Sun's mass $M_\odot$ = 1.989×10^{33} grams,
    with a mass-luminosity ratio of M/L_B = 4.2 $M_\odot/L_\odot$ to R_0 = 8.5 kpc.

   M = 4×10^{11} $M_\odot$ within R = 35 kpc = 114 000 light-years
    with a mass-luminosity ratio of M/L_B = 18 to R = 35 kpc.

   Total mass M = 2×10^{12} $M_\odot$ within R = 230 kpc or 750 000 light-years,
    with a total mass about ten times the mass of the visible stellar component, and a dark,
    invisible halo of about 90 percent of the total mass.

   Total mass-luminosity ratio $M/L_B \approx$ 100 $M_\odot/ L_\odot$ to R = 230 kpc.

---

^{a} Adapted from Lang, Kenneth R., *Astrophysical Formulae II. Space, Time, Matter and Cosmology,*
Third Edition, New York, Springer-Verlag 1999, pp. 10, 31, 42, 127.

named Omicron Ceti, and has been additionally proclaimed "the Amazing One" and
"the Wonderful". The German astronomer **David Fabricius (1564–1617)** discovered
the star's variation in 1596; when brightest, Mira is visible to the naked eye and when
dimmest it cannot be seen with the unaided eye. At its distance of 420 light-years, its
angular diameter of 0.048 seconds of arc corresponds to a physical radius of 3 AU or 3
times the average distance between the Earth and the Sun and about double the size of
the orbit of Mars. The absolute visual magnitude of Mira varies from −3 to −5, with an
absolute luminosity ranging from 1 to nearly 10 000 times the Sun's luminosity. Mira
is the prototype for thousands of long-period Mira variables, which vary in brightness
as the stars pulsate slowly.

Missing mass: Alternative term for dark matter. *See* dark matter.

MIT: Acronym for the Massachusetts Institute of Technology, located in Cambridge,
Massachusetts.

Mizar: A double star in Ursa Major, also known as Zeta Ursae Majoris, of apparent
visual magnitude 2.3. Its distant companion, Alcor, with an apparent visual magnitude

of 4.0, can be seen with the unaided, or naked, eye. Mizar, Alcor and the companion to Mizar are all spectroscopic binaries. *See* spectroscopic binary.

MMT: Acronym for **Multiple Mirror Telescope**. *See* Fred L. Whipple Observatory, and Multiple Mirror Telescope.

Molecular cloud: An unusually dense and large interstellar nebula of gas and dust whose mass of up to a million solar masses consists mainly of hydrogen molecules, also containing other molecules such as carbon monoxide, carbon dioxide and water. Regions of star formation are usually located near or within molecular clouds. *See* giant molecular cloud, interstellar molecules, and protostar.

Molecular hydrogen: A molecule denoted by H_2, consisting of two hydrogen atoms, each denoted by H. It is the most abundant molecule in interstellar space, with a distribution that can be mapped by observing carbon monoxide. *See* carbon monoxide and hydrogen molecule.

Molecule: The smallest unit of a chemical compound. A molecule is composed of two or more atoms, bound together by interaction of their electrons.

Month: The period of revolution of the Moon around the Earth. The sidereal month, reckoned from fixed star to fixed star, is 27.321 66 mean solar days, and this is equal to the rotation period of the Moon about its axis. The synodic month is the mean interval between conjunctions of the Moon and the Sun, and corresponds to the cycle of lunar phases. The synodic month, from one new Moon to the next new Moon, is equal to 29.530 59 mean solar days. The Gregorian calendar is divided into 12 months, with a Gregorian calendar year equal to 365.2425 days. *See* sidereal time.

Moon: A natural object that revolves around a planet. A moon is a natural satellite. The Earth has one moon, which is usually called the Moon with a capital M. The Moon is 3475 kilometers across, or roughly one quarter the size of the Earth, and it revolves around the Earth once every 27.32 days. Rocks returned from the Moon by the *Apollo* missions, from 1969 to 1972, indicate that most of the highland craters were created by an intense bombardment about 4.0 billion years ago and that the lunar maria were filled with lava from the Moon's interior between 3.9 and 3.2 billion years ago. The *Clementine* and/or *Lunar Prospector* spacecraft have provided information on the liquid core, surface composition, topography, and magnetism of the Moon, as well as possible evidence for water ice in permanently shaded regions at the lunar poles. The Moon's gravity draws the Earth's oceans into two tidal bulges, producing the tides, and the Moon acts as a brake on the Earth's rotation, causing the length of the day to steadily increase at the rate of 0.002 seconds per century and the Moon to move away from the Earth at the rate of 0.04 meters per year. The Moon was most likely formed by the giant impact of a Mars-sized body with the Earth early in its history. *See* ***Apollo Missions**, **Clementine spacecraft***, crater, ***Lunar Prospector**,* maria, and tides.

Morgan-Keenan classification: *See* stellar classification.

Mount Palomar Observatory: *See* Palomar Observatory.

TABLE 37   Moon[a]

Mass	7.349×10^{22} kilograms $= 0.00123$ M_E
Mean Radius	1737.5 kilometers $= 0.2725$ R_E
Mean Mass Density	3344 kilograms per cubic meter
Rotation Period	27.322 days $=$ sidereal month
Orbital Period, the Sidereal Month	27.322 days $=$ fixed star to fixed star
Synodic Month	29.53 days $=$ new Moon to new Moon
Mean Distance from Earth	3.844×10^8 meters
Increase in Mean Distance	0.0382 ± 0.0007 meters per year
Mean Orbital Speed	1023 meters per second
Angular Radius at Mean Distance (Geocentric)	15 minutes 32.6 seconds of arc
Angular Radius at Mean Distance (Topocentric)	15 minutes 48.3 seconds of arc
Age	4.6×10^9 years

[a] Here M_E and R_E respectively denote the mass and radius of the Earth.

Mount Stromlo Observatory: Located on Mount Stromlo, near Canberra, Australia, the observatory was established in 1924 as The Commonwealth Solar Observatory and operated since 1957 by the Australian National University as the Mount Stromlo Observatory. A devastating wildfire on 18 January 2003 destroyed all of its major telescopes, but redevelopment is now underway.

Mount Wilson Observatory: Located on Mount Wilson, a peak in the San Gabriel mountains near Pasadena, California, the Mount Wilson Observatory is the site of the Mount Wilson Solar Observatory, founded in 1904, the 60-inch (1.5-meter) reflecting telescope completed in 1908, and the 100-inch (2.5-meter) reflecting telescope, which was funded by the American businessman **John D. Hooker (1838–1911)** and completed in 1917. The American astronomer **Edwin Hubble (1889–1953)** used the 100-inch Hooker telescope to show that spiral and elliptical nebulae are galaxies, and to obtain the first evidence for the expansion of the Universe.

MSFC: Acronym for the Marshall Space Flight Center. *See* Marshall Space Flight Center.

M star: A red star of spectral type M, whose visible disk is colder and redder than the Sun. M-type stars on the main sequence are called red dwarfs, and they are the most common class of star in the Milky Way, accounting for about 80 percent of all stars in our Galaxy. The visible spectrum of an M star is characterized by absorption bands of titanium oxide molecules. M stars on the main sequence are smaller, less massive and cooler than the Sun. They have masses less than 0.5 solar masses, burn hydrogen in their core by the proton-proton chain, and have convective zones that can extend throughout the entire star for the lowest mass. They also emit less light than the Sun, and are so dim that not a single one is visible to the unaided, naked eye. Red, M-type giant or supergiant stars are hundreds or thousands of times brighter than the Sun. Proxima Centauri is a red dwarf star, Mira is a red giant star, and Antares and Betelgeuse are red supergiant stars. *See* Antares, Betelgeuse, dM star, flare star, giant

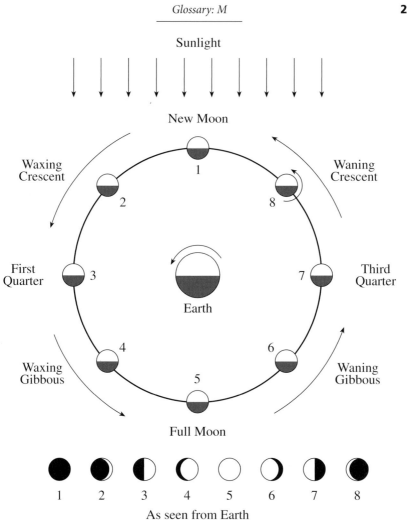

FIG. 40 Moon phases Light from the Sun illuminates one half of the Moon, while the other half is dark. As the Moon orbits the Earth, we see varying amounts of its illuminated surface. The phases seen by an observer on Earth (*bottom*) correspond to the numbered points along the lunar orbit. The period from new Moon to new Moon is 29.53 days, the length of the month. As the Earth completes its daily rotation, all nighttime observers see the same phase of the Moon.

star, Hertzsprung-Russell diagram, main sequence, Mira, M star, proton-proton chain, Proxima Centauri, red dwarf star, stellar classification, and supergiant star.

MSW effect: The transformation of a neutrino of one type or flavor into a neutrino of another kind while traveling through matter, named after **Lincoln Wolfenstein (1923–)** who originated the theory in 1978 and **Stanislav Mikheyev (1940–)** and **Alexei Y. Smirnov (1951–)** who further developed it about a decade later. The MSW effect could explain the solar neutrino problem if some of the electron neutrinos produced in the core of the Sun oscillate into muon or tau neutrinos on their way out of the Sun, thereby becoming invisible to most neutrino detectors. *See* neutrino oscillation, and solar neutrino problem.

Mullard Radio Astronomy Observatory: A radio astronomy observatory of the University of Cambridge at Lords Bridge near Cambridge, England, renowned for its discovery of pulsars and its pioneering work in aperture synthesis. The **Ryle Telescope**, named for the English radio astronomer **Martin Ryle (1918–1984)**, is an array of eight parabolic dishes each 13 meters (43 feet) in diameter, on a 5-kilometer, east-west baseline, originally built in 1971 for high-resolution imaging of radio galaxies and quasars and upgraded for mapping the microwave background. The name acknowledges funding by Mullard Limited. *See* aperture synthesis and pulsar.

Multiple Mirror Telescope: Abbreviated **MMT**, a visible-light telesocpe with a primary mirror of 6.5-meter (256-inch) diameter located on Mt. Hopkins, Arizona. It originally consisted of several smaller mirrors, but is now a singe mirror. The Fred L. Whipple Observatory and the University of Arizona operate the MMT. *See* Fred L. Whipple Observatory.

Muon: An elementary particle, denoted by the symbol μ. A muon is similar to an electron except that the muon has a mass that is 207 times that of an electron, and a brief life. A muon decays into neutrinos and an electron in just 2.197×10^{-6} seconds. Muons may have a positive or a negative charge. Muon neutrinos interact with muons, and the anti-muon is the antiparticle of the muon.

Naked-eye observation: Visual observation of cosmic objects without a telescope or other optical instrument, also called unaided observations.

NanoTesla: A unit of magnetic field strength, abbreviated nT. It is equal to 10^{-9} Tesla, to 10^{-5} Gauss, and to 1 gamma. *See* Gauss and Tesla.

NASA: Acronym for the National Aeronautics and Space Administration, United States. *See* National Aeronautics and Space Administration.

National Aeronautics and Space Administration: Abbreviated NASA, the United States government space agency formed in 1958. It is responsible for civilian manned and robotic activities in space, including launch vehicles, scientific satellites, and space probes. NASA Headquarters are in Washington, DC. NASA operates the Ames Research Center near San Francisco, California; the Goddard Space Flight Center, abbreviated GSFC, in Greenbelt Maryland; the Jet Propulsion Laboratory, or JPL for short, near Pasadena, California; the John F. Kennedy Space Center at Cape Canaveral, Florida; the Lyndon B. Johnson Space Center, in Houston, Texas; and the Marshall Space Flight Center, abbreviated MSFC, in Huntsville, Alabama. The Ames Research Center specializes in research that spans the spectrum of NASA interests, including aeronautics and planetary science; GSFC focuses on observations from space related to the Earth and its environment, the Solar System and the Universe; JPL includes robotic exploration of the Solar System and deep-space tracking; the Johnson Space Center includes mission control for manned space missions and specializes in human space exploration; the Kennedy Space Center prepares and launches space missions, and the Marshall Space Flight Center develops space transportation and propulsion technologies and includes the use of space research and development to benefit humanity. *See* Goddard Space Flight Center and Jet Propulsion Laboratory.

National Optical Astronomy Observatory: Abbreviated NOAO, it was formed in 1982 to consolidate under one director several ground-based astronomical optical observatories: the Kitt Peak National Observatory, the Cerro Tololo Inter-American Observatory, and the National Solar Observatory, with facilities at Sacramento Peak, New Mexico and Kitt Peak, Arizona. Today the National Solar Observatory has its own director. NOAO has its headquarters in Tucson, Arizona, is funded by the United States National Science Foundation and operated by the Association of Universities for Research in Astronomy.

National Radio Astronomy Observatory: Abbreviated NRAO, with its headquarters at Green Bank, West Virginia, the National Radio Astronomy Observatory operates the Very Large Array in New Mexico, and other powerful, advanced radio telescope facilities in the western hemisphere such as the **Very Long Baseline Array** and a 12-meter (39-feet) dish for millimeter-wavelength investigations located at Kitt Peak, Arizona. NRAO's **Robert C. Byrd Green Bank Radio Telescope** has dimensions of

TABLE 38   NASA and ESA space missions[a]

Apollo Missions: Six *Apollo* spacecraft transported twelve astronauts to the Moon and back between 1969 and 1972. They returned a total of 382 kilograms of rocks that were used to decipher the Moon's history and provide clues to its origin. The *Apollo 11* astronaut, **Neil A. Armstrong (1930–)**, was the first human to step on the Moon, from the *Lunar Module Eagle* on 20 July 1969.

Cassini-Huygens Mission: NASA launched the *Cassini-Huygens* spacecraft on 15 October 1997. Instruments aboard the spacecraft, which reached Saturn in July 2004, will improve our understanding of the planet Saturn, its rings, its magnetosphere, and its satellites. NASA's *Cassini Orbiter* will orbit Saturn and its moons for four years. On 14 January 2005, ESA's *Huygens Probe* dove into the thick, hazy atmosphere of Saturn's moon Titan and landed on its surface, discovering signs of flowing liquid, probably methane. The spacecraft is named for the Italian-born French astronomer **Gian (Giovanni) Domenico (Jean Domenique) Cassini (1625–1712)**, who discovered Cassini's gap in Saturn's rings, and the Dutch astronomer **Christiaan Huygens (1629–1695)**, who discovered Titan, Saturn's largest moon.

Chandra X-ray Observatory: NASA deployed the *Chandra X-ray Observatory,* abbreviated *CXO,* on 23 July 1999, from the *Space Shuttle Columbia.* It investigates the high-energy, X-ray regions of the Universe from cosmic objects such as active galactic nuclei, black holes, clusters of galaxies, dark matter, galaxies, neutron stars, pulsars, quasars, supernova remnants, supernovae, and white dwarfs. The X-ray observatory was named after the Indian-born American astrophysicist and Nobel laureate, **Subrahmanyan Chandrasekhar (1910–1995)**.

Compton Gamma Ray Observatory: NASA launched the *Compton Gamma Ray Observatory*, abbreviated *CGRO*, on 5 April 1991, from the *Space Shuttle Atlantis*, and it re-entered the Earth's atmosphere on 4 June 2000. It was the second of NASA's four great observatories in space, designed to study the Universe in an invisible, high-energy form of radiation known as gamma rays. It detected more than 2600 gamma-ray bursts, as well as gamma rays from black holes, pulsars, quasars and supernovae. The gamma ray observatory was named after the American physicist **Arthur Compton (1892–1962),** who is remembered for discovering the Compton effect, a phenomenon in which electromagnetic waves such as X-rays undergo an increase in wavelength after having been scattered by electrons.

COsmic Background Explorer: NASA launched the *COsmic Background Explorer*, abbreviated *COBE*, on 18 November 1989; instrument operations were terminated on 23 December 1993. These instruments measured the angular distribution and spectrum of the cosmic microwave background radiation, showing that it has a faint anisotropy and a blackbody spectrum, with a temperature of 2.725 kelvin.

Einstein Observatory: NASA launched the *Einstein Observatory,* the first fully imaging X-ray telescope put in space, on 12 November 1978, where it orbited Earth and obtained data until April 1981. It provided the first high-resolution X-ray studies of supernova remnants, discovered unexpectedly intense coronal emission from main-sequence stars, resolved numerous X-ray sources in nearby galaxies, and detected X-ray jets from Centaurus A and M87, aligned with radio jets. The observatory is named after the German-born American physicist **Albert Einstein (1879–1955)** who revolutionized our understanding of space, matter and time.

Galileo Mission: The *Galileo* spacecraft was launched by NASA onboard the *Space Shuttle Atlantis* on 18 October 1989, and arrived at Jupiter on 7 December 1995, orbiting the giant planet and vastly increasing our understanding of Jupiter and the four Galilean satellites, Io, Europa, Ganymede and Callisto. The mission is named after the Italian astronomer **Galileo Galilei (1564–1643)** who discovered Jupiter's four largest moons, now known as the Galilean satellites. The mission was terminated on 21 September 2003.

(continued)

TABLE 38   NASA and ESA space missions

Giotto Mission: ESA launched the *Giotto* spacecraft on 2 July 1985. It encountered Comet Halley on 13 March 1986, obtaining images of its dark, irregular jet-spewing nucleus and measuring the amount of water and dust emitted. The mission is named after the Florentine painter **Giotto di Bondone (c. 1267–c. 1337)**.

HIgh Precision PARallax COllecting Satellite: ESA launched its *HIgh Precision PARallax COllecting Satellite,* abbreviated *HIPPARCOS,* on 8 August 1989 and operated it until August 1993. It was the first space mission dedicated to astrometry, measuring the accurate positions, distances, motions, brightness and colors of stars. *HIPPARCOS* measured the parallax and proper motions of 120 000 stars with a precision of 0.001 to 0.002 seconds of arc, up to 100 times better than possible from the ground. The acronym also alludes to **Hipparchus of Nicaea (c. 190 BC– c. 120 BC)** who recorded accurate star positions more than two millennia previously.

Hubble Space Telescope: The world's first space-based optical, or visible-light, astronomi- cal telescope, the *Hubble Space Telescope*, with a 2.4-meter (94.5-inch) primary mirror, was launched on 24 April 1990 from the *Space Shuttle Discovery*, with servicing missions by shuttle astronauts in December 1993, February 1997, December 1999, and February 2002. It has observed newly formed galaxies when the Universe was half its present age, contributing to our understanding of the age and evolution of the Cosmos, watched super-massive black holes consuming the material around them, and helped astronomers determine how a mysterious "dark energy" has taken over the expansion of the Universe. The telescope was named for the American astronomer **Edwin Hubble (1889–1953)** who demonstrated that spiral nebulae are galaxies like our own Milky Way and found the first evidence for the expansion of the Universe.

Huygens Probe: See *Cassini-Huygens Mission*.

Magellan Mission: NASA launched its *Magellan* spacecraft to Venus on 4 May 1989. It ar- rived at the planet on 10 August 1990 and orbited the planet for four years, using radar to map 98 percent of the planet's surface with 100-meter vertical resolution and 50-kilometer spatial resolution. The *Magellan* radar images revealed rugged highlands, plains smoothed out by global lava flows, ubiquitous volcanoes, and sparse, pristine impact craters. It is named for the Portuguese explorer **Ferdinand Magellan (1480–1521)** whose ship Victoria completed the first circumnavigation of the Earth in 1522.

Mars Exploration Rover Missions: NASA's twin rovers, *Spirit* and *Opportunity*, launched on 10 June and 7 July 2003, respectively. *Spirit* landed in Gusev Crater on 4 January 2004 and *Opportunity* landed in Meridiani Planum on 25 January 2004, obtaining evidence that liquid water once flowed across Mars.

Mars Global Surveyor: NASA launched its *Mars Global Surveyor* on 7 November 1996, placing it into orbit around the red planet on 12 September 1997. High resolution images, beginning on 4 April 1999, obtained abundant evidence for past and recent water flow, dust storms and volcanic activity. The global topography was determined with a laser altimeter, and a magnetometer deter- mined the global distribution of remnant magnetic fields preserved in the ancient highland crust.

Mars 2001 Odyssey: NASA launched its *Mars 2001 Odyssey* spacecraft on 7 April 2001, placing it into orbit around the red planet on 24 October 2001. It has mapped the amount and distribution of chemical elements and minerals on the Martian surface, and provided evidence for substantial subsurface water ice.

Near Earth Asteroid Rendezvous Shoemaker Mission: NASA launched its *Near Earth Asteroid Rendezvous*, abbreviated *NEAR*, mission on 17 February 1996, and it arrived at the near-Earth asteroid 433 Eros on Valentine's day 14 February 2000. It was renamed *NEAR Shoemaker* after launch, in honor of the astronomer-geologist **Eugene M. Shoemaker (1928–1997)**, a pioneering expert on asteroid and comet impact. *NEAR Shoemaker* circled the asteroid 433 Eros for a year, determining its detailed size, shape, mass and mass density, and showing that it has a battered, homogeneous structure.

(continued)

TABLE 38   NASA and ESA space missions

Röntgen mission: An X-ray observatory developed through a cooperative program among Germany, the United States, and the United Kingdom, launched by NASA on 1 June 1990 and ending almost nine years later, on 12 February 1999. The spacecraft, abbreviated ***ROSAT*** for ***ROntgen SATellite,*** carried out an all-sky survey of cosmic X-rays, and obtained detailed information about the X-ray emission from comets, T Tauri stars, globular clusters, neutron stars, supernova remnants, planetary nebulae, nearby normal galaxies, active galactic nuclei, and clusters of galaxies. The mission was named after **Wilhelm Konrad Röntgen (1845−1923)**, the German physicist who discovered X-rays in 1895.

SOlar and Heliospheric Observatory: NASA launched the *SOlar and Heliospheric Observatory*, abbreviated *SOHO*, on 2 December 1995 with operations continuing into 2006. It is a cooperative project between ESA and NASA to study the Sun from its deep core, through its outer atmosphere and the solar wind out to a distance ten times beyond the Earth's orbit. It has obtained important new information about the structure and dynamics of the solar interior, the heating mechanism of the Sun's million-degree outer atmosphere, the solar corona, and the origin and acceleration of the solar wind.

Spitzer Space Telescope: NASA launched the *Spitzer Space Telescope*, abbreviated *SST*, on 25 August 2003, with an expected minimum 2.5-year mission lifetime. It is an infrared telescope with an 0.85-meter (33.5-inch) primary mirror, obtaining images and spectra of the infrared energy, or heat, radiated by cosmic objects at wavelengths between 3 and 180 microns, where 1 micron is one-millionth of a meter, and providing information on the formation, composition and evolution of planets, stars and galaxies. The mission was initially designated the *Space Infra-Red Telescope Facility*, or *SIRTF* for short, and renamed for the American astrophysicist **Lyman Spitzer, Jr. (1914−1997)**.

Uhuru Mission: The *Uhuru* satellite was launched from Kenya on 12 December 1970, on the seventh anniversary of Kenyan independence, when it was named *Uhuru,* the Swahili word for "freedom". It provided a comprehensive, all-sky survey of cosmic X-ray sources, including black holes and neutron stars in binary stellar systems, supernova remnants, Seyfert galaxies, and clusters of galaxies. It discovered X-ray emission from black holes and pulsars, and the diffuse X-ray emission from clusters of galaxies, which accounts for some of the so-called dark matter in the Universe. The mission ended in March 1973.

Ulysses Mission: The *Ulysses* spacecraft was a joint undertaking of ESA and NASA, launched by the *Space Shuttle Discovery* on 6 October 1990. During an encounter with Jupiter on 8 February 1992, the giant planet's gravitational field was used to accelerate the spacecraft and hurl it above the Sun's poles, but never at a distance closer to the Sun than the Earth is. *Ulysses* passed over the solar south pole in September 1994 and the north pole in July 1995, and then again over the south pole in November 2000 and the north pole in December 2001. Comparisons of *Ulysses* observations with those from other spacecraft, such as *Yohkoh* and *SOHO,* showed that much, if not all, of the high-speed solar wind comes from open magnetic fields in polar coronal holes, at least during the minimum of the Sun's 11-year cycle of magnetic activity.

Viking Mission: NASA's *Viking Mission* to Mars was composed of two spacecraft, *Viking 1* and *Viking 2,* each consisting of an orbiter and a lander. The orbiters imaged the entire surface of Mars at a resolution of 150 to 300 meters, showing volcanoes, lava plains, immense canyons, cratered areas, wind-formed features and evidence for catastrophic floods of water across the Martian surface in the past. The landers found no detectable organic molecules on the Martian surface, which means that this part of the surface now contains no cells, living, dormant or dead.

(*continued*)

TABLE 38   NASA and ESA space missions

Wilkinson Microwave Anisotropy Probe: NASA launched its *Wilkinson Microwave Anisotropy Probe,* abbreviated *WMAP,* on 30 June 2001, publishing definitive full sky maps of the faint anisotropy, or temperature variations, of the cosmic microwave background radiation in early 2003. When combined with the results of other diverse cosmic measurements, the *WMAP* data suggest that the Universe is 13.7 billion years old, that the first stars began to shine about 200 million years after the Big Bang, that the radiation in the *WMAP* images originated at the decoupling time 379,000 years after the Big Bang, that the Hubble constant has a value of 71 kilometers per second per Megaparsec, and that the Universe is composed of 4 percent atoms, 23 percent cold dark matter and 73 percent dark energy. The probe is named in honor of **David T. Wilkinson (1935–2002)** of Princeton University, a pioneer in cosmic microwave background research and the *MAP* instrument scientist.

Yohkoh Mission: The Japanese Institute of Space and Astronautical Science, abbreviated ISAS, launched its *Solar-A* spacecraft on 30 August 1991, renaming it *Yohkoh,* which means "sunbeam" in English. *Yohkoh* collected high-energy radiation from the Sun at soft X-ray and hard X-ray wavelengths, with high angular and spectral resolution, for more than a decade, providing new insights to the mechanisms of solar flare energy release and the heating of the million-degree solar corona. It was a collaborative effort of Japan, the United States and the United Kingdom.

---

[a]This abbreviated list does not include all NASA or ESA Missions, but these past, current and future spacecraft are listed alphabetically in our *Glossary.*

---

100 × 110 meters (330 × 360 feet), and began operation in August 2000. It is the world's largest fully steerable radio telescope and is named for a United States Senator from West Virginia, **Robert C. Byrd (1917–)**. *See* Effelsberg Radio Telescope, radio interferometer, radio telescope, Very Large Array, and Very Long Baseline Array.

National Solar Observatory: Abbreviated NSO, the United States National Solar Observatory includes three facilities dedicated to observing the Sun with high resolution at optical, or visible-light, wavelengths – the Sacramento Peak Observatory in New Mexico, with its 0.79-meter (30-inch) Richard B. Dunn Vacuum Tower Telescope, the Kitt Peak National Observatory's 1.5-meter (60-inch) McMath-Pierce Solar Telescope, also in Arizona, and the Global Oscillation Network Group of six solar telescopes that provide nearly continuous worldwide observations of the Sun. *See* Global Oscillation Network Group, Kitt Peak National Observatory, and National Optical Astronomy Observatory.

Near-Earth asteroid: Although the vast majority of asteroids travel in the main belt lying between the orbits of Mars and Jupiter, the near-Earth asteroids move closer to planet Earth. There are three populations of near-Earth asteroids, the Atens, Apollos and Amors, named for a representative asteroid in each class, 1221 Amor, 1862 Apollo and 2062 Aten. Both the Aten and Apollo asteroids move on eccentric orbits that can cross the Earth's path in space. The Atens are always close to the Sun, never moving out as far as the orbit of Mars. The elongated orbits of the Apollo objects loop from the main belt to within the Earth's orbit. The Amors come closest to the Sun, at their orbit perihelion, between the orbits of Mars and the Earth. *See* asteroid.

Near Earth Asteroid Rendezvous Shoemaker Mission: NASA launched its *Near Earth Asteroid Rendezvous*, abbreviated *NEAR*, spacecraft on 17 February 1996, and it arrived at the near-Earth asteroid 433 Eros on Valentine's Day 14 February 2000. It was the first spacecraft to orbit an asteroid, and the first to land on one. The spacecraft was renamed *NEAR Shoemaker* after launch, in honor of the astronomer-geologist **Eugene M. Shoemaker (1928–1997)**, a pioneering expert on asteroid and comet impact. *NEAR Shoemaker* circled the asteroid 433 Eros for a year, determining its detailed size, shape, mass and mass density, and showing that it has a battered, homogeneous structure. *See* Eros.

NEAR Shoemaker: Acronym for *Near Earth Asteroid Rendezvous Shoemaker*. See *Near Earth Asteroid Rendezvous Shoemaker Mission*.

Nebula: A region of gas and dust in interstellar space, including the emission nebulae, the planetary nebulae and the reflection nebulae. The word *nebula* is Latin for "cloud." An emission nebula emits bright spectral lines at visible wavelengths, planetary nebulae are shells of gas ejected by stars, and a reflection nebula shines by reflecting the light of nearby stars. Emission nebulae, such as the Orion Nebula, are the sites of current star and planet formation. Dark nebulae are not illuminated either externally or internally. The remnants of exploding stars, called supernova remnants, are also known as nebulae. The Crab Nebula is a famous example of a supernova remnant. Galaxies were also once called spiral nebulae or elliptical nebulae, before their enormous sizes and distances were discovered, and then for a brief time known as extra-galactic nebulae. *See* Crab Nebula, emission nebula, H II region, Orion Nebula, planetary nebula, reflection nebula, and supernova remnant.

Neptune: The eighth major planet from the Sun, and the smallest and most distant of the four giant planets; it marks the inner edge of the Kuiper Belt. Neptune is invisible to the naked eye, and can only be seen through a telescope. In 1989, the *Voyager 2* spacecraft flew by Neptune, discovering the planet's excess heat, raging winds, great dark spot, and thin, dark, widely-spaced rings. Neptune radiates 2.7 times the energy it receives from the Sun, and this internal heat drives an active atmosphere with high-speed winds and a short-lived Great Dark Spot. Most of the interior of Neptune probably consists of a vast internal ocean of water, methane and ammonia. The planet's magnetic axis is tilted by 46.8 degrees from the rotational axis, perhaps because it is generated in an internal shell of ionized water. Neptune's largest satellite, Triton, revolves about the planet in a direction opposite to that in which Neptune spins, and might have been formed elsewhere before being captured by Neptune. Triton has bright polar caps of nitrogen and methane ice, and geysers or volcanoes of ice that may now be erupting on its surface. Neptune was discovered in 1846 when the German astronomer **Johann Gottfried Galle (1812–1910)** and his student **Heinrich Louis d'Arrest (1822–1875)** located it near the position predicted by the French astronomer **Urbain Jean Joseph Leverrier (1811–1877)** on the basis of its gravitational perturbation of the orbit of the planet Uranus. The English astronomer **John Couch Adams (1819–1892)** performed similar calculations, also predicting the existence of the unknown planet at about the same time as Leverrier. *See* Kuiper belt and Triton.

TABLE 39   Neptune[a]

Mass	1024.3×10^{23} kilograms $= 17.14$ M_E
Radius	24 766 kilometers $= 3.883$ R_E
Mean Mass Density	1638 kilograms per cubic meter
Rotation Period	16.11 hours
Orbital Period	165.8 Earth years
Mean Distance from Sun	30.06 AU
Atmosphere	79 percent hydrogen 18 percent helium
Energy Balance	2.7 ± 0.3
Effective Temperature	59.3 kelvin
Temperature at 1-bar level	73 kelvin
Central Temperature	5000 kelvin
Magnetic Dipole Moment	25 D_E
Equatorial Magnetic Field Strength	0.14×10^{-4} tesla

[a]The Earth's mass, M_E, is 59.743×10^{23} kilograms, the Earth's equatorial radius, R_E, is 6378 kilometers, the astronomical unit, denoted AU, is the mean distance between the Earth and the Sun with a value of 1.496×10^{11} meters. The energy balance is the ratio of total radiated energy to the total energy absorbed from sunlight, the effective temperature is the temperature of a blackbody that would radiate the same amount of energy per unit area, a pressure of one bar is equal to the atmospheric pressure at sea level on Earth, and D_E is the magnetic dipole moment of the Earth.

Neutral hydrogen: An electrically neutral hydrogen atom, which has a proton for a nucleus and one surrounding electron. H I regions are composed of neutral hydrogen gas, and they emit radio waves that are 21 centimeters long. *See* H I region and twenty-one centimeter line.

Neutrino: A subatomic particle with no electric charge and very little mass that responds to the weak nuclear force but not to the strong nuclear and electromagnetic forces. In 1930 the Austrian physicist **Wolfgang Pauli (1900–1958)** proposed an uncharged and unseen particle to remove energy during radioactive beta decay, and the Italian physicist **Enrico Fermi (1901–1954)** developed the idea in 1934, giving the putative particle its name *neutrino*, Italian for "little neutral one." Neutrinos travel at very near the velocity of light and interact very weakly with other matter. They are emitted during radioactivity and generated during thermonuclear reactions in the Sun and other stars. There are three types, or flavors of neutrinos, named the electron, muon and tau neutrinos after the particles they interact with. Vast quantities of electron neutrinos are created as the result of nuclear reactions in the Sun's core. Other types of neutrinos are produced in man-made particle accelerators, nuclear reactors and by cosmic rays entering the Earth's atmosphere. *See* beta decay, hot dark matter, neutrino astronomy, proton-proton chain, radioactivity, solar neutrino problem, and weak nuclear force.

Neutrino oscillation: The change of one type or flavor of neutrino to another while traveling through matter or a vacuum. Neutrinos come in three flavors, the electron, muon and tau neutrinos. *See* MSW effect and solar neutrino problem.

Neutron: A subatomic particle with no electric charge found in all atomic nuclei except that of hydrogen. The English physicist **James Chadwick (1891–1974)** discovered the neutron in 1932 when bombarding atoms with energetic particles. The neutron has slightly more mass than a proton. The mass of the neutron is 1.008 665 atomic mass units and 1.6748×10^{-27} kilograms. The neutron is 1839 times heavier than an electron. When set free from an atomic nucleus, or created outside one, a neutron decays into a proton, an electron and an anti-electron neutrino with a neutron half-life of only 614 seconds or 10.25 minutes, and a mean life of just 887 seconds. *See* mass number and nucleon.

Neutron star: A very dense star with a radius of about 10 kilometers and a typical mass of 1.4 times the mass of the Sun, consisting of neutrons and supported by degenerate neutron pressure. Neutron stars are the collapsed cores of stars more massive than the Sun, produced in the final evolutionary stages of these stars and associated with supernova explosions. Rotating neutron stars are observed as radio pulsars, and are also thought to be components of some X-ray binary stars. *See* Crab Nebula, pulsar, and supernova.

New General Catalogue: *See* NGC.

New Technology Telescope: Abbreviated NTT, the New Technology Telescope is an optical/infrared telescope run by the European Southern Observatory at La Silla, Chile. It has a 3.58-meter (142-inch) mirror whose shape is controlled by active optics. *See* European Southern Observatory.

Newtonian telescope: A reflecting telescope in which the incoming light is reflected and focused by a mirror. This type of telescope is named after the English physicist **Isaac Newton (1643–1727)** who first built one in 1668. Newton's telescope had a primary concave spherical mirror and a flat secondary mirror, angled at 45 degrees, to redirect light to one side, at a place now called the Newtonian focus. Most modern telescopes use a parabolic mirror to gather and focus light. *See* reflecting telescope.

Next Generation Space Telescope: *See* **James Webb Space Telescope**.

NGC: Acronym for the *New General Catalogue of Nebulae and Clusters of Stars,* a catalogue of 7840 objects that was published in 1888 by the Danish-born Irish astronomer **John Louis Emil "J. L. E." Dreyer (1852–1926)**, with two supplementary *Index Catalogues,* abbreviated IC, containing another 5000 objects published in 1895 and 1908. NGC or IC numbers are now used to designate cosmic objects, including galaxies that were once thought to be nebulae. *See* IC.

NGST: Acronym for ***Next Generation Space Telescope***. *See* **James Webb Space Telescope**.

Nitrogen: The element with atomic number seven, and the dominant ingredient of the atmosphere of the Earth and Saturn's satellite Titan. Nitrogen is produced in giant stars, and cast into interstellar space when red giant stars expand and form planetary nebulae. *See* giant star and planetary nebula.

NOAA: Acronym for the National Oceanic and Atmospheric Administration, United States.

NOAO: Acronym for the National Optical Astronomy Observatory, United States. *See* National Optical Astronomy Observatory.

Nobel Prize: An international award given yearly since 1901 for achievements in physics, chemistry, physiology or medicine, literature and peace. Significant accomplishments in astronomy and astrophysics have been awarded the Nobel Prize in Physics. The awards are bestowed by the Nobel Foundation, which was endowed on 27 November 1895 in the last will of the Swedish industrial chemist and philanthropist **Alfred Nobel (1833–1896)**, who invented dynamite.

Noble gases: The rare and inert gases, or elements, helium, neon, argon, krypton, xenon, and radon, which are almost always mono-atomic and rarely undergo chemical reactions.

Non-baryonic matter: Hypothetical material that consists of exotic subatomic particles that are not baryons. Dark matter might be composed of non-baryonic particles that could be cold and slow or hot and fast. Inflation requires substantial quantities of non-baryonic matter. About 23 percent of the energy-matter content of the Universe could consist of cold, non-baryonic dark matter. *See* baryon, cold dark matter, dark matter, inflation, and hot dark matter.

Non-Euclidean geometry: Geometry that describes curved space, the three-dimensional analog of a sphere or a hyperbola. *See* Euclidean geometry.

Non-thermal particle: A particle that is not part of a thermal gas. A single temperature cannot describe these particles. *See* thermal gas.

Non-thermal radiation: The electromagnetic radiation produced by a non-thermal electron traveling at a speed close to that of light in the presence of a magnetic field. Such radiation is called synchrotron radiation after the man-made particle accelerator where it was first seen. More generally, the term non-thermal radiation denotes any electromagnetic radiation from an astronomical body that is not thermal in origin. *See* blackbody radiation and thermal radiation.

North America Nebula: An emission nebula, designated NGC 7000, in the constellation Cygnus, and named for the resemblance of its shape to the continent of North America. The North America Nebula measures 100×120 minutes of arc across, and is about 3500 light-years distant. An associated emission nebula is the Pelican Nebula. *See* emission nebula and Pelican Nebula.

Northern lights: A popular name for an aurora when observed from northern latitudes.

North Star: The bright star toward which the Earth's northern rotational axis points. The North Star is now Polaris, but the Earth's axis points in other directions during its slow 23 000-year precession. *See* Polaris and precession.

TABLE 40   Nobel Prize Winners

Year	Person	Accomplishment
1901[a]	**Wilhelm Röntgen (1845–1923)** (German physicist)	For the discovery of the remarkable rays subsequently named after him (and now called X-rays).
1902[a]	**Hendrik Lorentz (1853–1928) Pieter Zeeman (1865–1943)** (Dutch physicists)	For their research into the influence of magnetism upon radiation phenomena.
1903[a]	**Antoine Becquerel (1852–1908) Pierre Curie (1859–1906) Marie Curie (1867–1934)** (French physicists)	For his discovery of spontaneous radioactivity. For their investigations of radioactivity
1904[b]	**William Ramsay (1852–1916)** (English chemist)	In recognition of his services in the discovery of the inert gaseous elements in air, and his determination of their place in the periodic system.
1907[a]	**Albert Michelson (1852–1931)** (American physicist)	For his optical precision instruments and the spectroscopic and metrological investigations carried out with their aid.
1908[b]	**Ernest Rutherford (1871–1937)** (English physicist)	For his investigations into the disintegration of the elements, and the chemistry of radioactive substances.
1909[a]	**Guglielmo Marconi (1874–1937)** (Italian electrical engineer) **Karl Ferdinand Braun (1850–1918)** (German physicist)	For their contributions to the development of wireless telegraphy
1911[a]	**Wilhelm Wien (1864–1928)** (German physicist)	For his discovery of the laws governing heat radiation.
1911[b]	**Marie Curie (1867–1934)** (French physicist and chemist)	In recognition of her services by the discovery of the elements radium and polonium, by the isolation of radium and the study of the nature and compounds of this remarkable element.
1918[a]	**Max Planck (1858–1947)** (German physicist)	For his discovery of energy quanta.
1921[a]	**Albert Einstein (1879–1955)** (German physicist)	For his services to theoretical physics, and especially for his discovery of the law of the photoelectric effect.
1921[b]	**Fredrick Soddy (1877–1956)** (English chemist)	For his contributions to our knowledge of radioactive substances and his investigations into the origin and nature of isotopes.
1921[b]	**Francis Aston (1877–1945)** (English chemist)	For his discovery, by means of his mass spectrograph, of isotopes, in a large number of non-radioactive elements, and his enunciation of the whole-number rule.

(continued)

TABLE 40   Nobel Prize Winners

Year	Person	Accomplishment
1922[a]	**Niels Bohr (1885–1962)** (Dutch physicist)	For his investigation of the structure of atoms and of the radiation emanating from them.
1927[a]	**Arthur H. Compton (1892–1962)** (American physicist)	For his discovery of the effect named after him.
1935[a]	**James Chadwick (1891–1974)** (English physicist)	For the discovery of the neutron.
1936[a]	**Victor Franz Hess (1883–1964)** (Austrian physicist)	For his discovery of cosmic radiation (now called cosmic rays).
	Carl Anderson (1905–1991) (American physicist)	For his discovery of the positron.
1938[a]	**Enrico Fermi (1901–1954)** (Italian physicist)	For his demonstrations of the existence of new radioactive elements produced by neutron irradiation, and for his related discovery of nuclear reactions brought about by slow neutrons.
1939[a]	**Ernest Lawrence (1901–1958)** (American physicist)	For the invention and development of the cyclotron and for results obtained with it, especially with regard to artificial radioactive elements.
1945[a]	**Wolfgang Pauli (1900–1958)** (Austrian-born Swiss physicist)	For the discovery of the Exclusion Principle, also called the Pauli Principle.
1947[a]	**Edward Appleton (1892–1965)** (English physicist)	For his investigations of the physics of the upper atmosphere, especially for the discovery of the so-called Appleton layer (now called the ionosphere).
1959[a]	**Emilio Segrè (1905–1989)** (Italian-born American physicist) **Owen Chamberlain (1920–)** (American physicist)	For their discovery of the antiproton.
1961[a]	**Rudolf Mössbauer (1929–)** (German physicist)	For his researches concerning the resonance absorption of gamma radiation and his discovery in this connection of the effect which bears his name.
1964[a]	**Charles H. Townes (1915–)** (American physicist) **Nicolay Basov (1922–2001) Aleksandr Prokhorov (1916–2002)** (Russian scientists)	For fundamental work in the field of quantum electronics, which has led to the construction of oscillators and amplifiers based on the maser-laser principle.
1967[a]	**Hans Bethe (1906–2005)** (American physicist)	For his contributions to the theory of nuclear reactions, especially his discoveries concerning the energy production in stars.
1970[a]	**Hannes Alfvén (1908–1995)** (Swedish astrophysicist)	For fundamental work and discoveries in magnetohydrodynamics with fruitful applications in different parts of plasma physics.

(continued)

TABLE 40   Nobel Prize Winners

Year	Person	Accomplishment
1974[a]	**Martin Ryle (1918–1984)** (English radio astronomer)	For his pioneering observations and inventions in radio astrophysics, particularly the aperture synthesis technique.
1974[a]	**Antony Hewish (1924–)** (English radio astronomer)	For his pioneering research in radio astrophysics, particularly for his decisive role in the discovery of pulsars.
1978[a]	**Arno A. Penzias (1933–)** (American radio engineer) **Robert W. Wilson (1936–)** (American radio astronomer)	For their discovery of the cosmic microwave background radiation.
1983[a]	**Subramanyan Chandrasekhar (1910–1995)** (Indian-born American astrophysicist)	For his theoretical studies of the physical processes of importance to the structure and evolution of stars.
1983[a]	**William A. Fowler (1911–1995)** (American astrophysicist)	For his theoretical and experimental studies of the nuclear reactions of importance in the formation of the chemical elements in the Universe.
1989[a]	**Norman F. Ramsey (1915–)** (American physicist)	For the invention of the separated oscillatory fields method and its use in the hydrogen maser and other atomic clocks
1993[a]	**Russell A. Hulse (1950–) Joseph H. Taylor Jr. (1941–)** (American radio astronomers)	For their discovery of a new type of pulsar, a discovery that has opened up new possibilities for the study of gravitation.
1995[a]	**Frederick Reines (1918–1998)** (American physicist)	For the detection of the neutrino.
1995[b]	**Mario J. Molina (1943–) F. Sherwood Rowland (1927–)** (American chemists) **Paul J. Crutzen (1933–)** (Dutch chemist)	For their work in atmospheric chemistry, particularly concerning the formation and decomposition of ozone.
2002[a]	**Raymond Davis Jr. (1914–)** (American astrophysicist) **Masatoshi Koshiba (1926–)** (Japanese astrophysicist)	For their pioneering contributions to astrophysics, in particular for the detection of cosmic neutrinos.
2002[a]	**Riccardo Giacconi (1931–)** (Italian-born American astronomer)	For pioneering contributions to astrophysics, which have led to the discovery of cosmic X-ray sources.

[a] Nobel Prize in Physics with special significance for Astronomy and Astrophysics.

[b] Nobel Prize in Chemistry with special significance for Astronomy and Astrophysics.

Nova: An existing star, often very faint, that suddenly and unpredictably brightens by about 10 magnitudes and then gradually fades to its previous dim condition. A nova is a member of a close, double-star system in which mass flows from one star, a main-sequence star of late spectral type G, K or M, to another one, usually a white dwarf, producing an explosion on that star that makes it shine brightly for months. The explosion does not destroy either star, and the nova is never as intrinsically luminous as a supernova. Since the white dwarf can then suddenly become bright enough to be seen with the unaided eye, such events gave the impression that a new star had appeared where none was before. This explains the name *nova*, the Latin word for "new", but they are really old stars. *See* cataclysmic binary.

NRAO: Acronym for the National Radio Astronomy Observatory, United States. *See* National Radio Astronomy Observatory.

NRL: Acronym for the Naval Research Laboratory, United States.

NSF: Acronym for the National Science Foundation, United States.

Nuclear energy: The energy obtained by nuclear reactions, the source of the Sun's luminosity.

Nuclear fission: A reaction in which a heavy atomic nucleus is broken apart into two or more lighter nuclei, releasing energy. This process powers man-made nuclear reactors on Earth. *See* fission.

Nuclear force: The force that binds protons and neutrons within atomic nuclei, and which is effective only at distances less than 10^{-15} meters.

Nuclear fusion: The process by which two or more light atomic nuclei fuse together to make a heavier atomic nucleus, releasing energy. This process occurs at the high temperatures and pressures found near the centers of stars, and it makes most stars shine, including the Sun. All main-sequence stars shine by the nuclear fusion of light hydrogen nuclei into heavier helium nuclei, with the mass difference released as energy. *See* carbon-nitrogen-oxygen cycle, fusion, and proton-proton chain.

Nucleon: A particle in or from the nucleus of an atom, either a proton or a neutron. *See* atomic mass number.

Nucleosynthesis: The transformation of one chemical element or isotope into another by naturally occurring nuclear reactions. Heavier elements are built up from lighter ones by fusion reactions, while fission reactions break up heavy elements into lighter ones. Fusion reactions inside stars have created elements heavier than helium. Although helium is also synthesized from hydrogen inside stars, most of the helium in the Universe was created by nucleosynthesis during the first few minutes following the Big Bang. *See* Big-Bang nucleosynthesis and primordial nucleosynthesis.

Nucleus: The central, massive part of an atom. The atomic nucleus is composed of protons and (except for hydrogen) neutrons bound together by the strong nuclear force, and containing nearly all of an atom's mass. An atom's negatively charged electrons surround the nucleus, which has a positive electrical charge. The central region of a galaxy is also known as the nucleus of the galaxy.

Nulling: The use of interferometry to remove the bright light of a stellar disk in order to look for possible planets close to it. *See* interferometer and interferometry.

Nutation: In 1748 the English astronomer **James Bradley (1693–1762)** announced his discovery of a periodic oscillation of the Earth's rotational axis, which is superimposed on the precession of the axis. Bradley called the oscillation a nutation and attributed it to the gravitational interaction of the Moon with the Earth's equatorial bulge. Nutation has a variety of periods up to 18.6 years, and the 18.6-year term has a value of about 9.20 seconds of arc. Shorter periodicities of smaller amplitude are attributed to the gravitational interaction of the Sun with the Earth's equatorial bulge. The celestial coordinates of an object have to be corrected for nutation as well as precession. *See* celestial coordinates and precession.

O

OB association: A group of massive, luminous O and B stars. In 1949 the Armenian astronomer **Viktor Ambartsumian (1908–1996)** noticed that these associations cannot be bound together by their own gravitation, and that they must be slowly dispersing and have a relatively young age. The Dutch astronomer **Adriann Blaauw (1914–)** subsequently measured the expansion velocities of some OB associations, inferring that it took about a million years for them to disperse to their present size. Since this is comparable to the age of the individual stars, the associations and their component stars must have formed together.

Obliquity of the ecliptic: The angle between the plane of the ecliptic and the celestial equator, about 23.5 degrees. *See* celestial coordinates and ecliptic.

Occultation: The temporary disappearance, either complete or partial, of a celestial body when another nearer one moves between it and an observer and cuts off its light. A planet may occult a star or its own satellites, and the Moon occults nebulae, stars and cosmic radio sources. In 1679 the Danish astronomer **Ole Römer (1644–1710)** used the inconstant interval between Jupiter's occultation of its satellite Io to estimate the velocity of light. In 1977 the American astronomer **James L. Elliot (1943–)** unexpectedly discovered the rings of Uranus just before and after observing the planet's occultation of a star.

Olbers' paradox: According to Olbers' paradox, the sky in an infinite, uniform Universe should be covered by stars shining as brightly as our Sun, so the dark night sky should be as bright as the day. The paradox is named for the German astronomer **Heinrich Wilhelm Olbers (1758–1840)** who discussed it in 1823. It is resolved by the expansion of the Universe, which redshifts the most intense light of distant galaxies out of the visible part of the spectrum. *See* Doppler effect.

Olympus Mons: The largest volcano on Mars, rising more than 24 kilometers above its surroundings, almost three times the elevation of Mount Everest and five times higher than Hawaii's Mauna Loa volcano. The word *Mons* is Latin for "mountain", and Olympus was the abode of the Gods in Greek mythology.

Omega: A parameter, designated by the capital Greek letter omega, or Ω, that can describe the cosmic matter density and the shape of space in some situations. Omega is the ratio between the actual density of matter in the Universe and the critical mass density needed to close a homogeneous, isotropic Universe and eventually halt its expansion if the cosmological constant is zero and there is no dark energy. An omega equal to one or more corresponds to an open, ever-expanding Universe, in which the Universe does not have enough mass to reverse its expansion and will expand forever. If omega is exactly equal to one and there is no cosmological constant, then cosmic space is described by non-curved Euclidean geometry. Space curvature is needed for other values of omega with no dark energy or cosmological constant. *See* closed Universe, cosmological constant, critical mass density, and open Universe.

Omega Centauri: A globular star cluster about 17 000 light-years away in the direction of the constellation Centaurus. With an apparent visual magnitude of 3.7, it is the brightest globular star cluster in the sky, containing perhaps a million stars in a volume 200 light-years across.

Omega Nebula: An emission nebula, designated NGC 6618 or M17, located in the constellation Sagittarius. Other names for the Omega Nebula are the Horsehead Nebula and the Swan Nebula, all interpretations of its shape. The Omega Nebula measures 37 by 47 minutes of arc across and lies at a distance of about 5000 light-years. The Swiss astronomer **Philippe Loys de Chéseaux (1718–1751)** discovered it in 1746, and the French astronomer **Charles Messier (1730–1817)** rediscovered it in 1764. *See* emission nebula and Horsehead Nebula.

Oort comet cloud: A vast spherical cloud of comets belonging to the outer parts of the Solar System. The Oort comet cloud contains a million million, or 10^{12}, to ten million million, or 10^{13}, comets and extends out to roughly 100 000 astronomical units, or 0.5 parsecs, from the Sun, far beyond the Kuiper belt. The Oort comet cloud supplies the Solar System with long-period comets, which revolve around the Sun with periods greater than 200 Earth years. It is named for the Dutch astrophysicist **Jan Oort (1900–1992)**, who used the orbits of long-period comets to demonstrate, in 1950, that they originated in such a cloud. The Estonian astronomer **Ernst Öpik (1893–1985)** had previously shown, in 1932, that such a remote comet cloud is statistically stable against stellar perturbations. *See* comet, Kuiper belt, and long-period comet.

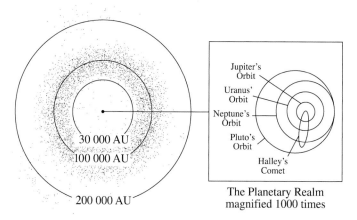

FIG. 41 **Oort comet cloud** More than 200 billion comets hibernate in the remote Oort comet cloud, shown here in cross section. It is located in the outer fringes of the solar system, at distances of about 100,000 AU from the Sun. By comparison, the distance to the nearest star, Proxima Centauri, is 0.27 million AU, while Neptune orbits the Sun at a mere 30 AU. The planetary realm therefore appears as an insignificant dot when compared to the comet cloud, and has to be magnified by a factor of 1,000 in order to be seen. This comet reservoir is named after the Dutch astronomer **Jan H. Oort (1900–1992)** who, in 1950, first postulated its existence.

Opacity: A measure of the ability of a gaseous atmosphere to absorb radiation and become opaque to it. A transparent gas has little or no opacity. Opacity also denotes the mean absorption of radiation, which is used with a diffusion equation to calculate the photon luminosity flow in the Sun or other stars.

Open cluster: A group of young stars in the spiral arms of our Milky Way Galaxy, which typically contains several hundred Population I stars in a volume that is several light-years across. An open cluster is also known as a galactic cluster, since it is located in the galactic plane. The Hyades, Pleiades and Praesepe are open clusters. In 1930 the Swiss-born American astronomer **Robert J. Trumpler (1886–1956)** used observations of the distances, sizes and reddening of open star clusters to convincingly demonstrate the interstellar absorption of starlight throughout the galactic plane. *See* Hyades, Pleiades, Population I star, Praesepe, and stellar populations.

Open Universe: An ever-expanding Universe in which the mass density of the Universe is less than the critical density, or omega is less than one. Space has hyperbolic curvature in an open Universe. *See* critical mass density and omega.

Opportunity Rover: *See Mars Exploration Rover mission.*

Opposition: A major planet is in opposition when it is aligned with the Earth and the Sun, and the direction of the planet is exactly opposite to the direction of the Sun. Venus and Mercury cannot be in opposition, because their orbits are closer to the Sun than the Earth's orbit. The other planets, such as Mars, are closest to the Earth at opposition, making it a favorable time for observations. *See* syzygy.

Optical astronomy: The study of visible light from objects in space. *See* visible light.

Optical depth: A measure of the radiation absorbed as it passes through a medium. A transparent medium has an optical depth of zero. The medium is optically thin when the optical depth is less than one, and optically thick when greater than one.

Optical radiation: Electromagnetic radiation that is visible to the human eye, with wavelengths of approximately 385 to 700 nanometers. *See* optical spectrum, optics, photosphere, and visible light.

Optical spectrum: Spectrum of a source that spans the visible wavelength range, approximately from 385 to 700 nanometers. *See* visible light.

Optics: The science that describes the lenses and mirrors that are used to focus and collect visible light.

Orbit: The path of a celestial body moving under the gravitational influence of a larger celestial body. Satellites orbit their planet, the planets orbit the Sun, and the two components of a binary star system orbit their common center of mass.

Orbital period: The time a body takes to revolve around another one. The Earth's orbital period around the Sun is one year, and more distant planets revolve with longer orbital periods in accordance with Kepler's third law. *See* Kepler's laws.

Orbiting Solar Observatory: Abbreviated *OSO*, a series of eight orbiting observatories launched between 1962 and 1971 by NASA to study the Sun in the ultraviolet and X-ray wavelengths from above the atmosphere.

Orion Nebula: An emission nebula visible to the naked eye as a diffuse glow marking the sword in the constellation Orion, and designated M42 or NGC 1976. It is a cloud of interstellar gas and dust illuminated by the ultraviolet light of bright young stars in the Trapezium. The Orion Nebula is located at a distance of about 1600 light-years. The northwestern part of the Orion Nebula is designated M43 or NGC 1982, and is separated from M42 by a dark lane of dust named the Fish Mouth from its shape. Star formation is still underway in the Orion nebula and in the dense molecular cloud in which it is embedded. *See* emission nebula, H II region, nebula, protoplanetary disk, and Trapezium.

Oscillating Universe: A closed Universe, whose expansion will eventually stop, followed by collapse and perhaps a Big-Bang rebound into another expansion. *See* closed Universe.

O star: A luminous, blue star of spectral type O, with a disk temperature that is much hotter than the Sun, or even a B star. Absorption lines of ionized helium characterize the visible spectrum of an O star. They are the hottest, brightest and most massive stars on the main sequence, with an effective temperature of between 32,000 and 50,000 degrees kelvin, an absolute luminosity of 50 000 to a million times that of the Sun, and a mass between 20 and 50 solar masses. They are relatively rare and short-lived, remaining on the main sequence for less than a million years. *See* Hertzsprung-Russell diagram and stellar classification.

Owl Nebula: A planetary nebula, designated M97 and NGC 3587, located in the constellation Ursa Major. The Owl Nebula is 3 minutes of arc across and gets its name from two adjacent dark patches, like eyes, that have the appearance of an owl's face. It has an apparent visual magnitude of 11 and a lies at a distance of about 1300 light-years. The central star has a magnitude of 16. *See* planetary nebula.

Oxygen: Element of atomic number eight and atomic mass number sixteen, denoted by the symbol O and the third most abundant element in the Universe after hydrogen and helium. Oxygen is synthesized within high mass, giant stars, which shine by converting helium into oxygen within their cores, and generated and dispersed into interstellar space during supernova explosions. Each molecule of water, or H_2O, consists of two atoms of hydrogen, H, and one of oxygen, O. Since humans breathe molecular oxygen, denoted O_2, it is a vital ingredient of the Earth's atmosphere, second only to nitrogen in abundance within our air. Plants produce molecular oxygen during photosynthesis, and the detection of substantial amounts of molecular oxygen in a planet's atmosphere would suggest the presence of life.

Ozone: A form of molecular oxygen containing three atoms of oxygen instead of the normal two, and hence denoted O_3, where O is an oxygen atom. It is created by the action of ultraviolet sunlight on the Earth's atmosphere, and this ozone layer shields the Earth's surface from deadly ultraviolet radiation from the Sun. The detection of substantial amounts of ozone in a planet's atmosphere would suggest the presence of life, which might be needed to produce the oxygen. *See* ozone hole and ozone layer.

Ozone depletion: *See* ozone hole.

Ozone hole: An exceptionally low concentration of ozone that forms above the South Pole in the local winter. The ozone hole was conclusively demonstrated in 1985, as the result of 27 years of continuous measurements of the ozone content above Halley Bay, Antarctica, begun by the British scientist **Gordon Miller Bourne "G. M. B." Dobson (1889–1976)** in 1957. Two American chemists, **Mario J. Molina (1943–)** and **F. Sherwood Rowland (1927–)** showed in 1974 that the chorine in man-made chemicals called chlorofluorocarbons, or CFCs for short, would destroy enormous amounts of ozone in the atmosphere. The CFCs were limited and eventually banned by an international treaty known as the *Montreal Protocol,* first signed in 1987 and strengthened in later years. *See* ozone layer.

Ozone layer: A region in the lower part of Earth's stratosphere, located at about 20 to 60 kilometers above sea level, where the greatest concentration of ozone appears. The ozone layer shields the Earth's surface from the Sun's energetic ultraviolet rays.

Pair annihilation: Mutual destruction of an electron and positron with the formation of energetic gamma rays that have a photon energy of 511 keV.

Pallas: The second asteroid to be discovered, by the German astronomer **Heinrich Wilhelm Olbers (1758–1840)**, with dimensions of 559 × 525 × 532 kilometers, a mean radius of 269 kilometers, and a mass density of 2710 kilograms per cubic meter. *See* asteroid.

Palomar Observatory: An observatory located on Mount Palomar, 90 miles southeast of the Mount Wilson Observatory, which is owned and operated by the California Institute of Technology. The largest telescope at the Palomar Observatory is the 5-meter (200-inch) **Hale telescope**, completed in 1949 and named after the American astronomer **George Ellery Hale (1868–1938)**. The 200-inch telescope played an important role in the discovery of quasars. A wide-field 1.2-meter (48-inch) **Schmidt telescope** on Mount Palomar was used for the all-sky Palomar Observatory Sky Survey, sponsored by the National Geographic Institute and completed in 1954; its red and blue plates were important in identifying the visible-light counterparts of the first radio galaxies and quasars. *See* quasars and Schmidt telescope.

Parallax: The small angular displacement of the apparent position of a planet or star when viewed from different locations, often designated by the symbol π. The distance of a planet can be inferred from parallax observations at widely separated locations on the Earth. The geocentric parallax is defined as the angular size of the Earth's equatorial radius at the distance of the object under consideration. The solar parallax is the Sun's geocentric parallax, defined as the angular size of the Earth's equatorial radius from a distance of 1 astronomical unit, with a value of 8.794 148 seconds of arc. Stellar parallax is determined from the angular displacement of a nearby star with respect to distant ones when observed at intervals of six months on opposite sides of the Earth's orbit. Such a parallax is known as the annual parallax. The distance of the star can be inferred from the measured annual parallax and the astronomical unit. The larger the parallax, the closer the star lies to Earth. The parallax, or distance, of a star can also be inferred from its spectral type, which indicates its absolute or intrinsic luminosity, and apparent luminosity using the inverse square law. The parallax of about 120 000 stars has been measured with instruments aboard the *HIPPARCOS* satellite, with a precision of 0.001 seconds of arc. *See* absolute luminosity, Alpha Centauri, annual parallax, apparent luminosity, astronomical unit, Barnard's Star, *HIPPARCOS*, inverse square law, Kapteyn's Star, Proxima Centauri, 61 Cygni, solar parallax, and spectroscopic parallax.

Parkes Observatory: A radio astronomy observatory in New South Wales, whose main telescope, a 64-meter (210-feet) dish, was completed in 1961 and used to receive radio communications during the historic *Apollo 11* landing on the Moon, on 21 July

1969, and the **Huygens Probe** landing on Titan on 14 January 2005. The radio telescope has also been active in pulsar research, discovering about 1000 of them, and as a component in Very Long Baseline Interferometry. The Australia Telescope National Facility, a division of the Commonwealth Scientific and Industrial Research Organization, operates the Parkes Observatory. *See* **Apollo Missions**, Australia Telescope National Facility, **Huygens Probe**, pulsar, radio interferometer, radio telescope, and Very Long Baseline Interferometry.

Parsec: A unit of distance used in stellar astronomy, defined as the distance at which one astronomical unit, or the mean Earth-Sun separation, subtends an angle of one second of arc. The distance of a nearby star in parsecs is given by the reciprocal of its annual parallax in seconds of arc. One parsec is equivalent to 3.0857×10^{16} meters, 3.261 633 light years, and 206 265 astronomical units. The nearest star other than the Sun is Proxima Centauri with a distance of 1.316 parsecs. Distances comparable to the extent of our Galaxy are measured in units of kiloparsecs, abbreviated kpc, where 1 kpc = one thousand parsecs = 1000 parsecs, and intergalactic distances are specified in units of Megaparsecs, abbreviated Mpc, with 1 Mpc = 1 000 000 parsecs = 1 million parsecs. *See* annual parallax.

Particle: Fundamental unit of matter and energy, including subatomic particles that are smaller than the atom.

Particle accelerator: A device that uses electric fields to propel charged, sub-atomic particles to great energies. In 1932 the English physicist **John Cockcroft (1897–1967)** and the Irish physicist **Ernest Walton (1903–1995)** used a particle accelerator to produce the first transmutation of an atomic nucleus with artificially accelerated atomic particles, for which they shared the 1951 Nobel Prize in Physics. In 1929 the American physicist **Ernest O. Lawrence (1901–1958)** invented the first circular particle accelerator, known as the cyclotron, capable of accelerating particles to very high energy, and leading to the discovery of numerous radioactive isotopes. At velocities near the velocity of light, the particles increase their mass and lose synchronization with the acceleration process, which is corrected in synchrotron particle accelerators independently invented by **Vladimer Veksler (1907–1966)** and **Edwin McMillan (1907–1991)**. *See* synchrotron radiation.

Particle physics: The branch of physics that deals with the smallest known structures of matter and energy.

Pathfinder: *See* **Mars Pathfinder**.

Pauli exclusion principle: *See* exclusion principle.

Pelican Nebula: An emission nebula, designated IC 5070, in the constellation Cygnus, adjacent to the North America Nebula, designated NGC 7000. *See* emission nebula and North America Nebula.

Penumbra: The lighter periphery of a sunspot seen in white light, surrounding the darker umbra. Also the outer part of a shadow cast during an eclipse, where only a partial eclipse is seen. *See* lunar eclipse, solar eclipse, sunspot and umbra.

Perfect cosmological principle: On a sufficiently large scale the Universe is the same in all directions and for all times. This perfect cosmological principle was introduced in 1948 by **Hermann Bondi (1919–2005)** and **Thomas Gold (1920–2004)**, two Austrian-born astrophysicists then at Cambridge University, when proposing the Steady State cosmology, which has since been discounted. *See* cosmological principle and Steady State.

Periastron: *See* perihelion.

Perihelion: For a planet, comet or other object orbiting the Sun, the point in the orbit that is closest to the Sun. The periastron is the closest point for a body orbiting a star other than the Sun. *See* aphelion.

Period: The time to complete one cycle of any periodic phenomenon, such as one orbit of a planet or one pulsation of a variable star.

Period-luminosity relation: A relationship between the period of light variation and the absolute luminosity of a variable star. Cepheid variable stars have a period-luminosity relation, in which the longer the pulsation period the more intrinsically luminous the star. The American astronomer **Henrietta Swan Leavitt (1868–1921)** discovered this period-luminosity relation in 1912. A Cepheid star's distance can be determined using this relation with its observed apparent luminosity and its variation period. *See* Cepheid variable star.

Perseids: One of the best meteor showers, occurring between 25 July and 20 August, with a maximum on 12 August, popularly known as the night of the shooting stars. In the Middle Ages, the shower was known as the tears of St Lawrence, who was martyred on 10 August 258. The radiant of the Perseids meteor shower occurs in the constellation Perseus, and it is associated with the orbital debris of Comet 109P/Swift-Tuttle. *See* meteor shower.

Phase: The illuminated portion of the Moon or the planets, as seen from the Earth. The phases of the Moon occur in a repeating pattern that depends on its orbital position, and they are known as the new, crescent, half, gibbous and full phases. *See* Moon.

Phobos: The largest and innermost of the two tiny satellites of Mars, discovered by the American astronomer **Asaph Hall (1829–1907)** in 1877. *Phobos* is the Greek word for "fear", and the name of one of the attendants of the Greek god of war, Ares, in Homer's *Illiad*. The satellite Phobos has dimensions of $13.3 \times 11.1 \times 9.3$ kilometers and a mean mass density of 1905 kilograms per cubic meter. Phobos moves around Mars three times faster than the planet rotates, and it is now moving toward Mars. In about 50 million years, Phobos will either crash into the surface of Mars or be torn apart by the planet's gravity to make a ring around Mars. Phobos could be a captured asteroid.

Photodissociation: The breakdown of molecules into their component atoms due to the absorption of light, especially ultraviolet sunlight.

Photoionization: The ionization of an atom by the absorption of a photon of electromagnetic radiation. Ionization can take place only if the photon carries at least the energy corresponding to the ionization potential of the atom. The ionization potential is the minimum energy required to overcome the force binding the electron within the atom. *See* ionization and ionization potential.

Photon: A corpuscle, particle, or quantum, of light or other electromagnetic radiation. A photon is a discrete unit or quantity of electromagnetic energy associated with waves of electromagnetic radiation. Photons have no electric charge and travel at the speed of light. Short wavelength, or high frequency, photons carry more energy than long wavelength, or low frequency, photons. *See* photon energy and Planck's constant.

Photon decoupling: *See* decoupling.

Photon energy: The energy of radiation of a particular frequency or wavelength. The photon energy is equal to the product of Planck's constant, 6.626×10^{-34} J s, and the frequency of the radiation. Short wavelength, or high frequency, photons have more energy than long wavelength, or low frequency, photons. *See* Planck's constant.

Photosphere: That part of the Sun from which visible light originates. The photosphere emits white light, or all the colors combined. It is the intensely bright, visible portion of the Sun, and the place where most of the Sun's energy escapes into space, the lowest layer of the Sun's atmosphere viewed in white light. The photosphere of any star is the region that gives rise to its visible continuum radiation. About 500 kilometers thick, the solar photosphere is a zone where the gaseous layers change from being completely opaque to radiation to being transparent. The effective temperature of the Sun's photosphere is 5780 kelvin. The continuum radiation of the photosphere is absorbed at certain wavelengths by slightly cooler gas just above it, producing the dark Fraunhofer lines. Sunspots, faculae, granules and the supergranulation are observed in the photosphere, and magnetograms describe the magnetic field in the photosphere. The in and out heaving motions, or oscillations, of the photosphere are used to decipher the internal dynamics and structure of the Sun. The transparent chromosphere lies just above the photosphere, and the transparent corona is found above the chromosphere. *See* chromosphere, corona, Fraunhofer line, granule, helioseismology, magnetogram, sunspot, supergranulation, and white light.

Photosynthesis: The chemical process whereby green plants use sunlight to manufacture carbohydrates from carbon dioxide and water and release oxygen as a by-product.

Pic du Midi Observatory: An observatory located in the French Pyrénées at an altitude of 2860 meters (9300 feet), and operated by the University of Tolouse. It has a 2-meter (78-inch) reflector, which was used for important early investigations of the Moon and planets.

Pinwheel Galaxy: A face-on spiral galaxy designated M101 and NGC 5457, located at a distance of about 25 million light-years. The Triangulum spiral galaxy, designated M33, is also sometimes called the Pinwheel Galaxy. *See* Messier Objects and Triangulum Galaxy.

Plage: From the French word for "beach", a plage is a bright, dense region in the chromosphere found above sunspots or other active areas of the solar photosphere. It always accompanies and outlives sunspot groups. Plages appear much brighter in the mono-chromatic light of the hydrogen-alpha line or the calcium H and K lines than the surrounding parts of the chromosphere. Plages are associated with faculae, which occur just below them in the photosphere, and are found in regions of enhanced magnetic field. *See* calcium H and K lines, chromosphere, faculae, H and K lines, and hydrogen-alpha line.

Planck epoch: Also called the Planck time, the first instant following the Big Bang when the cosmic matter density was so high that the gravitational force acted as strongly as the other fundamental forces.

Planck Mission: A collaborative ESA and NASA mission that is in development, with a projected launch in the year 2007. ***Planck*** will survey the entire sky and produce maps of it at nine different frequencies, providing detailed information on the three-degree cosmic microwave background radiation as well as the origin and evolution of the observable Universe. The mission is named after the German theoretical physicist **Max Planck (1858–1947)**, who derived the spectrum of blackbody radiation, introducing the quanta of energy; he was awarded the 1918 Nobel Prize in Physics for this achievement.

Planck time: *See* Planck epoch.

Planck's constant: A fundamental constant giving the ratio between the energy and frequency of an electromagnetic photon, or quantum of light, first used by the German physicist **Max Planck (1858–1947)** in deriving the formula for the radiation spectrum of a perfect absorber, or blackbody. Planck's constant is denoted by the symbol h, and has a value of $h = 6.626 \times 10^{-34}$ Joule second or $h = 6.626 \times 10^{-27}$ erg second.

Planet: An object orbiting the Sun or another star, which reflects the light that strikes it. A planet is not large enough to generate energy through nuclear fusion at its core, which means that it has to be less massive than about ten times Jupiter's mass. To be

TABLE 41   Planets – orbits

Planet	Semi-major Axis, a_p (AU)	Orbital Period, P_p (years)[a]	Mean Orbital Velocity (km s^{-1})[b]
Mercury	0.387 099	0.2409	47.89
Venus	0.723 332	0.6152	35.03
Earth	1.000 000	1.0000	29.79
Mars	1.523 688	1.8809	24.13
Jupiter	5.202 834	11.8622	13.06
Saturn	9.538 762	29.4577	9.64
Uranus	19.191 391	84.0139	6.81
Neptune	30.061 069	164.793	5.43
Pluto	39.529 402	247.7	4.74

[a] One tropical year $= 3.155\,692\,597\,47 \times 10^7$ seconds $= 365.24219$ days.
[b] 1.0 km s$^{-1} = 1,000$ meters per second.

TABLE 42 Planets – physical properties[a]

Terrestrial Planets	Mercury	Venus	Earth	Mars
Angular Diameter (seconds of arc)[b]	10.9	61.0		17.88
Equatorial Radius, R_p (km)	2440	6052	6378	3396
Equatorial Radius, R_p (R_E = 1.0)	0.382	0.949	1.000	0.532
Mass, M_p (10^{23} kg)	3.302	48.685	59.736	6.4185
Mass, M_p (M_E = 1.0)	0.0553	0.8150	1.00000	0.1074
Mean Mass Density (kg m^{-3})	5427	5204	5515	3934

Giant Planets	Jupiter	Saturn	Uranus	Neptune
Angular Diameter (seconds of arc)[b]	46.86	19.52	3.60	2.12
Equatorial Radius, R_p (km)	71 492	60 268	25 559	24 766
Equatorial Radius, R_p (R_E = 1.0)	11.209	9.449	4.007	3.883
Mass, M_p (10^{23} kg)	18 986	5684.6	868.32	1024.3
Mass, M_p (M_E = 1.0)	317.710	95.162	14.535	17.141
Mean Mass Density (kg m^{-3})	1326	687	1318	1638

[a] Adapted from Lang, Kenneth R., *The Cambridge Guide to the Solar System*, New York, Cambridge University Press 2003, p. 24.

[b] The largest angular diameter seen from the Earth when the planet is at its closest approach to Earth.

called a planet, the object must have a large enough mass to be forced by its gravity into becoming more or less spherical in shape, which means the object must be more massive than the largest asteroid; but asteroids are sometimes called minor planets. Some definitions demand that a planet should have an atmosphere. A planet is formed by the accretion of planetesimals within a protoplanetary disc, the rotating debris left after a star has formed. *See* asteroid, planetesimal, and protoplanetary disk.

Planetary nebula: A shell of illuminated gas that surrounds old and dying stars. The first planetary nebula, designated NGC 7009, was discovered in 1785 by the German-born English astronomer **William Herschel (1738–1822)**, describing it as "planetary" because its round, disk-like shape resembled a planet as seen in a small telescope, but there is no physical connection with planets. In 1864 the English astronomer **William Huggins (1824–1910)** discovered three emission lines radiated from the planetary nebula now known as NGC 6543, showing that it is made of low-density gas. One of lines was attributed to hydrogen, but the other two suggested the existence of an undiscovered element that became known as "nebulium". It wasn't until 1927 that the Dutch astronomer **Herman Zanstra (1894–1972)** used quantum theory to explain the hydrogen lines of diffuse nebulae, including the planetary nebulae, whose hydrogen atoms are ionized by the ultraviolet light of bright, central stars followed by the recombination of free electrons and protons to radiate hydrogen lines and make hydrogen atoms. In 1928 the American astronomer **Ira S. Bowen (1893–1973)** showed that the nebulium lines arise from ionized atoms of oxygen and nitrogen undergoing "forbidden" transitions in the low-density gas. A planetary nebula is usually ejected from a dying red giant star. Radiation from the newly exposed, hot blue core excites the gas in the planetary nebula, causing it to glow. The central stars of planetary nebulae are hot and blue, in transition between a red giant star and a white dwarf. Commonly known planetary nebulae are the Dumbbell Nebula – M27 and NGC 6853, the Helix Nebula – NGC 7293, the Owl Nebula – M97 and NGC 3587, the Ring Nebula in Lyra – M57 and NGC 6720 and the Saturn Nebula – NGC 7009. *See* Cat's Eye Nebula, Dumbbell Nebula, Helix Nebula, Hourglass Nebula, Owl Nebula, Ring Nebula and Saturn Nebula.

Planetesimal: A small body from a few meters to several kilometers in size, formed from dust and ice, if present, contained in a protoplanetary disk, the rotating debris left after a star has formed. Planets form from the accretion or gathering together of planetesimals and by collision and gravitational attraction of protoplanets formed by the planetesimals. *See* protoplanet, protoplanetary disk, and solar nebula.

Plasma: A completely ionized gas, consisting of electrons that have been pulled free of atoms, and ions, in which the temperature is too high for neutral, un-ionized atoms to exist. The high temperatures result in atoms losing their normal complement of electrons to leave positively charged ions behind, and it is too hot for the free electrons and ions to join together and form permanent atoms. The interior of the Sun and other main-sequence stars is a plasma consisting mainly of electrons and protons, the nuclei of hydrogen atoms. Plasma has been called the fourth state of matter, in addition to solid, liquid, and gas. Since the total negative electrical charge of the free electrons is equal to the total positive charge of the ions, plasma is electrically neutral over a sufficiently large volume. Most of the matter in the Universe is in the plasma state.

Plasma beta: The ratio of gas pressure to the pressure of the magnetic field within a plasma. The gas pressure increases with temperature, and the magnetic pressure is proportional to the square of the magnetic field strength.

Plasmasphere: A region of high-density, cold plasma, consisting of protons and electrons, which surrounds the Earth above the ionosphere at altitudes greater than 1000 kilometers. The plasmasphere extends out to between three and seven Earth radii.

Plate tectonics: According to this theory, the Earth's outer shell is subdivided into a mosaic of large rigid plates that are in continual motion, producing earthquakes and building mountains at their boundaries by *tectonics,* from the Greek word for "carpenter" or "building". The moving plates carry continents on their backs, accounting for continental drift, and are pushed along by the spreading sea floor. The Canadian geologist **John Tuzo Wilson (1908–1993)** introduced the concept of plate tectonics in 1965. *See* continental drift and sea-floor spreading.

Pleiades: A nearby open cluster of stars in the constellation Taurus that lies about 385 light-years away. The Pleiades is also known as the Seven Sisters whose seven bright blue-white B stars are named after the seven daughters of the god Atlas and the mortal Pleione in Greek mythology. In order of brightness, they are Alcyone, Electra, Maia, Merope, Taygeta, Celaeno, and Serope. The bright stars are relatively young, with an

TABLE 43 Planetary nebulae[a]

NGC/IC	Name	RA (2000) h m s	Dec. (2000) ° ′ ″	m_v	V_{exp} (km s^{-1})	Diam. (″)	Dist. (kpc)
Hb 5	Double Bubble	17 47 56.2	−29 59 42	—	—/—	15.	0.69
He3-1357	Stingray	17 16 21.1	−59 29 21	10.8	—/—	10.	—
IC 3568		12 31 46.6	+82 50 22	13.5	08/—	10.	2.7
IC 418	Spirograph	05 25 09.5	−12 44 15	10.2	06/12	12.	0.6
IC 4406	Retina	14 19 15.5	−43 55 27	12.5	06/14	35.	2.4
MyCn18	Hourglass	13 39 35.1	−67 22 52	13.0	10/—	4	1.60
Mz 3	Ant	16 17 17.4	−51 59 10	17.6	—/—	25.	1.27
NGC 2392	Eskimo	07 26 13.2	+21 00 51	10.5	70/90	19.5	1.2
NGC 2440		07 41 55.4	−18 12 33	17.5	22/—	16	0.74
NGC 3242	Eye of Jupiter	10 24 45.8	−18 38 37	12.3	20/28	25	1.1
NGC 3587	Owl	11 14 47.7	+55 01 09	16.0	23/40	170.	0.62
NGC 3918	Blue Planetary	11 50 17.8	−57 10 57	15.7	24/26	19	0.54
NGC 5307		13 51 03.3	−51 12 21	10.	11/—	12	1.60
NGC 6210		16 44 29.5	+23 47 59	12.7	21/36	6.2	2.0
NGC 6369	Little Ghost	11 21 35.8	+18 27 27	14.	42/—	38.	0.66
NGC 6543	Cat's Eye	17 58 33.3	+66 37 59	11.2	20/20	19	0.55
NGC 6720	Ring	18 51 43.6	+32 57 55	15.0	27/—	76.	0.9
NGC 6751		19 03 15.0	−06 04 07	15.5	40/38	20.5	2.6
NGC 6826	Blinking Eye	19 44 48.3	+50 31 30	10.6	11/—	25	0.7
NGC 6853	Dumbbell	19 59 36.3	+22 43 16	10.5	15/32	402.	0.26
NGC 7009	Saturn	21 04 10.9	−11 21 48	12.8	21/20	28	0.42
NGC 7293	Helix	22 29 38.6	−20 50 14	13.5	14/24	980.	0.16

[a]Courtesy of Arsen R. Hajian, US Naval Observatory. The central star has magnitude, m_v. The expansion velocity, V_{exp}, is given for the light of [OIII]/[NII] in kilometers per second, abbreviated km s^{-1}, the angular diameter is in seconds of arc, denoted ″, and the distance is in units of kiloparsec, or kpc, where 1 kpc = 3260 light-years.

estimated age of about 70 million years, still enmeshed in a cloud of dusty gas that reflects the starlight and creates the Pleiades or Merope reflection nebula. *See* open cluster and reflection nebula.

Plurality of worlds: Hypothesis that the Universe contains inhabited planets other than the Earth.

Pluto: A small planet discovered at the outer edge of the planetary system on 18 February 1930 by **Clyde William Tombaugh (1906–1997)** using the 0.33-meter (13-inch) photographic telescope at the Lowell Observatory. Pluto moves in a highly elongated orbit that carries it between 29.7 and 49.3 AU from the Sun, within and outside the orbit of Neptune. The American astronomer **William Henry Pickering (1858–1938)**

predicted the existence of such a planet in 1909, calling it Planet O, and it was also predicted by the wealthy American **Percival Lowell (1855–1916)** in 1915, dubbing it Planet X. Both Pickering and Lowell suggested that such a planet might be producing noticeable perturbations in the apparent motion of Uranus beyond those caused by Neptune. However, after accounting for the gravitational effects of Neptune, using a precise mass obtained when *Voyager 2* encountered the planet in 1989, the small unexplained differences between the predicted and observed locations of Uranus disappeared. This means that Neptune is the outermost major planet in the Solar System. In 1978 **James W. Christy (1938–)** of the United States Naval Observatory, noticed a varying elongation on images taken to refine the orbit of Pluto, realizing that the changing orientation of the elongation implied that Pluto has a satellite and announcing the discovery on 7 July 1978. **Robert S. Harrington (1942–1993)**, another astronomer at the Naval Observatory, calculated that Pluto and its moon would undergo a series of mutual eclipses and occultations, beginning in early 1985. The first successful observation of one of these events, on 17 February 1985, fully confirmed the existence of the satellite, which was officially named Charon. It moves around Pluto at a distance of 19 640 kilometers, once every 6.387 Earth days. Charon's slow revolution about Pluto is a result of Pluto's small mass – only 0.2 percent (0.002) of the Earth's mass. Pluto is also relatively small, with a radius of about 1160 kilometers or about one-fifth the size of the Earth. So Pluto is most likely just one of many similar small objects located in the Kuiper belt just outside the orbit of Neptune. *See* Kuiper belt.

Polaris: The bright star toward which the northern end of the Earth's rotation axis now points, also called the North Star. It is located at the end of the handle of the Little Dipper in Ursa Minor, and is hence also designated as Alpha Ursae Minoris. Polaris is a yellow-white F5 supergiant star with an apparent visual magnitude of 1.98. At a distance of 425 light-years, Polaris has an absolute magnitude of –3.59, shining 2200 times more brightly then the Sun. Polaris is the sky's brightest Cepheid variable star, but it changes by a mere 0.03 magnitudes, with a period of 3.97 days, an effect that is invisible to the unaided eye. Polaris is about half a degree away from the north celestial pole, to which the rotation axis points, but precession is bringing Polaris closer to the pole, and it will be closest around the year 2100. Also as the result of the 26 000-year precession, the Earth's rotation axis is slowly moving across the sky, and in 13 000 years the celestial north pole will be located near the bright star Vega and 47 degrees away from Polaris. *See* North Star, pole star, and precession.

Polarity: The directionality of a magnet or magnetic field, being north- or south-seeking. According to one convention, magnetic lines of force emerge from regions of positive north polarity and re-enter regions of negative south polarity. However, a magnet in a compass is marked so that it points toward the Earth's magnetic North Pole.

Polarization: When radiation is polarized its electromagnetic vibrations are not randomly oriented, but instead have a preferred direction, oscillating in a plane. Synchrotron radiation is polarized.

Pole star: The nearest naked-eye star to the projection of the Earth's rotational axis on the celestial sphere, or the closest bright star to the north or south celestial pole.

Polaris is now the North Pole star, but in about 13 000 years it will be Vega. *See* Polaris and precession.

Pollux: A giant K0 star 34 light-years away in the direction of the constellation Gemini. Pollux, also named Beta Geminorum, has an apparent visual magnitude of +1.16, an absolute visual magnitude of +1.1, and has an intrinsic luminosity 40 times that of the Sun.

Population I star: A relatively young star containing a high abundance of metals that tends to be located in the arms and disk of spiral galaxies. The Sun is a Population I star. *See* stellar populations.

Population II star: A relatively old star whose abundance of heavy elements is much less than that of the Sun and other Population I stars, consisting almost solely of hydrogen and helium. They tend to be located in elliptical galaxies, the central regions of spiral galaxies and globular star clusters. *See* globular cluster, and stellar populations.

Positron: A positively charged anti-particle of the electron, named a positron for a "positive electron". The positron is a subatomic particle with the mass of the electron but an equal positive electric charge. *See* antimatter and beta decay.

Potential energy: The energy that an object possesses as a result of its position.

Poynting-Robertson effect: In 1903 the English physicist **John Poynting (1852–1914)** suggested that the Sun's radiation causes small particles orbiting the Sun to gradually approach it and eventually plunge into the Sun. When dust particles between a micron and a centimeter, or between 0.000001 and 0.01 meters, in size absorb energy from the Sun and re-radiate it, they lose kinetic energy and their orbital radius becomes slightly smaller. The American physicist **Howard Robertson (1903–1961)** developed this idea in 1937, deriving this so-called Poynting-Robertson effect from the theory of relativity.

Praesepe: An open star cluster, designated M44, which is about 577 light-years away in the constellation Cancer. It is also known as the Beehive or manger. *See* open cluster.

Precession: The gyration of the Earth's rotational axis in space with a period of about 26 000 years, discovered in the second century BC by the Greek astronomer **Hipparchus (c. 190–c. 120 BC)**. The precession is produced by the tidal action of the Moon and Sun on the equatorial bulge of the spinning Earth, causing it to wobble like a top. As a result, the Earth's axis of rotation sweeps out a cone in space, with an opening angle equal to the obliquity of the ecliptic, or about 23.5 degrees, completing one circuit in about 26 000 years. This precession causes an east to west movement of the equinoxes along the ecliptic, at the rate of about 50.3 seconds of arc per year, or 3.35 seconds of time per year, and it is therefore sometimes called the precession of the equinoxes. As the result of precession, the celestial coordinates of cosmic objects are slowly changing, so the stars that we now see in mid-winter are not the ones seen by the ancients in mid-winter, although the shift is too slow to be noticed by the naked eye in an individual's lifetime. A periodic oscillation of the Earth's rotational axis, known as nutation, is superimposed upon its precession. *See* celestial coordinate, equinox, nutation, Polaris, right ascension, and Vernal Equinox.

FIG. 42 Precession The pole of the Earth's rotation traces out a circle on the sky once every 26 000 years, sweeping out a cone with an angular radius of about 23.5 degrees. The Greek astronomer **Hipparchus (lived c. 146 BC)** discovered this precession in the 2nd century BC. The north celestial pole now lies near the bright star Polaris, but as the result of precession the rotational axis will point towards another bright north star, Vega, in roughly 13 000 years. This motion of the Earth's rotational axis also causes a precession of the equinoxes, and therefore a slow change in the celestial coordinates of any cosmic object. The Vernal Equinox will move through all the signs of the zodiac in about 26 000 years.

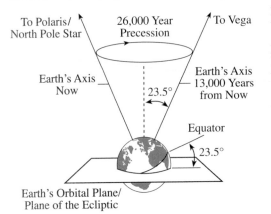

Prime focus: The focal point of the objective lens or primary mirror of a telescope, without any other optical counterpart such as a secondary mirror.

Primordial nucleosynthesis: The creation of light elements just minutes after the Big Bang, including all of the hydrogen and deuterium now present in the Universe and most of the helium. *See* nucleosynthesis.

Procyon: The eighth-brightest star in the sky, with an apparent visual magnitude of 0.38, lying about 11.4 light-years in the direction of the constellation Canis Minor. The name *Procyon* is derived from the ancient Greek for "before the dog" owing to the fact that Procyon rises before the dog star Sirius. Procyon is also known as Alpha Canis Minoris and the "Little Dog Star" or "Lesser Dog Star" from the constellation name *Canis Minor,* meaning "Little Dog." Procyon is a subgiant star of spectral type F5, an absolute magnitude of +2.71, and an intrinsic luminosity seven times that of the Sun. Procyon is a spectroscopic binary star with a period of 40.4 years and a white dwarf companion.

Prominence: A region of high-density, cool, 10 000-kelvin gas embedded in the lower part of the low-density, hot, million-degree solar corona. A prominence is apparently suspended above the photosphere by magnetic fields. It can be seen at the limb of the Sun during a solar eclipse or with a coronagraph. A prominence is a filament viewed on the limb of the Sun in the light of the hydrogen-alpha line. A quiescent prominence occurs away from active regions and can last for weeks or many months. An active prominence is a short-lived, high-speed eruption associated with active regions, sunspots and flares. An eruptive prominence is often associated with a coronal mass ejection. *See* active region, coronagraph, coronal mass ejection, filament, and hydrogen-alpha line.

Proper motion: The apparent motion of a star across the line of sight, and on the celestial sphere, often designated with the symbol μ. Proper motion is perpendicular to the line of sight to the star. When a star's distance is known, the proper motion can be converted into a transverse velocity. The two stars with the largest proper motions are Barnard's star and Kapteyn's star, both red dwarf stars, with proper motions of 10.3 and 8.67 seconds of arc per year, respectively. A few hundred stars have proper motions greater than 1 second of arc per year, but most of them have proper motions of less than 0.1 seconds of arc per year. Instruments aboard the *HIPPARCOS* satellite have determined the proper motions of 120 000 stars with a precision of 0.002 seconds of arc. *See* Barnard's Star, *HIPPARCOS*, and Kapteyn's Star.

Protogalaxy: Matter that is beginning to come together to form a galaxy.

Proton: A subatomic particle with positive electric charge located in the nucleus of every atom or set free from it. In 1920 the New-Zealand born British physicist **Ernest Rutherford (1871–1937)** announced that the massive nuclei of all atoms are composed of the positively charged hydrogen nuclei, which he named protons. The nucleus of a hydrogen atom is a proton. The atomic number, denoted by Z, specifies the number of protons in the nucleus of any atom. The proton has a mass of 1.672623×10^{-27} kilograms, and a mass-energy equivalent to 938.3 MeV. A proton is 1836 times more massive than an electron. *See* atomic number and nucleon.

Proton-proton chain: Abbreviated p-p chain, a sequence of thermonuclear reactions in which hydrogen nuclei, or protons, are transformed into helium nuclei. The German-born American physicist **Hans Bethe (1906–2005)** proposed the details of the proton-proton chain of thermonuclear reactions in 1938–39. It is the main source of energy in the Sun and of all main-sequence stars with a mass less than 1.5 times the Sun's mass. In the first stage of the proton-proton chain, two protons combine to form a deuterium nucleus, releasing a positron, a neutrino and radiation. In the second stage the deuterium nucleus combines with a proton to form an isotope of helium, denoted ^{3}He, again releasing radiation. In the last stage, two ^{3}He nuclei combine to form the normal helium nucleus, denoted ^{4}He, releasing two protons and radiation. Overall, four protons are converted into one helium nucleus, with the mass difference released as energy. *See* hydrogen burning.

Proton-proton reaction: *See* proton-proton chain.

Protoplanet: A body formed by the accretion of planetesimals within a protoplanetary disk. The collision and gravitational coalescence of protoplanets resulted in the formation of planets, within our Solar System and around other stars. *See* planetesimal and protoplanetary disk.

Protoplanetary disk: A disk of dust and gas rotating about a young star, representing the early stages of planet formation through the gathering together of planetesimals and protoplanets. The protoplanetary disk of the Solar System is known as the solar nebula. Protoplanetary disks have been detected at infrared wavelengths, first in 1983 with the *InfraRed Astronomical Satellite*, and since 2003 with the *Spitzer Space Telescope*. Instruments aboard the *Hubble Space Telescope* have discovered numerous protoplanetary disks at visible-light wavelengths, particularly surrounding young stars in the Orion Nebula. *See* Orion Nebula, planetesimal, protoplanet, solar nebula and Vega.

Protostar: An embryonic star in the process of formation, which is luminous owing to the release of gravitational potential energy from the infall of nebula material. A protostar has not yet begun to shine by nuclear fusion in its core. Protostars are formed when dense clumps of gas and dust collapse within massive molecular clouds. They are exceptionally bright at infrared wavelengths. Observations at infrared wavelengths can detect protostars within dark clouds. *See* molecular cloud.

Proxima Centauri: The closest star to the Earth other than the Sun, with a parallax of 0.772 seconds of arc and located at a distance of 4.22 light-years from the Earth. Proxima Centauri is a red M dwarf star, with an apparent visual magnitude of 11.01. Also known as Alpha Centauri C, the star Proxima Centauri is the nearest and faintest of the three stars in the Alpha Centauri system, located just 0.17 light-years closer than the central, brighter pair. *See* Alpha Centauri and parallax.

PSR: Acronym for pulsar.

PSR 0531+21: A pulsar in the Crab Nebula supernova remnant, spinning 30 times a second, discovered in 1968.

PSR 0833–45: A pulsar with a period of 89 milliseconds, discovered in 1968 in the Vela supernova remnant.

PSR 1257+12: A pulsar in the constellation Virgo that has a least two Earth-sized planets discovered in 1992 by the American radio astronomers **Aleksander Wolszczan (1946–)** and **Dale A. Frail (1961–)**. The pulsar is located at a distance of about 1300 light-years from the Earth, and was the first star, other than the Sun, to be known to have a planet orbiting it. *See* Arecibo Observatory.

PSR 1913+16: A binary pulsar discovered in 1974 by the American radio astronomers **Russell A. Hulse (1950–)** and **Joseph H. Taylor Jr. (1941–)**, consisting of one pulsar and another neutron star revolving about each other. It has been used to test the *General Theory of Relativity*, and to provide evidence for the existence of gravitational radiation. *See* Arecibo Observatory and binary pulsar.

PSR 1937+21: The first millisecond pulsar to be discovered, in 1982 by the American radio astronomer **Donald C. Backer (1943–)** using the Arecibo Observatory. The radio pulses repeat with a period of 1.56 milliseconds. *See* Arecibo Observatory, and millisecond pulsar.

PSR 1919+21: The first pulsar to be discovered, by the Irish astronomer **Jocelyn Bell (1943–)** in 1967 while a graduate student working under the direction of the English radio astronomer **Antony Hewish (1924–)**.

Ptolemaic cosmology: A description of the apparent motions of the planets, Moon and Sun around the Earth. *See* epicycle.

Pulsar: An object emitting periodic pulses of radio radiation, believed to be a rotating neutron star formed as the aftermath of a supernova explosion. A pulsar is not pulsating, but it instead acts like a lighthouse that sends a beam of radiation sweeping past Earth every time it spins. The repetition period of pulsars is extremely regular,

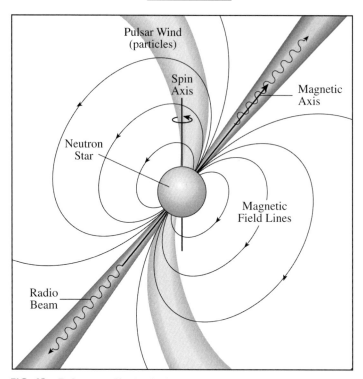

FIG. 43   Pulsar – radio A spinning neutron star has a powerful magnetic field whose axis intersects the north and south magnetic poles. The rotating fields generate strong electric currents and accelerate electrons, which emit an intense, narrow beam of radio radiation from each magnetic pole. Because the magnetic field axis is inclined to the neutron star's rotation axis, these beams wheel around the sky as the neutron star rotates, once per revolution. If one of the beams sweeps across the Earth, a bright pulse of radio emission, called a pulsar, is observed once per rotation of the neutron star. A wind of charged particles also streams from the poles but gets swept around to spray in all directions as the field rotates.

requiring atomic clocks to detect any changes. The Irish astronomer and graduate student **Jocelyn Bell (1943–)** and her research advisor, the English radio astronomer **Antony Hewish (1924–)** discovered the first pulsar in 1967, at the Mullard Radio Astronomy Observatory in Cambridge, England. They were studying rapid fluctuations of extra-galactic radio sources; this twinkling or scintillation is produced by the solar wind. The first radio pulsar to be discovered is designated PSR 1919+21 for its celestial coordinates, and it had a period of 1.337 279 5 seconds. In 1968 American astrophysicist **Thomas Gold (1920–2004)** proposed that pulsars are highly magnetized, rapidly rotating neutron stars, predicting that pulsars with much shorter periods would be found and that their pulsar periods will lengthen as the neutron star loses rotational energy. His predictions were confirmed in 1969 when a pulsar was discovered spinning 30 times a second in the Crab Nebula supernova remnant, and subse-

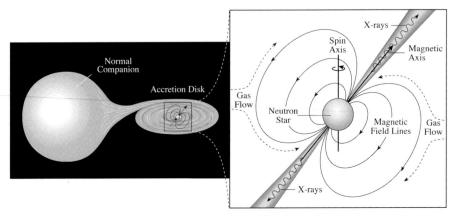

FIG. 44   Pulsar – X-ray Material from a normal visible-light star spills over onto its companion, an invisible neutron star. The flow of gas is diverted by the powerful magnetic fields of the neutron star, which channels the in-falling material onto the magnetic polar caps. The impact of the gas on the star's surface creates a pair of X-ray hotspots aligned along the magnetic axis at each magnetic cap. Because the magnetic field axis is inclined to the neutron star's rotation axis, the X-ray radiation from the hot spots sweeps across the sky once per revolution, producing the appearance of periodic X-ray pulsations if one of the hot spots intersects the observer's line of sight.

quently found to have an increasing period. In 1974 the American radio astronomers **Russell A. Hulse (1950–)** and **Joseph H. Taylor Jr. (1941–)** discovered the binary radio pulsar PSR 1913+16, which consists of two neutron stars, one radio loud and the other radio quiet, in orbit about a common center of mass. Accurate timing observations of the radio pulsar indicted that the two neutron stars are slowly approaching each other, losing orbital energy at the rate expected if they are emitting gravitational waves. In 1982 the American radio astronomer **Donald C. Backer (1943–)** discovered the first millisecond pulsar, designated PSR 1937+21, with a period of 1.56 milliseconds. Another millisecond pulsar, PSR 1257+12 with a period of 6.2 milliseconds, was discovered in 1990 by the American astronomer **Aleksander Wolszczan (1946–)**; accurate timing observations of this pulsar resulted in the 1992 discovery of two Earth-sized planets orbiting PSR 1257+12. Instruments aboard the *Uhuru* satellite, launched in 1970, discovered repeating pulses of X-ray radiation, with periods of about a second. These X-ray pulsars, such as Centaurus X-3 and Hercules X-1, are binary stars with the neutron-star component pulling in matter from a nearby visible-light companion; X-rays are emitted by the gas falling into the neutron star. As the result of this accretion of matter, the rotating neutron stars are gaining rotational energy and speeding up, with shorter pulsation period as time goes on. Individual radio pulsars such as the Crab Nebula and Veil Nebula pulsars have also been detected at X-ray wavelengths with the *Chandra X-ray Observatory*. *See* Arecibo Observatory, binary pulsar, Centaurus X-3, Crab Nebula, extrasolar planet, gravitational radiation, Hercules X-1, millisecond pulsar, Mullard Radio Astronomy Observatory, neutron star, PSR 1257+12, PSR 1919+21, PSR 1937+21, and *Uhuru Mission*.

Pulsating star: A variable star that pulsates, brightening and dimming as it moves in and out. The Cepheid and RR Lyrae stars are examples of pulsating stars. The bright Cepheid variable stars have a period-luminosity relation that is useful for determining the distances of the stars, and the galaxies or globular star clusters they are located in. *See* Cepheid variable, Delta Cephei, period-luminosity relation, and RR Lyrae.

Quandrantids: A strong meteor shower occurring during the first week of January, with a maximum on 3 to 4 January and a radiant in the constellation Boötes. The meteor shower's name is derived from an old constellation Quadrans Muralis, which is no longer officially recognized. *See* meteor shower.

Quantum mechanics: The theory of atomic and subatomic systems based on the notion of quantized energy in radiation, the photons, and in the angular momentum and energy levels of electrons in atoms, arising from the failure of classical concepts of waves or particles individually to describe such systems. *See* uncertainty principle.

Quantum tunneling: *See* tunnel effect.

Quasar: A relatively compact, very distant, and extremely luminous object that emits radiation over a wide range of wavelengths from X-ray to optical and radio wavelengths. A quasar can generate over a trillion times as much radiation as the Sun, from a region comparable in size to the Solar System. The word "quasar" is derived from quasi-stellar radio sources, because this type of object was discovered by determining the distances, or redshifts, to two powerful radio sources, denoted 3C 273 and 3C 48, and the small angular extent of their star-like images at first suggested that they were stars. Quasars are actually very compact extra-galactic objects with exceptionally large redshifts and distances of billions of light-years, sometimes emitting no detectable radio radiation and thought to be the nuclei of young galaxies. The first quasar to be discovered was 3C 273, whose large redshift was discovered by the Dutch-born American astronomer **Maarten Schmidt (1929–)**, using an accurate radio position obtained by the English radio astronomer **Cyril Hazard (1928–)** and his colleagues during a lunar occultation of 3C 273. Quasars are the brightest type of active galactic nucleus and are most likely powered by a supermassive black hole. The radiation we now detect from the distant quasars was emitted long ago when the Universe was young. *See* active galactic nucleus, gamma-ray burst, and supermassive black hole.

Quintessence: The fifth, heavenly element in ancient Greek cosmology, which supplements the four terrestrial elements: earth, air, fire and water. Quintessence is also the name given to a hypothetical anti-gravity force that may accelerate the Universe. *See* dark energy.

Radar: A form of radio observation in which a signal is emitted and the reflected signal is received and studied. Acronym for RAdio Detection And Ranging. The Scottish engineer **Robert Watson-Watt (1892–1973)** patented his first "radio locator" in 1919, and participated in the subsequent development of radar, which became fully realized by American and British scientists during World War II (1939–1945). After the war, radar was used to measure the distance, rotation and topography of the Moon and nearby planets. *See* Mercury, radar astronomy, and Venus.

Radar astronomy: The use of radar to determine the distance, rotation, and radio reflection characteristics of cosmic objects. Distance can be determined from the time delay between the transmission of a radar pulse and the reception of the "echo"; just multiply the time delay by the velocity of light and divide by two. Rotation can be inferred from the Doppler effect broadening of the wavelength of the returned radar signal. Radio engineers with the United States Army Signal Corps Laboratory first recorded radar echo signals from the Moon on 10 January 1946, approximately 2.5 seconds after transmission; the Hungarian electrical engineer **Zoltan Bay (1900–1992)** repeated the accomplishment a month later. Radar measurements of the planets Venus and Mercury from the **Arecibo Observatory**, first reported by the American scientist **Irwin I. Shapiro (1929–)** in 1964 for Venus and by the American scientists **Gordon H. Pettengill (1926–)** and **Rolf B. Dyce (1929–)** in 1965 for Mercury, indicate that Mercury rotates with a period of 58.646 Earth days, just 2/3 of its orbital period of 87.969 Earth days and that Venus rotates with a period of 243.025 Earth days, in the retrograde direction, longer than Venus' orbital period of 224.7 Earth days. The distance of Venus, and hence the astronomical unit or mean distance from the Earth to the Sun, was also accurately determined by radar measurements in the 1960s from the Arecibo Observatory and from NASA's Goldstone tracking station and the Jodrell Bank Observatory. Beginning in 1990, the radar instrument aboard the *Magellan* spacecraft spent more than four years mapping out the surface of Venus in great detail. *See* Arecibo Observatory, astronomical unit, *Magellan Mission*, Mercury, radar, and Venus.

Radial velocity: The speed at which an object is apparently moving either toward or away from an observer along the line of sight. The Doppler effect measures the radial velocity, and the larger the Doppler shift in wavelength the greater the radial velocity. The radial velocity is a relative velocity, consisting of both the observer's motion along the line of sight and the object's motion along the line of sight. To determine an object's true velocity in space, it is necessary to know the observer's motion along the line of sight, and also to know the object's transverse velocity, which is across the line of sight. *See* Doppler effect, Doppler shift, proper motion, and redshift.

Radian: A dimensionless unit of angular measure equal to $2.062\,65 \times 10^5$ seconds of arc. There are 2π radians in a full circle of 360 degrees, where $\pi = 3.141\,59$.

TABLE 44   Radioactive Isotopes

Radioactive Parent [Name (Symbol) Mass No.]	Stable Daughter [Name (Symbol) Mass No.]	Half-life (10^6 years)
Rubidium (Rb) 187	Strontium (Sr) 87	48 800
Rhenium (Re) 187	Osmium (Os) 187	44 000
Lutetium (Lu) 176	Hafnium (Hf) 176	35 700
Thorium (Th) 232	Lead (Pb) 208	14 050
Uranium (U) 238	Lead (Pb) 206	4 470
Potassium (K) 40	Argon (Ar) 40	1 270
Uranium (U) 235	Lead (Pb) 207	704
Samarium (Sm) 146	Neodymium (Nd) 142	100
Plutonium (Pu) 244	Thorium (Th) 232	83
Iodine (I) 129	Xenon (Xe) 129	16
Palladium (Pd) 107	Silver (Ag) 107	6.5
Manganese (Mn) 53	Chromium (Cr) 53	3.7
Aluminum (Al) 26	Magnesium (Mg) 26	0.72

Radiant: The point on the celestial sphere from which a meteor shower seems to come, designated by the constellation that point is in.

Radiation: A process that carries energy through space. *See* electromagnetic radiation.

Radiation belt: A ring-shaped region around a planet in which electrically charged particles – electrons, protons, and other ions – are trapped, following spiral trajectories around the direction of the magnetic field of the planet. The main radiation belts surrounding the Earth are known as the Van Allen belts, containing electrons or protons that are mainly from the solar wind. Similar regions exist around other planets with magnetic fields, such as Jupiter. The Earth also has an inner radiation belt containing ions of material from interstellar space. *See* Van Allen belts and South Atlantic Anomaly.

Radiation pressure: The pressure exerted by electromagnetic radiation or photons. Radiation pressure can compete with gas pressure in supporting giant stars, and it blows the dust tails of comets away from the Sun. In 1916 the English astronomer **Arthur Eddington (1882–1944)** showed that radiation pressure must stand with gravitation and gas pressure as the third major factor in maintaining the equilibrium of a star. In 1924 he derived the mass-luminosity relation of stars assuming that they were supported by gas and radiation pressure. *See* gas pressure and mass-luminosity relation.

Radiative zone: An interior layer of the Sun, lying between the energy-generating core and the convective zone, where energy travels outward by radiation.

Radio: *See* radio radiation.

Radioactive decay: *See* radioactivity.

Radioactivity: The spontaneous decay of certain rare, unstable, heavy nuclei into more stable light nuclei, with the release of energy and the emission of particles, including electrons, helium nuclei and neutrinos. The process by which certain kinds of atomic nuclei naturally decompose or decay, with the spontaneous emission of subatomic particles and gamma rays, and governed by the weak nuclear force. *See* alpha decay, beta decay, gamma-ray radiation, isotope, neutrino, and weak nuclear force.

Radio astronomy: The investigation of the radio radiation of cosmic objects. *See* cosmic microwave background radiation, quasar, radio galaxy, radio radiation, synchrotron radiation, and twenty-one centimeter line.

Radio galaxy: A galaxy that emits intense radio radiation, often from two lobes apparently hurled out from a central elliptical galaxy. The biggest radio galaxies have lobes separated by about 15 million light-years, comparable in size to a typical cluster of galaxies. A radio galaxy shines by the synchrotron radiation of energetic electrons spiraling about a magnetic field and traveling near the speed of light. In 1974 the American astronomer **Roger D. Blandford (1949–)** and English astronomer **Martin Rees (1942–)** proposed that the radio lobes of radio galaxies are fed by jets of high-energy electrons sent out along the rotation axis of a rotating, supermassive black hole in the elliptical galaxy. *See* black hole, Cygnus A, supermassive black hole, and synchrotron radiation.

Radio interferometer: The electronic combination of radio signals detected by two or more radio telescopes from the same cosmic object at the same time, using the interference of the signals to create the angular resolution of a radio telescope as large as the separation of the two or more component radio telescopes. *See* Australia Telescope, interferometry, Jodrell Bank Observatory, MERLIN, Mullard Radio Astronomy Observatory, National Radio Astronomy Observatory, Parkes Observatory, Very Large Array, Very Long Baseline Array, and Very Long Baseline Interferometry.

Radio radiation: The part of the electromagnetic spectrum whose radiation has the longest wavelengths and smallest frequencies of all types, with wavelengths ranging from about 0.001 meter to 30 meters and frequencies ranging between 10 MHz and 300 GHz. The German physicist **Heinrich Rudolf Hertz (1857–1894)** discovered long-wavelength radio waves in 1887, generating them with an oscillating electric current and demonstrating that they can be sent through the air traveling at the velocity of light. The American radio engineer **Karl Jansky (1905–1950)** discovered cosmic radio emission in 1933, and the American radio engineer **Grote Reber (1911–2002)** reported his discovery of discrete, cosmic radio sources in 1944. The radio wavelength spectrum includes millimeter waves, from 0.001 to 0.010 meters in wavelength, and microwaves with wavelengths from 0.01 to 0.06 meters. Unlike visible light, radio waves penetrate interstellar dust and can be detected from throughout our Galaxy, especially at the 21-centimeter line of neutral hydrogen. Distant galaxies also emit intense radio radiation. *See* cosmic microwave background radiation, quasar, radio galaxy, radio radiation, synchrotron radiation, and twenty-one centimeter line.

Radio telescope: A large radio antenna designed to concentrate radio waves and permit detection of faint radio signals reaching us from emission nebulae, galaxies, interstellar hydrogen, planets, pulsars, quasars, radio galaxies and the Sun and other stars. A large single dish or parabolic reflector sends the incoming radio waves to a focus where they are converted into electronic signals. A larger radio telescope collects more energy, detecting fainter sources, and has greater angular resolution, permitting the detection of finer detail. Better resolution can be obtained by linking the electronic signals of two or more dishes to form a radio interferometer with an angular resolution comparable to a single antenna as big as the largest separation of the dishes. An example is the Very Large Array. *See* Effelsberg Radio Telescope, interferometry, Jodrell Bank Observatory, Mullard Radio Astronomy Observatory, National Radio Astronomy Observatory, Parkes Observatory, and Very Large Array.

Ramaty High Energy Solar Spectroscopic Imager: NASA launched the **Ramaty High Energy Solar Spectroscopic Imager**, abbreviated **RHESSI**, on 5 February 2002, with an anticipated two-year lifetime. It obtained high-resolution imaging spectroscopy of solar flares at hard X-rays and gamma-ray wavelengths, in order to study solar flare particle acceleration and flare energy release. It determined the frequency, location and evolution of impulsive flare energy release in the corona and located the sites of particle acceleration and energy deposition at all phases of solar flares. The spacecraft was named after **Reuven Ramaty (1937–2001)**, a pioneer in the fields of solar physics, gamma-ray astronomy, nuclear physics and cosmic rays.

Recombination: The capture of an electron by a proton or any other ion with a positive electrical charge. It is the inverse process to ionization.

Reddening: The scattering away of blue light that occurs when light passes through dust, causing the observed light to be redder than that emitted. Reddening by the Earth's atmosphere is responsible for the red color of a sunset, as well as the full red Moon seen during a total lunar eclipse, and reddening by interstellar dust makes more distant stars appear redder. When radiation passes through a longer path, containing a greater amount of dust, the amount of reddening is increased.

Red dwarf: *See* red dwarf star.

Red dwarf star: A cool, low mass star that lies near the lower end of the main sequence in the Hertzsprung-Russell diagram, typically of spectral class M. They have the lowest mass and temperature of all stars that shine by thermonuclear reactions, with a mass less than about 0.5 solar masses, a radius less than about half the radius of the Sun, and an effective temperature less than 4000 kelvin. The red dwarf, or dwarf M, stars are the most common class of star in our Galaxy, accounting for about 70 percent of all stars. *See* dM star, Hertzsprung-Russell diagram, M star, and spectral classification.

Red giant: A star with a vast, expanded low-density atmosphere, often a hundred times as large as the Sun, with a relatively cool disk temperature when compared with other giant stars. Red giants lie in the upper right hand part of the Hertzsprung-Russell diagram. Red giant stars have a spectral type of M, and hence a red color. Mira is a red giant star, and the Sun will eventually become one. *See* giant star, Hertzsprung-Russell diagram, Mira, and M-type star.

Red planet: Popular name for the planet Mars. The ancients associated the blood-red color with warfare, and the Greeks and Romans named the planet after their gods of war – Ares and Mars, respectively. *See* Mars.

Redshift: An increase in wavelength of electromagnetic radiation; at visible-light, or optical, wavelengths the shift is toward longer, redder wavelengths. The redshift of a star, galaxy, or other cosmic object is attributed to the Doppler shift of the radiation from a source that is moving away from the observer along the line of sight, and the larger the redshift, the faster the star is moving away from us. The redshift is given the symbol z, and is equal to the change in wavelength of a particular spectral line divided by its wavelength when not moving or at rest. The radial velocity, along the line of sight, is given by the product of the redshift z and the velocity of light c = 299 792.46 kilometers per second, provided that the radial velocity is much less than the velocity of light. Redshifts greater than 1.0 are possible, but the radial velocities are always less than the velocity of light. The expanding Universe was discovered by noticing the increase of galaxy redshifts with distance, now known as the Hubble law. The expansion of space can produce a redshift by stretching light waves from a distant galaxy while they travel to Earth. This type of redshift is sometimes called a cosmological redshift to distinguish it from one caused by movement. The farther a galaxy, the longer its light waves have traveled through space and the more stretched and redshifted they have become. *See* Doppler effect, Doppler shift, expanding Universe, Hubble diagram, Hubble law, and radial velocity.

Red supergiant: A supergiant star with spectral type M, and the largest star in the Universe. Antares and Betelgeuse are red supergiants.

Reflecting telescope: Also known as a reflector, a telescope that gathers visible-light or infrared radiation and forms an image by the reflection of light from a primary concave mirror, usually parabolic in shape. Reflecting telescopes differ in the arrangement of their secondary mirrors. *See* Cassegrain telescope, Coudé focus, and Newtonian telescope.

Reflection nebula: A nebula in which the starlight is reflected off the dust particles. As discovered in 1913 by the American astronomer **Vesto M. Slipher (1875–1969)**, the visible-light spectra of what we now call "reflection" nebulae exhibit absorption lines identical to those found in their central stars, indicating that they are reflecting the starlight. An example of a bright reflection nebula is found around the star Merope in the Pleiades open star cluster. *See* Pleiades.

Refracting telescope: Also known as a refractor, a telescope that gathers radiation and forms an image by the refraction of light through a lens, called the objective. The first telescopes were refracting telescopes. *See* telescope.

Regolith: A layer of dust, fine-grained material, pebbles and rock fragments formed on the Moon, asteroids, and other bodies in the Solar System by eons of bombardment by small meteorites that break apart the surface rock. The term *regolith* is from the Greek word for "rock layer."

Regulus: A blue-white main sequence, or dwarf, star of apparent visual magnitude 1.36 and spectral type B7. It is also named Alpha Leonis, being the brightest star in the direction of the constellation Leo. The name Regulus means "Little King", a designation apparently given by the Polish astronomer **Nicolaus Copernicus (1473–1543)**. At a distance of 78 light-years, it has a parallax of 0.042 seconds of arc, an absolute magnitude of –0.5 and an intrinsic luminosity about 135 times that of the Sun.

Relativistic: Approaching the velocity of light. Particles moving at relativistic speeds demonstrate effects predicted by the *Special Theory of Relativity,* such as length contraction, mass increase and time dilation. *See Special Theory of Relativity.*

Relativity: *See General Theory of Relativity* and *Special Theory of Relativity.*

Resolution: The degree to which the fine details in an image are separated and detected or seen. Also known as angular resolution and resolving power, the smallest angular size distinguishable by a telescope, which determines the ability of a telescope to discern fine detail of a celestial object. The angular resolution, denoted by θ, in radians is given by the ratio of the wavelength, λ, to the diameter, D, of the primary mirror or reflector, with $\theta = \lambda/D$ radians. To convert to seconds of arc, or arc seconds, 1 radian $= 2.062\,65 \times 10^5$ seconds of arc. The human eye has an angular resolution, or resolving power, of about 1 minute of arc. The resolving power of an optical telescope, operating at visible-light wavelengths, is found by dividing 110 by the aperture in millimeters, or by dividing 4.56 by the aperture in inches; the aperture is the diameter of the objective lens or the primary mirror. However, the resolution of such an optical telescope is limited to about 1 second of arc due to atmospheric turbulence, and this means that a telescope with a mirror or lens larger than about 0.12 meters in diameter has an effective angular resolution of 1 second of arc when observing from the ground, but the full angular resolution is possible if the telescope is in space. The atmosphere does not limit the resolution of a radio telescope. *See* scintillation and seeing.

Resolving power: *See* resolution.

Retrograde loop: The temporary apparent motion of a planet in a direction opposite to its normal progress across the sky. This reversal is called a retrograde loop. It is explained by the Earth catching up and overtaking the slower orbital motion of a planet that is more distant from the Sun.

Retrograde orbital motion: An orbital motion in the direction opposite to that of the Earth's orbit around the Sun, or clockwise as opposed to anti-clockwise when viewed from above the Sun's north pole. Some of the small, outer satellites of Jupiter, some comets such as Halley's Comet, and Triton, the largest satellite of Neptune, have retrograde orbits.

Retrograde rotation: A rotational motion in the direction opposite to that of the Earth's orbit around the Sun, or clockwise as opposed to anti-clockwise when viewed from above the Sun's north pole. Venus and Uranus have retrograde rotation, but all the other planets rotate in the same direction that they orbit the Sun and all the major planets orbit in the same direction as the Earth.

Revolve: The orbital motion of one object around another. The Moon revolves around the Earth, and the Earth revolves around the Sun.

RHESSI: Acronym for *Ramaty High Energy Spectroscopic Imager*. *See* **Ramaty High Energy Spectroscopic Imager.**

Rigel: A blue-white B supergiant star, the seventh brightest star in the sky and the brightest star in the constellation Orion. Rigel is also known as Beta Orionis, although it is brighter than Betelgeuse, designated Alpha Orionis, the second brightest star in Orion. Rigel is 770 light-years distant, with a parallax of 0.004 seconds of arc. It has an apparent visual magnitude of 0.18, and an absolute visual magnitude of –6.7, with an intrinsic luminosity of about 40 000 times that of the Sun and the most intrinsically luminous star known. *See* B star and stellar classification.

Right ascension: The celestial longitude of a cosmic object measured eastward along the celestial equator from the Vernal Equinox, also known as the First Point of Aries, to the foot of the great circle that intersects the object. The right ascension is often denoted by α, the lower case Greek letter alpha, or by R.A. The right ascension is expressed in hours, minutes and seconds of time, with twenty-four hours in the complete circle of 360 degrees. The local sidereal time is equal to the right ascension of an object on the observer's meridian. As the result of precession, the right ascension increases with time at the rate of about 3.35 seconds of time per year. *See* celestial coordinates, day, precession, and sidereal time.

Rigil Kentaurus: *See* Alpha Centauri.

Ring: A collection of small particles in orbital motion about a planet, forming either a thin ring or a wider disk that nowhere touches the planet. In 1646 the Dutch astronomer **Christiaan Huygens (1629–1695)** identified the first planetary ring, circling Saturn. Most planetary rings are inside the Roche limit, located at about 2.5 times its radius. In 1847 the French mathematician **Edouard Roche (1820–1883)** showed that a planet's tidal forces will tear any large satellite into pieces if it is within the Roche limit, while also preventing small bodies from coalescing to form a larger moon. In 1857 the Scottish physicist **James Clerk Maxwell (1831–1879)** showed that the rings of Saturn remain suspended in space because they are not solid, but instead composed

TABLE 45   Rings – Saturn

Ring	Width (km)	Closest Distance (km)	Distance Range[a] (R_S)	Particle Size (m)	Optical Depth	Mass (kg)
D	7 540	66 970	1.11 to 1.235	$<10^{-6}$	0.0001	
C	17 490	74 510	1.235 to 1.525	0.01 to 3.0	0.05 to 0.35	1×10^{17}
B	25 580	92 000	1.525 to 1.949	0.01 to 5.0	0.4 to 2.5	2×10^{19}
A	14 610	122 170	1.949 to 2.025	0.01 to 7.5	0.05 to 0.15	4×10^{17}
F	50	140 180	2.324	10^{-7} to 10^{-5}	0.1	
G	500 to 3 000	170 180	2.82	3×10^{-8}	2×10^{-6}	
E	302 000	181 000	3 to 8	1×10^{-6}	1.5×10^{-5}	7×10^8

[a]The distance range is given in units of Saturn's apparent equatorial radius, R_S = 60 330 kilometers. At the 1-bar pressure level, the radius is 60 268 kilometers.

of a vast number of small particles, each pursuing its individual orbit around Saturn in the plane of the planet's equator. In 1895 the American astronomer **James Keeler (1857–1900)** used the Doppler effect to show that the innumerable particles that make up Saturn's rings act as tiny satellites, with the inner parts moving at a faster speed than the outer ones. The thin ring of Jupiter was discovered in 1979 when a camera aboard the retreating *Voyager 1* spacecraft looked back at the shadowed side of Jupiter. The American astronomer **James L. Elliot (1943–)** discovered the thin, widely spaced rings of Uranus in 1977, when a star passed behind the planet. Similar stellar occultations provided controversial evidence for thin, widely spaced rings of Neptune, which were conclusively discovered in 1989 by the cameras aboard the *Voyager 2* spacecraft. *See* Jupiter, Neptune, Roche limit, Saturn, and Uranus.

Ring Nebula: A planetary nebula designated M57 and NGC 6720, in the constellation Lyra. It has an apparent visual magnitude of 9, measures 70 by 150 seconds of arc, and has a distance of about 2000 light-years. The Ring Nebula derives its name from its hollow appearance like a smoke ring. Its blue central star is only fifteenth magnitude. *See* planetary nebula.

R Monocerotis: *See* Hubble's variable nebula.

Roche limit: As shown by the French mathematician **Edouard Roche (1820–1883)** in 1847, a large satellite will be torn to pieces by the differential gravitational forces of a planet if the satellite is within a distance of 2.5 times the planet's radius from the planet center, now known as the Roche limit. These forces also prevent small particles from coalescing to form a large moon within this distance.

Roche lobe: In 1847 the French mathematician **Edouard Roche (1820–1883)** investigated the gravitational interaction of binary star systems, showing that each star has its own zone of gravitational influence, which is now called a Roche lobe. When the two stars are close together, their Roche lobes touch at a point where their gravitational forces exactly cancel. If the largest companion star expands and fills its Roche lobe, matter is transferred to the other star; when the other companion is a neutron star or a black hole, the infalling matter can generate X-rays. *See* black hole, and pulsar.

Röntgen Mission: An X-ray observatory, abbreviated *ROSAT* for *ROntgen SATellite,* developed through a cooperative program among Germany, the United States, and the United Kingdom, launched by NASA on 1 June 1990 and ending after almost nine years, on 12 February 1999. It carried out an all-sky survey of cosmic X-rays, and obtained detailed information about the X-ray emission from comets, young T Tauri stars, globular clusters, neutron stars, supernova remnants, planetary nebulae, nearby normal galaxies, active galactic nuclei, and clusters of galaxies. The mission was named after **Wilhelm Konrad Röntgen (1845–1923)**, the German physicist who discovered X-rays in 1895, receiving the first Nobel Prize for Physics in 1901 for this achievement.

ROSAT: Acronym for the *ROnten SATellite*. *See Röntgen mission*.

Rosette Nebula: An emission nebula, designated NGC 2237–39 and NGC 2246, located about 5000 light-years away in the constellation Monoceros. The Rosette Nebula surrounds the star cluster known as NGC 2264, whose stars illuminate the nebula and have blasted out a cavity about 12 light-years across in the central part of the nebula. *See* emission nebula.

Rossi X-ray Timing Explorer: NASA launched the ***Rossi X-ray Timing Explorer***, abbreviated ***RXTE***, on 30 December 1995, obtaining about a decade of subsequent observations of variable cosmic X-ray sources with high time resolution and providing new insights to black holes, gamma-ray bursts, neutron stars and pulsars. The spacecraft is named for the Italian-born, United States physicist **Bruno B. Rossi (1905–1993)**, a pioneer in studies of cosmic rays, X-ray astronomy and space plasma physics.

Rotation: The spin of an object about its own axis. The Earth rotates once a day.

Royal Greenwich Observatory: The Royal Observatory of England, founded by **Charles II (1660–1683)** in 1675 at Greenwich, England, appointing **John Flamsteed (1646–1719)** as his first Astronomer Royal in March of that year. The Observatory was built to improve navigation at sea and to find the east-west position, or longitude, while at sea and out of sight of land, by astronomical means. In 1884, the Greenwich Meridian was chosen to be the Prime Meridian of the World, with exactly zero degrees longitude, dividing the eastern and western hemispheres of the Earth. It is defined by the position of the Observatory's large Transit Circle telescope built in 1850 by the seventh Astronomer Royal **George Biddell Airy (1801–1892)**. Since the late-19th century, the Prime Meridian at Greenwich has served as the co-ordinate base for the calculation of Greenwich Mean Time, also called Universal Time. After World War II (1939–1941), the Royal Observatory moved to Herstmonceux in Sussex; in 1990 it moved to Cambridge, and was closed in 1998 with its functions transferred to the Royal Observatory, Edinburgh. *See* chronometer, Greenwich Mean Time, longitude, and Universal Time.

Royal Observatory, Edinburgh: An observatory in Edinburgh, Scotland, which is an administrative center for astronomy in the United Kingdom.

RR Lyrae: A pulsating, helium-burning giant star of spectral type F5, whose apparent visual magnitude changes between 7.2 and 8.0 with a period of 13 hours 36 minutes and 14.9 seconds. At its distance of 750 light-years, RR Lyrae shines with an average luminosity 40 times that of the Sun. It is the prototype for the RR Lyrae variable stars.

Runaway Star: A star of spectral type O or B with an exceptionally high velocity. In 1961 the Dutch astronomer **Adriaan Blaauw (1914–)** proposed that these stars are escaped members of binary star systems in which one star has become a supernova. He noted that the runaway stars are massive stars whose high velocities are comparable to the orbital velocities expected from massive binary star systems, and that the more massive component would quickly evolve to the supernova stage and thereby release the other member as a high-velocity star.

RXTE: Acronym for ***Rossi X-ray Timing Explorer***. *See* ***Rossi X-ray Timing Explorer***.

Sagittarius A*: A strong radio source at the center of our Galaxy, and probably a massive black hole. As the name suggests, it is located in the constellation Sagittarius.

Satellite: A natural or artificial object orbiting another object of larger size and mass. A natural satellite that orbits a planet is also known as a moon, and man-made,

TABLE 46   Satellites – large[a]

Name	Mean Radius[b] (km)	Mass (10^{20} kg)	Mean Mass Density (kg m^{-3})	Distance from Planet Center (10^6 m)	Period of Revolution[c] (days)	Year of Discovery
EARTH						
Moon	1738	734.9	3344	384.4	27.3217	—
JUPITER						
Io	1822	894.0	3528	422	1.77	1610
Europa	1561	480.0	3014	671	3.55	1610
Ganymede	2631	1482.3	1942	1070	7.16	1610
Callisto	2410	1076.6	1834	1883	16.7	1610
SATURN						
Mimas	199	0.38	1142	186	0.94	1789
Enceladus	249	0.8	1000	238	1.37	1789
Tethys	530	7.6	1006	295	1.89	1684
Dione	559	10.5	1498	377	2.74	1684
Rhea	764	24.9	1236	527	4.52	1672
Titan	2575	1345.7	1881	1222	15.9	1655
Hyperion	142	—	1250	1464	21.3	1848
Iapetus	718	18.8	1025	3561	79.3	1671
Phoebe	110	—	2300	12944	550R	1898
URANUS						
Miranda	236	0.71	1201	130	1.41	1948
Ariel	579	14.4	1665	191	2.52	1851
Umbriel	585	11.8	1400	266	4.14	1851
Titania	789	34.3	1715	436	8.71	1787
Oberon	761	28.7	1630	583	13.5	1787
NEPTUNE						
Triton	1353	214.2	2061	354	5.88R	1846
Nereid	170		1500	5513	360	1949

[a] The physical parameters of the natural satellites are given on the Web at http://ssd.jpl.nasa.gov/sat.props.html
[b] The radii are given in units of kilometers, abbreviated km, the mass is in kilograms, abbreviated kg, and the mass density in kilograms per cubic meter, abbreviated kg m^{-3}. By way of comparison, the equatorial radius of the planet Mercury is 2440 kilometers, so Ganymede and Titan are both bigger than Mercury.
[c] The letter R following the period denotes a satellite revolving about its planet in the retrograde direction, opposite to that of the planet's rotation and orbital motion about the Sun.

Earth-orbiting satellites are called artificial satellites. A satellite galaxy orbits a larger galaxy. Mercury and Venus have no satellites, the Earth has one, the Moon with a capital M, and Mars has two small satellites, named Phobos and Deimos. The four largest satellites of Jupiter are known as the Galilean satellites, named Callisto, Europa, Ganymede and Io. Saturn's largest satellite is named Titan, and Neptune's largest satellite is called Triton. The giant planets have many small satellites. *See* Callisto, Deimos, Europa, Galilean satellites, Ganymede, Io, Moon, Phobos, Titan and Triton.

Saturn: The sixth major planet from the Sun, and the remotest planet known before the invention of the telescope. Saturn is the second largest of the four giant planets, and the only planet with such magnificent rings. Saturn has the lowest mass density of any planet in the Solar System, and it is primarily composed of the lightest element hydrogen, in liquid molecular and liquid metallic form. Saturn radiates almost twice the energy it receives from the Sun, and most of the planet's excess heat is generated by helium raining down into its inner metallic hydrogen shell. In 1655 the Dutch astronomer **Christiaan Huygens (1629–1695)**, discovered Titan, Saturn's largest moon, and in 1656 he realized that Saturn is surrounded by a ring, publishing in 1659 a Latin anagram that, when interpreted, explained Saturn's varying, handle-shaped appearance by "It is surrounded by a thin, flat ring, nowhere touching it and inclined to the ecliptic." In 1675 the Italian-born, French astronomer **Gian (Giovanni) Domenico (Jean Dominique) Cassini (1625–1712)** was the first to distinguish two zones within what was thought to be a single ring around Saturn, divided by a dark gap now known as

TABLE 47   Saturn[a]

Mass	5684.6×10^{23} kilograms $= 95.184 \, M_E$
Equatorial Radius at 1 bar	60 268 kilometers $= 9.449 \, R_E$
Polar Radius at 1 bar	54 364 kilometers
Mean Mass Density	687.3 kilograms per cubic meter
Rotation Period	10 hours 39 minutes 22.3 seconds $= 10.6562$ hours
Orbital Period	29.458 Earth years
Mean Distance from Sun	1.4294×10^{12} meters $= 9.539$ AU
Age	4.6×10^9 years
Atmosphere	97 percent molecular hydrogen, 3 percent helium
Energy Balance	1.79 ± 0.10
Effective Temperature	95.0 kelvin
Temperature at 1-bar level	134 kelvin
Central Temperature	13 000 kelvin
Magnetic Dipole Moment	$600 \, D_E$
Equatorial Magnetic Field Strength	0.22×10^{-4} tesla or $0.72 \, B_E$

[a] The symbols M_E, R_E, D_P, B_E denote respectively the mass, radius, magnetic dipole moment and magnetic field strength of the Earth. One bar is equivalent to the atmospheric pressure at sea level on Earth. The energy balance is the ratio of total radiated energy to the total energy absorbed from sunlight, and the effective temperature is the temperature of a blackbody that would radiate the same amount of energy per unit area.

Cassini's division. Cassini also correctly suggested that Saturn's rings are composed of myriad tiny satellites, although it was not until the mathematical work of **James Clerk Maxwell (1831–1879)** in 1857 that he was proved to be correct. Saturn and its rings were explored in close-up detail using instruments aboard the *Voyager 1* and *2* spacecraft, which respectively flew by the planet in 1980 and 1981. The *Cassini-Huygens* spacecraft arrived at Saturn in July 2004, to begin a four-year study with the *Cassini Orbiter*. The *Huygens Probe* landed on the surface of Titan on 14 January 2005. *See Cassini-Huygens Mission*, *Huygens Probe*, rings, and Titan.

Saturn Nebula: The planetary nebula designated NGC 7009 located in the constellation Aquarius, with handle-like protuberances that give it the appearance of Saturn. The Saturn Nebula has an apparent visual magnitude of 8. *See* planetary nebula.

SAX: Acronym for the *Satellite per Astronomia X*.

Schmidt telescope: A telescope with a wide field of view, useful for early supernova searches and for whole-sky photographic surveys, such as the Palomar Observatory Sky Survey with a 1.2-meter (48-inch) Schmidt telescope. The Estonian astronomer **Bernhard Schmidt (1879–1935)** invented it in 1930. *See* Anglo-Australian Observatory and Palomar Observatory.

Schwarzschild metric: The line element outside a massive, non-rotating, spherically symmetric body in a vacuum, describing the gravitational space curvature by that mass. It was derived in 1916 by the German astronomer **Karl Schwarzschild (1873–1916)**.

Schwarzschild radius: The critical radius that marks the event horizon of a black hole, from which nothing can escape, not even light. When a very massive body undergoes gravitational collapse to a size smaller than the Schwarzschild radius, it becomes a black hole. The Schwarzschild radius, $R_S = 2GM/c^2$, where G is the gravitation constant, M is the mass and c is the velocity of light. The radius is named for the German astronomer **Karl Schwarzschild (1873–1916)**, who in 1916 first derived the exact equations for the curved space outside an isolated, spherical, non-rotating mass. *See* black holes and event horizon.

Scintillation: The twinkling of the stars caused by wind-blown clouds or other non-uniform density distributions in our atmosphere, and the fluctuations of distant radio sources caused by the solar wind or winds in interstellar space. The stars twinkle, causing an apparent change in their color and brightness, but the Moon and planets like Venus do not scintillate because of their larger angular extent. Scintillation in the Earth's atmosphere limits the angular resolution of even the best ground-based optical, or visible-light, telescope to about one second of arc, but this does not affect the resolution of a radio telecope. *See* resolution and seeing.

Scorpius X-1: The first extra-solar X-ray source, discovered in 1962 by the Italian-born American astronomer **Riccardo Giacconi (1931–)** and his colleagues. It is a low-mass X-ray binary source, consisting of a visible-light companion and the X-ray source, presumably an accreting neutron star, in orbit around a common center of mass with a period of 0.787 days.

SDSS: Acronym for Sloan Digital Sky Survey. *See* Sloan Digital Sky Survey.

Sea-floor spreading: The sea floor is formed in a volcanic rift in a mid-ocean ridge, spreads sideways, and turns cold and heavy as it moves away from its source in two directions, eventually sinking and disappearing in a deep-ocean trench. The American geologist **Harry H. Hess (1906–1969)** first suggested the notion of sea-floor spreading in the early 1960s. The British geophysicists **Frederick Vine (1939–1988)** and **Drummond Matthews (1931–1997)** provided concrete evidence of sea-floor spreading when they discovered a symmetric pattern of magnetic field reversals in the rocks on the floor of the Atlantic Ocean on either side of the mid-ocean ridge.

Season: A periodic change in the weather conditions on a planet caused by the tilt of its rotational axis and its orbit around the Sun. The annual orbit of the Earth around the Sun produces summer and winter when the relevant hemisphere is pointed toward or away from the Sun. Mars has similar seasons, which last about twice as long as the Earth's seasons, since the Martian year, or the orbital period of Mars, is nearly two Earth years. On Earth we traditionally divide the year into four seasons, spring, summer, autumn and winter, each lasting about three months. In the northern hemisphere, the first days of spring, summer, autumn and winter respectively occur on 21 March the Vernal Equinox, 21 June the summer solstice, 23 September the autumnal equinox, and 22 December the winter solstice. *See* equinox and solstice.

Secondary cosmic rays: Subatomic particles produced by collisions between primary cosmic rays and atoms or molecules in Earth's atmosphere.

Second of arc: A unit of angular measure of which there are 60 in 1 arc minute and therefore 3600 in 1 arc degree. One arc second is equal to 725 kilometers on the visible disk of the Sun. A second of arc is denoted by the symbol ″, and it is also called an arc second. There are 2.06265×10^5 seconds of arc in one radian, and 2π radians in a full circle of 360 degrees, where the constant $\pi = 3.14159$.

Seeing: Fluctuations in a visible-light image due to refractive inhomogeneities in the Earth's atmosphere. Seeing is caused by the random turbulent motion in the Earth's atmosphere. In conditions of good seeing, images are sharp and steady; in poor seeing, they are extended and blurred and appear to be in constant motion. Seeing usually limits the angular resolution of ground-based optical telescopes to about one second of arc. *See* scintillation.

Seven Sisters: The seven brightest stars in the Pleiades, an open star cluster in the constellation Taurus. *See* Pleiades.

Sextant: Instrument employed to measure the elevation of astronomical objects above the horizon.

Seyfert galaxy: A spiral galaxy characterized by high-speed motions in its center. The first such galaxies were discovered in 1943 by the American astronomer **Carl K. Seyfert (1911–1960)**. A Seyfert galaxy has a moderately bright, compact active galactic nucleus, presumably powered by a supermassive black hole. An example of a Seyfert galaxy is the spiral galaxy designated M77 or NGC 1068. *See* active galactic nucleus and supermassive black hole.

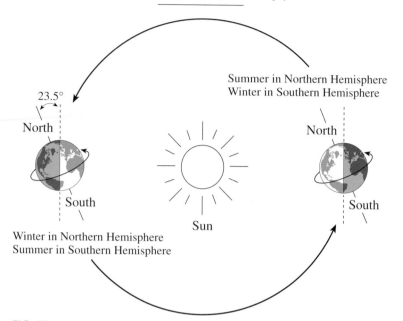

FIG. 45 Seasons As the Earth orbits the Sun, the Earth's rotational axis in a given hemisphere is tilted toward or away from the Sun. This variable tilt produces the seasons by changing the angle at which the Sun's rays strike different parts of the Earth's surface. The greatest sunward tilt occurs in the summer when the Sun's rays strike the surface most directly. In the winter, the relevant hemisphere is tilted away from the Sun and the Sun's rays obliquely strike the surface. When it is summer in the northern hemisphere, it is winter in the southern hemisphere and vice versa. (Notice that the radius of the Earth and Sun and the Earth's orbit are not drawn to scale.)

Shock wave: A sudden discontinuous change in density and pressure propagating in a gas or plasma at supersonic velocity. There is also a change in particle flow speed and in the magnetic and electric field strengths associated with a shock.

Shoemaker-Levy 9 Comet: A comet discovered on 23 March 1993 by **Carolyn Shoemaker (1929–)**, **Eugene Shoemaker (1928–1997)** and **David H. Levy (1948–)**, the ninth in a series of objects the trio found traveling around the Sun in short-period orbits. They were observing the fragments of a comet that passed too close to Jupiter on 7 July 1992, and was broken into pieces by the differential gravitational forces of the giant planet. Orbital calculations indicated that the train of comet fragments would plunge into Jupiter between 16 and 22 July 1994, and astronomers throughout the world witnessed the collisions, which created explosive fireballs and dark scars on Jupiter.

Short-period comet: A comet whose orbital period around the Sun is less than 200 years. Many short-period comets originate in the Kuiper belt. *See* comet and Kuiper belt.

Sidereal: Measured or determined with reference to the stars.

Sidereal period: The orbital or rotation period of a planet, the Sun, or other celestial body with respect to a background stars. *See* synodic period.

Sidereal time: Local time determined from the rotation of the Earth with respect to the fixed stars. Sidereal time is equal to the right ascension of an object on the observer's meridian. The local sidereal time is 0 hours when the Vernal Equinox crosses the observer's meridian. The sidereal day is equal to 23 hours 56 minutes 04.090524 seconds of mean solar time, or nearly 4 minutes shorter than the mean solar day. As a result, the stars rise and set four minutes earlier each day reckoned relative to the Sun. The sidereal month is the time for one revolution of the Moon around the Earth relative to a fixed star, equal to 27.32166 mean solar days. A sidereal year is the time required for the Earth to complete one revolution around the Sun relative to the fixed stars, and it is equal to 365.25636 mean solar days. *See* celestial coordinates, day, month, right ascension, and year.

Siding Spring Observatory: An observatory founded by the Australian National University in 1962 at Siding Spring, New South Wales, with a 2.3-meter (90-inch) telescope. The Anglo-Australian Observatory is also located at the site. *See* Anglo-Australian Observatory.

SIM: Acronym for *Space Interferometry Mission*. *See* ***Space Interferometry Mission***.

Singularity: A place in space or in time where the formula for some physical quantity becomes infinite. It was once thought that the Schwarzschild radius marked a singularity of a black hole, but it was removed to the center of the hole by a change in coordinate systems. *See* black hole and Schwarzschild radius.

Sirius: The apparently brightest star in the night sky, a white A1 dwarf, or main sequence, star with an apparent visual magnitude of –1.46. The Greek name *Sirius* means "searing", "scorching" and "sparkling." The other names for Sirius are Alpha Canis Majoris and the "Dog Star", due to its location in the constellation Canis Majoris, Orion's bigger Hunting Dog. The first glimpse of Sirius in the morning twilight foretold the rising of the Nile in Egypt's Old Kingdom, whose symbol was a dog. Sirius shines with an absolute visual magnitude of +1.42, or about 23 times more luminous than the Sun, so Sirius appears so bright because it is relatively nearby, located at a distance of just 8.6 light-years. Sirius is also a double star, whose faint component is a white dwarf star designated as Sirius B. In 1842 the German astronomer **Friedrich Wilhelm Bessel (1784–1846)** attributed periodic variation in the proper motion of the bright component, designated Sirius A, to the gravitational attraction of an unseen companion star, called Sirius B, which was first detected in 1862 by the American astronomer **Alvan Clark (1804–1887)**. It was so hard to find because Sirius B is 10 000 times fainter than nearby Sirius A. In 1915 the American astronomer **Walter S. Adams (1876–1956)** showed that Sirius B is a hot star of spectral type A0. The high temperature and low intrinsic luminosity meant that the faint companion star had to be very small, only about the size of the Earth. Such stars became known as white dwarf stars. *See* Stefan-Boltzmann law and white dwarf.

SIRTF: Acronym for the *Space Infra-Red Telescope Facility*. See *Spitzer Space Telescope*.

61 Cygni: The first star other than the Sun to have its distance measured. Known as the "Flying Star", due to its large proper motion of more than 5 seconds of arc per year, 61 Cygni was presumably nearby. The parallax, and hence the distance, of 61 Cygni was announced in 1838 by the German astronomer **Friedrich Wilhelm Bessel (1784–1846)**, who obtained a parallax of 0.31 seconds of arc, very close to the modern value of 0.287 seconds of arc, corresponding to the distance of 11.4 light-years. The star is a double orange dwarf that lies in the constellation Cygnus, whose component K stars have apparent visual magnitudes of 5.21 and 6.03.

Sloan Digital Sky Survey: The Sloan Digital Sky Survey, abbreviated SDSS, uses sophisticated electronic instruments with a dedicated 25-meter (100-inch) reflector, located at the Apache Point Observatory in the Sacramento Mountains of New Mexico, to map one quarter of the entire sky, determining the position and absolute brightness of more than 100 million celestial objects. The SDSS also determines the distances, or redshifts, to more than a million galaxies and quasars. The distribution of galaxies in different directions and depths, or redshifts, reveals a chain of galaxies, dubbed the Sloan Great Wall, that is 1.37 billion light-years across, together with huge voids, or seemingly empty places with few galaxies.

Small Magellanic Cloud: The second largest and the second nearest, of the irregular galaxies that orbit the Milky Way – the largest and nearest is the Large Magellanic Cloud. The Small Magellanic Cloud is located at a distance of 190 000 light-years in the southern sky. The Magellanic Clouds are named for the Portuguese explorer **Ferdinand Magellan (1480–1521)**, whose crew reported seeing them from the Southern Hemisphere during the first circumnavigation of the globe, completed in 1522. *See* Large Magellanic Cloud and Magellanic clouds.

Smithsonian Astrophysical Observatory: The Smithsonian Astrophysical Observatory, abbreviated SAO, is a research institute of the Smithsonian Institution, joined with the Harvard College Observatory to form the Harvard-Smithsonian Center for Astrophysics, abbreviated CfA. It is headquartered at Cambridge, Massachusetts. *See* Center for Astrophysics.

SMM: Acronym for *Solar Maximum Mission*. See *Solar Maximum Mission*.

SN: Acronym for SuperNova. *See* supernova.

SNC meteorite: About a dozen igneous meteorites identified to have come from Mars, on the basis of their element abundance and trapped gases. Most of them solidified from molten material less than 1.5 billion years ago. They have been collectively named the SNC meteorites after the initials of the locations where they were first observed to fall from the sky – near Shergotty, India in 1865, Nakhla, Egypt in 1911, and Chassigny, France in 1815.

SN 1987A: *See* Supernova 1987A.

SNO: Acronym for the Sudbury Neutrino Observatory, Canada. *See* Sudbury Neutrino Observatory.

SNR: Acronym for SuperNova Remnant. *See* supernova remnant.

Soft X-rays: Electromagnetic radiation with photon energies of 1 to 10 keV and wavelengths between about 10^{-9} and 10^{-10} meters.

SOHO: Acronym for the *SOlar and Heliospheric Observatory*. *See SOlar and Heliospheric Observatory*.

Solar activity cycle: A cyclical variation in solar activity with a period of about 11 years between maxima, or minima, of solar activity. Waxing and waning of various forms of solar activity, such as sunspots, flares, and coronal mass ejections, characterize the solar activity cycle. Over the course of an 11-year cycle, sunspots vary both in number and latitude. The complete cycle of the solar magnetic field is 22 years. A dynamo driven by differential rotation and convection may maintain the solar cycle. Activity cycles similar to the solar cycle are apparently typical of stars with convection zones. *See* butterfly diagram, Hale law, solar maximum, solar minimum, Spörer's law, and sunspot cycle.

SOlar and Heliospheric Observatory: NASA launched the *SOlar and Heliospheric Observatory*, abbreviated *SOHO*, on 2 December 1995. It reached its permanent position on 14 February 1996, with operations continuing into 2006. *SOHO* orbits the Sun at the first, L_1, Lagrangian point where the gravitational forces of the Earth and Sun are equal. It is a cooperative project between ESA and NASA to study the Sun from its deep core, through its outer atmosphere and the solar wind out to a distance ten times beyond the Earth's orbit. It has obtained important new information about helioseismology, the structure and dynamics of the solar interior, the heating mechanism of the Sun's million-degree outer atmosphere, the solar corona, and the origin and acceleration of the solar wind. *See* Lagrangian point.

Solar atmosphere: The outer layers of the Sun, from the photosphere through the chromosphere, transition region and corona. The term atmosphere is used to describe the outermost gaseous layers of the Sun because they are relatively transparent at visible wavelengths. *See* atmosphere.

Solar-B: A mission of the Japanese space agency, the Institute of Space and Astronautical Science abbreviated ISAS, proposed as a follow-on to the Japan/United States/United Kingdom *Yohkoh*, or *Solar-A*, collaboration. ISAS has scheduled the launch of *Solar-B* for September 2006. The spacecraft consists of a coordinated set of optical, extreme-ultraviolet and X-ray instruments to measure the detailed density, temperature and velocity structures in the photosphere, transition region and low corona with high spatial, spectral and temporal resolution, resulting in new information about the Sun's varying magnetic fields and their relationship to solar eruptions and atmosphere expansion.

Solar constant: The total amount of solar energy, integrated over all wavelengths, received per unit time and unit area at the mean Sun-Earth distance outside the Earth's atmosphere. Its value is 1366.2 Joule per second per square meter, which is equivalent to 1366.2 watt per square meter, with an uncertainty of ±1.0 in the same units. The solar constant is not constant, but instead varies in tandem with the 11-year solar activity cycle, by about 0.1 percent from maximum to minimum. Temporary dips of up to 0.3 percent and a few days duration are due to the presence of large sunspots on the visible solar disk. The large number of sunspots near the peak in the 11-year cycle is accompanied by a rise in magnetic activity, which creates an increase in the luminous output of plages that exceeds the cooling effects of sunspots. *See* plage, solar activity cycle, and sunspot.

Solar core: The region at the center of the Sun where nuclear reactions release vast quantities of energy.

Solar corona: *See* corona.

Solar cycle: The approximately 11-year variation in solar activity, as well as the number and position of sunspots. Taking Hale's law of sunspot magnetic polarity into account, it leads to a 22-year magnetic cycle. *See* Hale's law, solar activity cycle, and sunspot cycle.

Solar dynamo: *See* dynamo.

Solar eclipse: A blockage of light from the Sun when the Moon is positioned precisely between the Sun and the Earth. A total solar eclipse is seen at places where the umbra of the Moon's shadow cone falls on and moves over the Earth's surface. Although the total duration of a solar eclipse can be as much as four hours, the Sun is completely covered by the Moon, at totality, for at most 7.5 minutes. During the brief moments of a total solar eclipse, darkness falls, and the outer parts of the Sun, the chromosphere and the corona, are seen. At any given point of Earth's surface, a total solar eclipse occurs, on the average, once every 360 years. *See* Baily's beads, chromosphere, corona, and prominence.

Solar flare: A sudden and violent release of matter and energy within a solar active region in the form of electromagnetic radiation, energetic particles, wave motions and shock waves, lasting minutes to hours. A sudden brightening in an active region observed in chromospheric and coronal emissions that typically lasts tens of minutes. The frequency and intensity of solar flares increase near the maximum of the 11-year solar activity cycle. Solar flares accelerate charged particles into interplanetary space. The impulsive or flash phase of flares usually lasts for a few minutes, during which matter can reach temperatures of hundreds of millions of degrees. The flare subsequently fades during the gradual or decay phase lasting about an hour. Most of the radiation is emitted as X-rays but flares are also observed at visible hydrogen-alpha wavelengths and radio wavelengths. They are probably caused by the sudden release of large amounts (up to 10^{25} Joule) of magnetic energy in a relatively small volume in the solar corona. *See* active region, flare, and solar activity cycle.

Solar mass: The amount of mass in the Sun, equal to 1.989×10^{30} kilograms, and the unit in which stellar and galaxy masses are expressed.

Solar Maximum Mission: NASA launched the ***Solar Maximum Mission***, abbreviated ***SMM***, on 14 February 1980, with an in-orbit repair from the ***Space Shuttle Challenger***

TABLE 48 Solar eclipses – total[a]

Date	Maximum Duration (minutes)	Path of totality
21 June 2001	4.95	Atlantic Ocean, Angola, Zambia, Zimbabwe, Mozambique, Madagascar
04 December 2002	2.07	Angola, Botswana, Zambia, Zimbabwe, South Africa, Mozambique, south Indian Ocean, southern Australia
23 November 2003	1.99	Antarctica
08 April 2005	0.70	Eastern Pacific Ocean
29 March 2006	4.12	Eastern Brazil, Atlantic Ocean, Ghana, Togo, Benin, Nigeria, Niger, Chad, Libya, Egypt, Turkey, Russia
01 August 2008	2.45	Northern Canada, Greenland, Arctic Ocean, Russia, Mongolia, China
22 July 2009	6.65	India, Nepal, Bhutan, Burma, China, Pacific Ocean
11 July 2010	5.33	South Pacific Ocean, Easter Island, Chile, Argentina
13 November 2012	4.03	Northern Australia, south Pacific Ocean
03 November 2013	1.67	Atlantic Ocean, Gabon, Congo, Zaire, Uganda, Kenya
20 March 2015	2.78	North Atlantic Ocean, Faeroe Islands, Arctic Oceans, Svalbard
09 March 2016	4.17	Indonesia (Sumatra, Borneo, Sulawesi, Halmahera), Pacific Ocean
21 August 2017	2.67	Pacific Ocean, United States (Oregon, Idaho, Wyoming, Nebraska, Missouri, Illinois, Kentucky, Tennessee, North Carolina, South Carolina), Atlantic Ocean
02 July 2019	4.55	South Pacific Ocean, Chile, Argentina
14 December 2020	2.17	Pacific Ocean, Chile, Argentina, south Atlantic Ocean, South Africa
04 December 2021	1.92	Antarctica
20 April 2023	1.27	South Indian Ocean, Western Australia, Indonesia, Pacific Ocean
08 April 2024	4.47	Pacific Ocean, Mexico, United States (Texas, Oklahoma, Arkansas, Missouri, Kentucky, Illinois, Indiana, Ohio, Pennsylvania, New York, Vermont, New Hampshire, Maine), southeastern Canada, Atlantic Ocean
12 August 2026	2.32	Greenland, Iceland, Spain
02 August 2027	6.38	Atlantic Ocean, Morocco, Spain, Algeria, Libya, Egypt, Saudi Arabia, Yemen, Somalia
22 July 2028	5.17	South Indian Ocean, Australia, New Zealand
25 November 2030	3.73	South West Africa, Botswana, South Africa, south Indian Ocean, southeastern Australia
14 November 2031	1.13	Central Pacific Ocean
30 March 2033	2.63	Alaska, Arctic Ocean
20 March 2034	4.17	Atlantic Ocean, Nigeria, Cameroon, Chad, Sudan, Egypt, Saudi Arabia, Iran, Afghanistan, Pakistan, India, China

(continued)

TABLE 48   Solar eclipses – total[a]

Date	Maximum Duration (minutes)	Path of totality
02 September 2035	2.90	China, Korea, Japan, Pacific Ocean
13 July 2037	3.98	Australia, New Zealand, south Pacific Ocean
25–26 December 2038	2.30	Indian Ocean, western and southern Australia, New Zealand, Pacific Ocean
15 December 2039	1.85	Antarctica
30 April 2041	1.85	South Atlantic Ocean, Angola, Congo, Uganda, Kenya, Somalia
20 April 2042	4.85	Indonesia, Malaysia, Philippines, Pacific Ocean
23 August 2044	2.08	Greenland, Canada, United States (Montana, North Dakota)
12 August 2045	6.01	United States (California, Nevada, Utah, Colorado, Kansas, Oklahoma, Arkansas, Mississippi, Alabama, Georgia, Florida), Haiti, Dominican Republic, Venezuela, Guyana, Suriname, French Guyana, Brazil
02 August 2046	4.85	Eastern Brazil, Atlantic Ocean, Namibia, Botswana, South Africa, south Indian Ocean
05 December 2048	3.47	South Pacific Ocean, Chile, Argentina, south Atlantic Ocean, Namibia, Botswana
25 November 2049	0.63	Indian Ocean, Indonesia
20 May 2050	0.37	South Pacific Ocean
30 March 2052	4.13	Pacific Ocean, Mexico, United States (Louisiana, Alabama, Florida, Georgia, South Carolina), Atlantic Ocean

[a]Total eclipses of the Sun, adapted from *Totality: Eclipses of the Sun* by M. Littmann and K. Willcox, Honolulu: University of Hawaii Press 1991, Appendix 1, pp. 175–178.

on 6 April 1984 and a mission end on 17 November 1989. *SMM* obtained important new insights to powerful eruptions on the Sun known as solar flares and coronal mass ejections, during a maximum in the 11-year solar activity cycle.

Solar nebula: The disk-shaped cloud of gas and dust out of which the Sun and planetary system formed.

Solar neutrino problem: Massive, subterranean neutrino detectors find only one-third to one-half the number of solar neutrinos that theoretical calculations predict. The solar neutrino problem has been explained by a transformation of solar neutrinos into an undetectable form on their way out of the Sun. *See* MSW effect, neutrino oscillation, and Sudbury Neutrino Observatory.

Solar parallax: The angular size of the Earth's radius at a distance of one astronomical unit, amounting to 8.794 148 seconds of arc. *See* astronomical unit and parallax.

Solar System: The Solar System consists of the Sun, its planets, the interplanetary medium, the solar wind, and the smaller bodies orbiting the Sun, including asteroids,

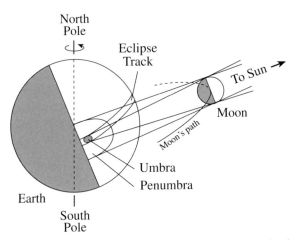

FIG. 46 Solar eclipse During a solar eclipse, the Moon casts its shadow upon the Earth. No portion of the Sun's photosphere can be seen from the umbral region of the Moon's shadow (*small gray spot*); but the Sun's light is only partially blocked in the penumbral region (*larger half circle*). A total solar eclipse, observable only from the umbral region, traces a narrow path across the Earth's surface.

FIG. 47 Solar flare model A solar flare is powered by magnetic energy released from a magnetic interaction site above the top of the coronal loop shown here. Electrons are accel-

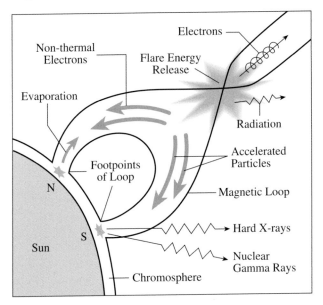

erated to high speed, generating a burst of radio energy as well as impulsive loop-top hard X-ray emission. Some of these non-thermal electrons are channeled down the magnetic loop and strike the chromosphere at nearly the speed of light, emitting hard X-rays at the loop footpoints. When beams of accelerated protons enter the dense, lower atmosphere, they cause nuclear reactions that result in gamma-ray spectral lines and energetic neutrons. Material in the chromosphere is heated very quickly and rises into the loop, accompanied by a slow, gradual increase in soft X-ray radiation. This upwelling of heated material is called chromospheric evaporation.

comets, and objects in the Kuiper belt and the Oort comet cloud. The planets and Sun were formed together from the collapse of a rotating, interstellar cloud of gas and dust known as the solar nebula. *See* asteroid, comet, Kuiper belt, Oort comet cloud, planet, solar nebula, and solar wind.

Solar TErrestrial RElations Observatory: NASA plans to launch the two identical spacecraft of the **Solar TErrestrial RElations Observatory**, abbreviated **STEREO**, in 2006. One spacecraft will lead Earth in its orbit, and one will be lagging; the combined images and radio tracking from the two spacecraft will provide a three-dimensional view of coronal mass ejections from their onset at the Sun to the orbit of the Earth, improving our understanding of these solar explosions and aiding space weather forecast capabilities.

Solar Time: Time kept by the rising and setting of the Sun viewed from the rotating Earth, more technically known as Greenwich Mean Time and Universal Time. *See* Greenwich Mean Time, Royal Greenwich Observatory, and Universal Time.

Solar wind: A steady flow of energetic charged particles, mainly electrons and protons, and entrained magnetic fields moving out from the Sun in all directions into interplanetary space at supersonic speed. In 1951 the German astronomer **Ludwig Biermann (1907–1986)** proposed that the straight ion tails of comets, which always point away from the Sun, are accelerated by a continuous, electrified flow of charged particles from the Sun. In 1958 the American physicist **Eugene N. Parker (1927–)** showed that the million-degree electrons and protons in the corona will overcome the Sun's gravity and accelerate to supersonic speeds, naming the resultant radial outflow the solar wind. He also proposed that the solar wind carries the Sun's magnetic field with it, forming a spiral magnetic pattern in interplanetary space as the Sun rotates. Measurements from the Soviet *Luna 2* spacecraft in 1959 confirmed the existence of the solar wind by direct measurements of its protons, and in 1962–63 the American *Mariner 2* spacecraft determined its density, about 5 million electrons and 5 million protons per cubic meter near Earth, and measured its speed. The solar wind has a fast component, or high-speed stream, with a velocity of about 800 kilometers per second, and a slow-speed component moving at about half this speed. In 2000, comparisons of observations with the *Ulysses* spacecraft, which passed over the Sun's poles, with those of the *Yohkoh* and **SOHO** solar spacecraft demonstrated that much, if not all, of the high-speed solar wind comes from open magnetic fields in polar coronal holes, at least during the minimum in the Sun's 11-year activity cycle, and that the slow wind is confined to low latitudes near an equatorial streamer belt. The solar wind carries away about 10^{-13} of the Sun's mass per year. Although the solar wind is diverted around the Earth's magnetosphere, some of its particles enter the magnetosphere. *See* aurora, corona, coronal hole, coronal streamer, heliosphere, high-speed stream, magnetosphere, and Van Allen belts.

Solstice: The summer and winter solstices are the extreme northern and southern position of the Sun during its yearly path among the stars, when the Sun reaches its greatest declination of 23.5 degrees north and south. The two dates when the Sun is farthest from the equator. On the summer solstice, occurring around 21 June, the Sun is farthest north and the daylight hours in the northern hemisphere are longest. On

TABLE 49   Solar wind[a]

Parameter	Mean Value
Particle Density, N	10 million particles per cubic meter (5 million electrons and 5 million protons)
Velocity, V	400 km s^{-1} and 800 km s^{-1}
Flux, F	10^{12} to 10^{13} particles per square meter per second.
Temperature, T	1.2×10^5 K (protons) to 1.4×10^5 K (electrons)
Particle Thermal Energy, kT	2×10^{-18} J $\approx$ 12 eV
Proton Kinetic Energy, $0.5\,m_p V^2$	10^{-16} J $\approx$ 1,000 eV = 1 keV
Thermal Energy Density, NkT	10^{-11} J m^{-3}
Proton Energy Density, $0.25\,N\,m_p\,V^2$	10^{-9} J m^{-3}
Magnetic Field Strength, B	6×10^{-9} T = 6 nT = 6×10^{-5} G

[a] Mean values of solar-wind parameters at the mean distance of the Earth from the Sun, or at one astronomical unit, 1 AU, where 1 AU = 1.496×10^{11} m; the Sun's radius, $R_\odot$, is $R_\odot = 6.955 \times 10^8$ m. Boltzmann's constant k = $1.380\,66 \times 10^{-23}$ J K^{-1} relates temperature and thermal energy. The proton mass m_p = 1.672 623 $\times 10^{-27}$ kg.

the winter solstice, occurring around December 22, the Sun is farthest south and the daylight hours in the northern hemisphere are shortest. *See* declination and season.

Sombrero Galaxy: The spiral galaxy about 30 million light-years away in the direction of the constellation Virgo, designated M104 and NGC 4594, seen nearly edge-on with a dark band of interstellar dust crossing the center of its large central bulge.

Sound speed: *See* velocity of sound.

Sound waves: *See* acoustic waves.

Southern cross: Popular name for the southern-hemisphere constellation Crux. *See* Acrux, and Crux.

Space Infra-Red Telescope Facility: Abbreviated ***SIRTF*** and renamed the ***Spitzer Space Telescope***. *See* ***Spitzer Space Telescope***.

Space Interferometry Mission: NASA's ***Space Interferometery Mission***, abbreviated ***SIM***, is in development, with a launch scheduled for 2009. It will use a space interferometer to probe nearby stars for Earth-sized planets and to measure the positions and distances of stars hundreds of times more accurately than any previous program.

Space Telescope Science Institute: The Space Telescope Science Institute, abbreviated STScI, is the astronomical research center responsible for operating the ***Hubble Space Telescope*** as an international observatory. The STScI is located at the Homewood campus of Johns Hopkins University.

Space-time: A four-dimensional description with three dimensions of space and one of time, employed in the *Special* and *General Theory of Relativity*. *See Special Theory of Relativity* and *General Theory of Relativity*.

Space weather: Varying conditions in interplanetary space and the Earth's magnetosphere controlled by the variable solar wind.

Special Theory of Relativity: Explanation of the properties of objects moving at high speeds, approaching the velocity of light, proposed in 1905 by the German physicist **Albert Einstein (1879–1955)**. The high-speed objects exhibit an increase in mass, a Lorentz contraction of length, and a slowing of time, or time dilation, when compared to non-moving objects at rest. The *Special Theory of Relativity* rests upon the postulates: 1. The Principle of Relativity, which states that the laws of nature and the results of experiments performed in an inertial frame are independent of the uniform velocity of the system. 2. There exists in nature a limiting, invariant speed, the speed of light, a universal constant. *See* Lorentz contraction, Michelson-Morley experiment, Minkowski metric, time dilation, and velocity of light.

Spectral classification: *See* stellar classification.

Spectral line: A feature observed in the spectra of stars and other luminous objects at a specific frequency or wavelength, either in bright emission or in dark absorption. A spectral line looks like a line in a display of radiation intensity as a function of wavelength or frequency. Atoms produce this spectral feature as they absorb or emit light, and it can be used to determine the chemical ingredients of the radiating source. Spectral lines are also used to infer the radial velocity and magnetic field of the radiator. *See* absorption line, Doppler effect, emission line, radial velocity, spectrum, and Zeeman effect.

Spectral type: *See* spectral classification.

Spectrograph: Also known as a spectrometer, a spectrograph spreads light or other electromagnetic radiation into its component wavelengths, collectively known as a spectrum, and records the result electronically or photographically. Spectra are now often recorded with Charge Coupled Devices, or CCDs, from which a computer can analyze information in digital form. High-dispersion instruments spread the spectral lines widely so that a particular wavelength can be studied in detail. An example of such a high-dispersion device is the spectroheliograph. *See* Charge Coupled Device and spectroheliograph.

Spectroheliograph: A type of spectrograph used to image the Sun in the light of one particular wavelength only, invented independently in 1891 by the French astronomer **Henri Deslandres (1853–1948)** and by the American astronomer **George Ellery Hale (1868–1938)**.

Spectrometer: *See* spectrograph.

Spectroscope: An instrument that spreads electromagnetic radiation into its various wavelengths, especially the component colors of visible light. The display, known as a spectrum, can be used to determine the ingredients, magnetic fields and motions of the cosmic source of the radiation. *See* spectral line.

Spectroscopic binary: A double star whose two components are inferred from the periodic Doppler shifts of the absorption lines in their combined spectrum. The German

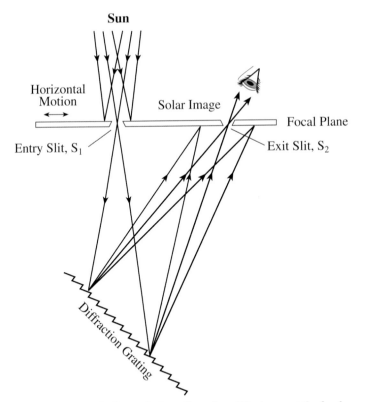

Sun

Horizontal
Motion

Solar Image

Focal Plane

Entry Slit, S_1

Exit Slit, S_2

Diffraction Grating

FIG. 48 **Spectroheliograph** A small section of the image at the focal plane of a telescope is selected with a narrow entry slit, S_1, and light passes to a diffraction grating, producing a spectrum. A second slit, S_2, at the focal plane selects a specific wavelength from the spectrum. If the plate containing the two slits is moved horizontally, then the entrance slit passes adjacent strips of the image. The light leaving the moving exit slit then builds up an image of the Sun at a specific wavelength.

astronomer **Hermann Carl Vogel (1841–1907)** discovered spectroscopic binary stars when studying the periodic displacements of the spectral lines of the eclipsing binary stars Algol and Spica. *See* Algol and Doppler effect.

Spectroscopic parallax: The determination of a star's parallax, or distance, from its spectral type and apparent luminosity. The absolute luminosity can be inferred from the spectral type, and it falls off as the inverse square of the distance to give the apparent luminosity. *See* absolute luminosity, apparent luminosity, inverse square law, parallax, and stellar classification.

Spectroscopy: The measurement and study of a spectrum, including the wavelength and intensity of emission and absorption lines, to determine the chemical composition, motion or magnetic field of the radiating source. *See* spectrum.

Spectrum: The distribution of the intensity of electromagnetic radiation with wavelength. Electromagnetic radiation, arranged in order of wavelength from long wavelength, low frequency, radio emissions to short wavelength, high frequency, gamma rays; also, a narrower band of wavelengths, called the visible spectrum, as when sunlight is dispersed by a prism or rainbow into its component colors. A continuous spectrum is an unbroken distribution of radiation over a broad range of wavelengths, such as the separation of white light into its component colors from red to violet. Continuous spectra are often punctuated with emission or absorption lines, called line spectra, which can be examined to reveal the composition, motion and magnetic field of the radiating source. *See* absorption line, Doppler effect, Doppler shift, emission line, Fraunhofer lines, spectral line, spectroheliograph, and Zeeman effect

Speed of light: *See* velocity of light.

Speed of sound: *See* velocity of sound.

Spica: A blue-white main-sequence, or dwarf, star of spectral type B1 and apparent visual magnitude 0.98, also known as Alpha Virginis. At a distance of 260 light-years, it has a parallax of 0.012 seconds of arc, an absolute magnitude of –3.5 and an intrinsic luminosity about 13 000 times that of the Sun. Spica also has a companion dwarf star of spectral type B4 that is less than 10 percent as luminous as Spica, and separated from Spica by only 0.122 AU or 26 solar radii. The name means "the Spike" or "Ear of Wheat", which the Virgin Mary holds in ancient portrayals of the constellation Virginis.

Spicule: Narrow, predominantly radial, spike-like structures extending from the solar chromosphere into the corona, observed in hydrogen-alpha lines and concentrated at the cell boundaries of the supergranulation. A spicule lasts between five to fifteen minutes and has a velocity of about 25 kilometers per second. They are about 1 kilometer thick and more than 10,000 kilometers long. *See* chromosphere, hydrogen-alpha line, and supergranulation.

Spiral arms: A pinwheel structure, composed of dust, gas and young stars, which winds its way out from the center of a spiral galaxy.

Spiral galaxy: A galaxy containing spiral arms coiling and winding out from a central bulge or nucleus, forming a flattened, disk-shaped region. A typical spiral galaxy has a spherical central bulge of older stars surrounded by a flattened galactic disk that contains a spiral pattern of young stars, as well as interstellar gas and dust. Our Milky Way Galaxy and the Andromeda galaxy, M31, are spiral galaxies, as are the Pinwheel and Sombrero galaxies.

Spiral nebula: Historical name for a type of object that became known as a spiral galaxy. *See* spiral galaxy.

Spirit: *See **Mars Exploration Rover Mission**.*

Spitzer Space Telescope: NASA launched the ***Spitzer Space Telescope***, abbreviated *SST*, on 25 August 2003, with an expected 2.5-year mission lifetime. It is an infrared telescope with an 0.85-meter (33.5-inch) primary mirror, obtaining images and spectra of the infrared energy, or heat, radiated by cosmic objects at wavelengths between 3 and 180 microns, where 1 micron is one-millionth of a meter. The *Spitzer Space Telescope* provides information on the formation, composition and evolution of planets, stars and galaxies. The mission was initially designated the *Space Infra-Red Telescope Facility*, or *SIRTF* for short, and renamed after the American astrophysicist **Lyman Spitzer, Jr. (1914–1997)**.

Spörer minimum: A period of low sunspot activity in the fifteenth century (about AD 1420–1570), named after the German astronomer **Gustav Friedrich Wilhelm Spörer (1822–1895)** who also called attention to the Maunder minimum as early as 1887. *See* Maunder minimum.

Spörer's law: The appearance of sunspots at lower solar latitudes over the course of the 11-year solar activity cycle, drifting from mid-latitudes of 30 to 40 degrees north and south towards the equator as the cycle progresses. The sunspot migration to lower average latitudes was first discovered in 1869 by the English astronomer **Richard Carrington (1826–1875)**, over an incomplete part of one solar cycle, and investigated in greater detail by the German astronomer **Gustav Friedrich Wilhelm Spörer (1822–1895)**. *See* solar activity cycle, sunspot, and sunspot cycle.

SS 433: An emission line star that exhibits varying wavelengths changes, with simultaneous redshifts and blueshifts, discovered by the American astronomer **Bruce Margon (1948–)** and his colleagues in 1979. If the wavelength changes are attributed to the Doppler effect of source motion, then the star seemed to be both coming and going with speeds of up to 50 thousand kilometers per second. But the star itself isn't moving; matter is instead being ejected in two narrow, collimated jets of matter from a central neutron star. A bright massive companion star, with a mass of 10 to 20 solar masses, is losing its outer layers in a stellar wind that flows down toward the neutron star, overloading an accretion disk that cannot digest all of the in-falling matter and hurls some of it out in opposite directions along the tightly collimated jets.

SST: Acronym for *Spitzer Space Telescope*. *See Spitzer Space Telescope*.

Star: A self-luminous ball of gas like the Sun whose radiant energy is produced by nuclear fusion reactions in the stellar core, or an astronomical object that once radiated this way, such as a white dwarf or neutron star. Most stars are found on the main sequence of the Hertzsprung-Russell diagram; they shine by the conversion of hydrogen nuclei into helium nuclei in either the carbon-nitrogen-oxygen cycle or the proton-proton chain. When the core hydrogen is depleted, a main-sequence star evolves into a giant star, obtaining its radiant energy by the nuclear burning of helium or heavier elements. The temperature and luminosity of a star is determined by its mass. The most massive stars are about 100 solar masses, or one hundred times heavier than the Sun. The minimum stellar mass is about 0.085 solar masses, or about 85 times the mass of Jupiter; less massive stars do not have enough mass and sufficient pressure and temperature to initiate nuclear reactions in their cores. *See* carbon-nitrogen-oxygen cycle, fusion, giant star, Hertzsprung-Russell diagram, main sequence, mass-luminosity relation, and proton-proton chain.

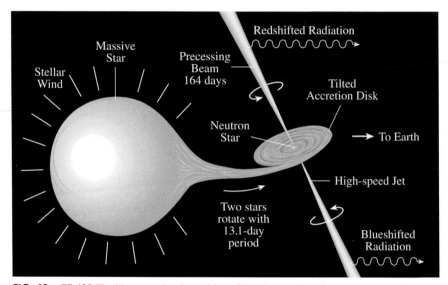

FIG. 49   SS 433 The bizarre and unique object, SS 433, consists of a neutron star and a massive visible-light star in orbit about each other with a 13.1-day period. The massive star is losing gas in a stellar wind, and the neutron star pulls some of this material into an accretion disk that emits X-rays. However, the in-falling matter cannot be accreted fast enough, and the excess material is shot out from the center of the accretion disk in two oppositely directed jets. Bullet-like concentrations of gas spew out along the jets, moving at a quarter the speed of light. The spectral lines of SS 433 come from the twin high-speed jets of gas, which create simultaneous red and blue wavelength shifts and make the star look like it is coming and going at the same time. The accretion disk is tipped slightly and precesses like a top, making the jets swing steadily around a cone every 164 days. The component of velocity directed along an observer's line of sight varies with the same period, so the observer sees both a redshifted and a blueshifted line, with wavelengths that change with time.

Starburst galaxy: A galaxy undergoing an exceptionally high rate of star formation. Discovered in 1983 with instruments aboard the *InfraRed Astronomical Satellite*, abbreviated *IRAS*, starburst galaxies are much brighter at infrared wavelengths, when compared to their visible-light output, than other galaxies.

Star cluster: Gravitationally bound aggregation of stars, much smaller and much less massive than a galaxy. Star clusters are of two types: the open clusters, containing hundreds of Populations I stars, and the more massive globular clusters consisting of between ten thousand and ten million older Populations II stars within a volume that is about 100 light-years across. Open clusters are found in the disk of our Galaxy and globular clusters in the galactic halo. *See* globular cluster and open cluster.

Stardust: NASA launched the *Stardust* spacecraft on 7 February 1999. After nearly four years of space travel, it encountered Comet Wild 2, on 02 January 2004, where it obtained close-up images of the comet nucleus and collected dust and carbon-based samples using a substance called aerogel. The cargo was parachuted in a reentry capsule to the Earth's surface on 15 January 2006.

TABLE 50 Stars – brightest[a]

| Star | R.A (2000) | | | Dec (2000) | | | D[b] (lyr) | Greek Letter Name | Spectral Class[c] | | Visual[d] Magnitude | | Absolute Magnitude | |
	h	m	s	°	′	″			A	B	A	B	A	B
Sirius	06	45	08.9	−16	42	58	8.6	Alpha CMa	A0 V	wd	−1.46	+08.7	+1.42	+11.6
Canopus	06	23	57.1	−52	41	45	326	Alpha Car	F0 II	—	−0.62	—	−5.63	—
Arcturus	14	15	39.7	+19	10	57	37	Alpha Boo	K1 III	—	−0.04	—	−0.30	—
Rigil Kentaurus	14	39	35.9	−60	50	07	4.4	Alpha Cen	G2 V	K1 V	−0.01	+1.33	+4.34	+5.68
Vega	18	36	56.3	+38	47	01	25.3	Alpha Lyr	A0 V	—	+0.03	—	+0.65	—
Capella [e]	05	16	41.4	+45	59	53	42	Alpha Aur	G8 III	G1III	+0.08	+10.2	−0.40	+9.5
Rigel	05	14	32.3	−08	12	06	770	Beta Ori	B8 Ia	B9	+0.18	+6.6	−6.7	−0.4
Procyon	07	39	18.1	+05	13	30	11.4	Alpha CMi	F5 IV	wd	+0.38	+10.7	+2.71	+13.0
Achernar	01	37	42.9	−57	14	12	142	Alpha Eri	B3 V	—	+0.45	—	−2.7	—
Betelgeuse	05	55	10.3	+07	24	25	427	Alpha Ori	M2 Ia	—	+0.45v	—	−5.1	—

TABLE 50   Stars – brightest[a]

Star	R.A (2000)			Dec (2000)			D[b] (lyr)	Greek Letter Name	Spectral Class[c]		Visual[d] Magnitude		Absolute Magnitude	
	h	m	s	°	′	″			A	B	A	B	A	B
Hadar	14	03	49.4	−60	22	23	320	Beta Cen	B1 III	—	+0.61	+4.0	−4.3	−0.8
Altair	19	50	47.0	+08	52	06	16.8	Alpha Aql	A7 IV-V	—	+0.76	—	+2.20	—
Aldebaran	04	35	55.2	+16	30	33	65	Alpha Tau	K5 III	M2V	+0.87	+13.0	−0.6	+12.0
Antares	16	29	24.4	−26	25	55	604	Alpha Sco	M1.5 Ib	B4 V	+0.90	+5.3	−5.3	−00.3
Spica	13	25	11.6	−11	09	41	260	Alpha Vir	B1 V	—	+0.98	—	−3.5	—
Pollux	07	45	18.9	+28	01	34	34	Beta Gem	K0 III	—	+1.16	—	+1.1	—
Fomalhaut	22	57	39.1	−29	37	20	25.1	Alpha PsA	A3 V	K4 V	+1.17	+6.5	+1.7	+7.3
Deneb	20	41	25.9	+45	16	49	3230	Alpha Cyg	A2 Ia	—	+1.25	—	−8.7	—
Mimosa	12	47	43.2	−59	41	19	490	Beta Cru	B0.5 III	—	+1.25	—	−4.6	—
Acrux	12	26	35.9	−63	05	57	320	Alpha Cru	B0.5 IV	B1 V	+1.58	+2.09	−3.8	−3.3

[a] Adapted from Kaler, James B., *Stars*, New York, Scientific American Library 1992, Appendix 3, p. 250, *The Hundred Greatest Stars*, New York, Copernicus Books, Springer-Verlag, 2002, Appendix C, p. 208, Lang, Kenneth R. *Astrophysical Data*, New York, Springer, 1992, p. 166, and Murdin, Paul (ed.), *Encyclopedia of Astronomy and Astrophysics*, New York, Nature Publishing Group, 2001. See also STARS at http://www.astro.uiuc.edu/~ kaler/sow/sowlist.html.

[b] The distances, D, are in light-years, abbreviated lyr.

[c] The luminosity classes are Ia = Supergiant of high luminosity, Ib = Supergiant of lower luminosity, II = bright giant, III = Normal giant, IV = Subgiant, V = Main-sequence star, or dwarf star, VI = Subdwarf.

[d] The stars are listed in order of increasing apparent visual magnitude or decreasing apparent brightness for the brightest component A; if the star is a binary star the other component is B.

[e] Capella has a third component C with a spectral class M5 V, a visual magnitude of +13.7 and an absolute visual magnitude of +13.

TABLE 51   Stars – nearest[a]

Star	Apparent magnitude[b]	Spectral type	Absolute magnitude	Distance (light-years)
Sun	−26.78	G2	4.82	—[c]
Proxima Centauri	11.01 v	M5	15.45	4.22
α Centauri A	−0.01	G2	4.34	4.39
α Centauri B	1.35	K1	5.70	4.39
Barnard's Star	9.54	M4	13.24	5.94
Wolf 359	13.46 v	M6.5	16.57	7.80
Lalande 21185	7.49	M2	10.46	8.31
Sirius A	−1.44	A0	1.45	8.60
Sirius B	8.44	DA2	11.34	8.60
UV Ceti A	12.56 v	M5.5	15.42	8.70
UV Ceti B	12.96 v	M6.5	15.81	8.70
Ross 154	10.37	M3.5	13.00	9.69
Ross 248	12.27	M5.5	14.77	10.30
ε Eridani	3.72	K2	6.18	10.50
HD 217987	7.35	M2	9.76	10.73
Ross 128	11.12 v	M4.5	13.50	10.89
L789–6	12.32	M5	14.63	11.20
61 Cygni A	5.20 v	K5	7.49	11.36
Procyon A	0.40	F5	2.68	11.41
Procyon B	10.7	DF	13.00	11.41
61 Cygni B	6.05 v	K7	8.33	11.42
HD 173740	9.70	K5	11.97	11.47
HD 173739	8.94	K5	11.18	11.64
GX Andromedae A	8.09 v	M1	10.33	11.64
GX Andromedae B	11.10	M4	13.35	11.64
G51-15	14.81	M6.5	17.01	11.80
ε Indi	4.69	K5	6.89	11.83
τ Ceti	3.49	G8	5.68	11.90
L372–58	13.03	M4.5	15.21	11.90

[a] Adapted from Ian Ridpath and John Woodruff, *Philip's Astronomy Dictionary*, London, George Philip Ltd, p. 203.

[b] The v means variable.

[c] The mean distance of the Sun is 0.0000158 light-years.

Star streaming: In 1905 the Dutch astronomer **Jacobus Kapteyn (1851–1922)** announced that the nearby stars are apparently moving in two large, intermingled and oppositely directed streams passing through each other at a relative velocity of about 40 kilometers per second. One star stream carries roughly half the nearby stars toward the constellation Sagittarius at a speed of about 20 kilometers per second, while the other stream carries the remaining closest stars away from this direction at a comparable speed. As explained by the Swedish astronomer **Bertil Lindblad (1895–1965)** in 1925, the two star streams are explained if all the stars in the Milky Way are whirling in differential rotation about a remote galactic center in Sagittarius. *See* differential rotation.

Steady State Cosmology: Theory of an expanding Universe whose overall properties are unchanging in space and time, and which was therefore never in a state of appreciably higher density and had no Big Bang. Although the Universe does not evolve and must remain forever in the same Steady State, without beginning or end, individual galaxies can evolve and the Universe can expand. Matter is constantly being created out of apparently empty space in order to maintain an unchanging mass density as the Universe expands. This theory, proposed in 1948 by two Austrian-born scientists, then at Cambridge University, **Hermann Bondi (1919–2005)** and **Thomas Gold (1920–2004),** and the English astrophysicist **Fred Hoyle (1915–2001)**, is not consistent with the observed three-degree cosmic microwave background radiation and the observed cosmic abundance of helium, which are both attributed to the Big Bang. *See* Big Bang, Big-Bang nucleosynthesis, and cosmic microwave background radiation.

Stefan-Boltzmann constant: The constant of proportionality, denoted by the symbol σ, relating the radiant flux per unit area from a blackbody to the fourth power of its effective temperature. The constant $\sigma = 5.670\ 51 \times 10^{-8}$ Joule m^{-2} °K^{-4} s^{-1}. *See* Stefan-Boltzmann law.

Stefan-Boltzmann law: The absolute luminosity of the thermal emission from a blackbody is given by the Stefan-Boltzmann law in which $L = 4\pi\sigma R^2 T_{eff}^4$, where $\pi = 3.14159$, the Stefan-Boltzmann constant, $\sigma = 5.67051 \times 10^{-8}$ Joule m^{-2} °K^{-4} s^{-1}, the radius is R and the effective temperature is T_{eff}. This law is named for the Austrian physicist **Josef Stefan (1835–1893)**, who showed in 1879 that the radiant energy of a thermal source increases with the fourth power of the temperature, and the Austrian physicist **Ludwig Boltzmann (1844–1906)** who derived the expression in 1884, based on the theory of a blackbody. *See* absolute luminosity, blackbody, and luminosity.

Stellar activity: Emission from a star in excess of that expected from a purely radiative atmosphere, suggesting additional heating sources related to magnetic fields. *See* flares, flare star, and solar activity cycle.

Stellar black hole: A black hole formed from the death of a massive star during a supernova explosion. *See* black hole.

Stellar classification: The sequence of stellar spectral types arranged according to the presence or absence of certain lines in the spectra, designated as O, B, A, F, G, K and M. It describes a sequence of temperature and color, from the hot, blue O and B stars to the cool, red M ones. In 1868 the Italian astronomer **Angelo Pietro Secchi**

TABLE 52 Stellar classification[a]

Class	Dominant Lines	Color	Color Index	Effective Temperature (K)	Examples
O	He II	Blue	−0.3	28 000−50 000	χ Per, ε Ori
B	He I	Blue-White	−0.2	9 900−28 000	Rigel, Spica
A	H	White	0.0	7400−9900	Vega, Sirius
F	Metals; H	Yellow-White	0.3	6000−7400	Procyon
G	Ca II; Metals	Yellow	0.7	4900−6000	Sun, α Cen A
K	Ca II; Ca I	Orange	1.2	3500−4900	Arcturus
M	TiO; Ca I	Orange-Red	1.4	2000−3500	Betelgeuse

[a] An H denotes hydrogen, He is helium, Ca is calcium, and TiO is a titanium oxide molecule. The Roman numeral I denotes a neutral, un-ionized atom, the number II describes an ionized atom missing one electron.

(1818–1878) classified about 4000 stars by their color and visual spectra, dividing them into four main groups with the colors white, yellow, orange and red, and with the respective spectral features of strong absorption lines of hydrogen, strong calcium lines, strong metallic lines and broad absorption lines in the violet. In 1918–24, the American astronomer **Annie Jump Cannon (1863–1961)**, working on photographic plates under the direction of **Edward C. Pickering (1846–1919)**, classified over 500 000 stars for the *Henry Draper Catalogue,* distinguishing the stars on the basis of the absorption lines in their photographic spectra, and arranging most of them in a smooth and continuous spectral sequence O, B, A, F, G, K, M. Her colleague, **Antonia C. Maury (1866–1952)** eventually examined about as many stars as Cannon, and in the process discovered that the spectral classification required a two-dimensional scheme. When the distances and absolute luminosities of stars were determined, the American astronomer **Henry Norris Russell (1877–1957)** discovered, in 1914, that two well-defined classes of stars, the dwarfs and giants, are described by a plot of the absolute luminosities of stars against spectral type. Such a plot is now known as the Hertzsprung-Russell diagram. Also in 1914, the American astronomer **Walter S. Adams (1876–1956)** and the German astronomer **Arnold Kohlschütter (1883–1969)** introduced spectral criteria for the determination of absolute stellar luminosity, making it possible to determine the luminosities of both dwarf, or main sequence, and giant stars, and to estimate the distances of millions of stars from their spectra and apparent brightness alone. In 1943 the American astronomers **William W. Morgan (1906–1994)**, **Philip C. Keenan (1908–2000)** and **Edith Kellman (1911–)** published *An Atlas of Stellar Spectra* that established formal rules for describing a star by spectral type, which defines the disk temperature, and luminosity class, with the designation I for supergiants, II for bright giants, III for normal giants, IV for subgiants, V for dwarf or main-sequence stars, and VI for subdwarfs. This system, known as the MK system, is now widely used. *See* A star, B star, F star, G star, Hertzsprung-Russell diagram, K star, M star and O star.

Stellar evolution: How a star changes with time, producing heavier atomic nuclei from light ones in the process. The present-day birth place of stars is dark, giant clouds of molecular hydrogen whose total mass is between 100 000 and one million times the mass of the Sun. If such a cloud becomes sufficiently massive and dense, the mutual gravitation of its parts will overcome the outward pressure from inside, and the cloud will start falling in on itself. Once collapse is underway, the giant cloud will fragment, and the pieces will themselves collapse until their cores become hot enough to ignite nuclear fusion, burning hydrogen to make the star shine. It has then settled down for a long, rather uneventful life as a main-sequence star, the longest stop in its life history. Both the total number of stars and their lifetime rapidly increase down the main sequence from the high mass, luminous upper end to the low-mass dimmer lower end. Ninety percent of all main-sequence stars have a mass below 0.8 solar masses, and have been on the main sequence ever since they were born. They thus provide us with no information about stellar evolution. Some of the massive stars, that were born long ago, have had enough time to burn up their available hydrogen fuel and advance to the next stage of the stellar life cycle. What happens next is described in the Hertzsprung-Russell diagrams of globular star clusters, whose various kinks and bends have been deciphered using complex theoretical models and high-speed computers. After spending most of its life on the main sequence, at essentially constant effective temperature and luminosity, an aging star consumes the available hydrogen in its core, converting it all into helium. The core then contracts, liberating gravitational energy, and the envelope expands into a giant star. The star's outer layers cool and become red or yellow, but its central temperature increases until it is hot enough to burn helium. Very massive stars near the top of the main sequence become supergiants; those with masses comparable to the Sun become red giants. The supergiants quickly consume their successive internal fuel sources at ever-increasing central temperatures and rates; the helium is converted into carbon, and then the carbon ash serves as fuel, and so on up the chain of the nuclear species. The giants are not sufficiently massive to burn anything but helium. Intense winds from a giant star blow some of its outer atmosphere back into interstellar space. The ultimate fate of a star also depends upon its mass. When giant stars with a moderate, Sun-like mass use up their helium fuel, they collapse until they are about the size of the Earth, or only about one hundredth, or 0.01, of the Sun's radius. The collapsed star is called a white dwarf star, for it has a small size and it is initially white in color. Dying stars with moderate mass can blow off a spherical puff of gas, known as a planetary nebula, before settling down to their final state. More massive stars have an explosive death. After their core has become hot enough to produce iron, there is nothing left to support the star. The core collapses to become a neutron star or a black hole, and the outer layers are blown outward during a supernova explosion, seeding interstellar space with heavy elements that are incorporated into the next generation of stars. *See* giant molecular cloud, giant star, globular cluster, gravitational collapse, Hayashi track, Hertzsprung-Russell diagram, main sequence, nucleosynthesis, planetary nebula, star, supergiant, supernova, and white dwarf.

Stellar parallax: *See* parallax.

Stellar populations: A classification of stars according to their age, composition, and locations in the Galaxy, discovered in 1943 by the German-born American astronomer **Walter Baade (1893–1960)**. Young bluish Population I stars are found in the spiral arms and plane of our Galaxy; older reddish Population II stars are predominantly visible in the halo, including globular clusters, and central nucleus of our Galaxy. Population I stars contain more heavy elements than Population II stars. The Population II stars date back to the formation of the Galaxy from gravitational collapse, while the Population I stars originated after the rotating galactic disk and first generation of stars were formed. The Population I stars contain heavy elements produced inside former stars and dispersed into interstellar space by stellar winds and supernova explosions. *See* Population I and Population II.

Stellar wind: The outward flow of charged particles, mostly protons and electrons, from a star. Young stars evolving toward the main sequence have powerful stellar winds, up to a thousand times stronger than the Sun's solar wind. Old stars evolving into red giants also have strong stellar winds. The velocity of stellar winds range from a few hundred to several thousand kilometers per second. *See* solar wind.

Stephan's Quintet: A group of five peculiar galaxies in the direction of the constellation Pegasus, designated NGC 7317, 7318 A, 7318 B, 7319, and 7320, the first four have the same redshift, or distance, suggesting a physical connection. The group is named for the French astronomer **Edouard Stephan (1837–1923)**, who discovered it in 1877.

STEREO: Acronym for ***Solar TErrestrial RElations Observatory***. *See **Solar TErrestrial RElations Observatory***.

Steward Observatory: The University of Arizona in Tucson operates the Steward Observatory, which includes the 2.3-meter (90-inch) Bok telescope at Kitt Peak. It participates in numerous collaborative projects on Mount Graham, near Safford, Arizona, including the 10-meter (400-inch) Heinrich Hertz Submillimeter Telescope, the 1.8-meter (72-inch) Vatican Advanced Technology Telescope, and the two 8.4-meter (336-inch) components of the Large Binocular Telescope that is under construction. The Steward Observatory also participates in the 6.5-meter (256-inch) Multiple Mirror Telescope on Mount Hopkins, near Amado, Arizona, and the twin 6.5-meter (256-inch) Magellan Telescopes at the Las Campanas Observatory in Chile. *See* Fred L. Whipple Observatory and Las Campanas Observatory.

Stony meteorite: About 94 percent of the meteorites that have been seen to fall and then recovered are stones, composed of 75 to 100 percent silicates and the rest, 0 to 25 percent, nickel and iron. *See* meteorites.

Streamer: *See* coronal streamer and helmet streamer.

String theory: Hypothesis that subatomic particles have extension along one axis, and that their properties are determined by the arrangement and vibration of these strings.

Strong nuclear force: Fundamental force of nature that holds neutrons and protons together, forming the nucleus of an atom.

STScI: Acronym for the Space Telescope Science Institute. *See* Space Telescope Science Institute.

Subaru telescope: A Japanese optical and infrared telescope with a primary mirror that is 8.2 meters (323 inches) in diameter, located near the summit of Mauna Kea, Hawaii. The Subaru telescope is a project of the Astronomical Observatory of Japan, and it is named for the young star cluster the Pleiades, whose Japanese name is "Subaru".

Subatomic: Smaller in size than an atom.

Subatomic particle: *See* particle and subatomic.

Submillimeter radiation: Electromagnetic radiation at wavelengths less than a millimeter, or 0.001 meters, and longer than 0.1 millimeter or 0.0001 meters.

Subsonic: Moving at a speed less than that of sound, or with a velocity that is slower than the velocity of sound. In air on Earth, the velocity of sound is about 340 meters per second, but the velocity of sound depends on both the temperature and composition of the gas. *See* velocity of sound.

Sudbury Neutrino Observatory: Abbreviated SNO, a massive underground neutrino detector in Canada filled with heavy water. Observations from SNO have confirmed that solar neutrinos change form on the way from the Sun to the Earth, and that the amount of neutrinos emitted by the Sun is in accordance with theoretical expectations. *See* MSW effect and solar neutrino problem.

Sun: The central star of the Solar System, around which all the planets, asteroids and comets revolve in their orbits, formed together with the planets 4.6 billion years ago. The Sun is a yellow dwarf star on the main sequence, of spectral type G2 with an effective temperature of 5780 kelvin. The absolute visual magnitude of the Sun is $+4.82$, and its apparent visual magnitude is -26.72. The Sun has a radius of 0.7 million kilometers, or 109 times the Earth's radius, a mass of 1.989×10^{30} kilograms, or 333 000 times the mass of the Earth, and a mean mass density of 1409 kilograms per cubic meter. The Sun has an absolute luminosity of 3.85×10^{26} Joule per second, and is much more luminous than the average star, outshining 95 percent of the stars in the Milky Way. The total amount of solar energy received per unit time and unit area at the mean Earth-Sun distance, or at one astronomical unit, is called the solar constant and amounts to 1366 Joule per second per square meter. The mean distance of the Sun from the Earth is one astronomical unit, which is 149.598 million kilometers. The visible-light spectrum of the photosphere exhibits dark absorption lines, also known as Fraunhofer lines, which can be examined to determine the ingredients of the Sun. Its principal chemical constituents are hydrogen and helium, which account for 92.1 and 7.8 percent of the atoms, respectively, and for respective mass fractions of 70.68 and 27.43 percent. The Sun's energy source is the nuclear fusion of hydrogen into helium by the proton-proton chain, taking place near the center of the Sun, which has a temperature of 15.6 million kelvin. Overlying this core is the radiative zone where the high-energy radiation, pro-

TABLE 53 Sun[a]

Mean distance, AU	$1.495\,978\,7 \times 10^{11}$ m
(from radar measurements of the distance to Venus and Kepler's third law)	
Light travel time from Sun to Earth	499.004 782 seconds
Radius, $R_\odot$	6.955×10^8 m (109 Earth radii)
(from distance and angular extent)	
Volume	1.412×10^{27} m^3 (1.3 million Earths)
Mass, $M_\odot$	1.989×10^{30} kg $=$ (332 946 Earth masses)
(from distance and Earth's orbital period using Kepler's third law)	
Escape velocity at photosphere	6.178×10^5 m s^{-1}
Mean density	1,409 kg m^{-3}
Solar constant, $f_\odot$	1366 J s^{-1} m^{-2} $=$ 1366 W m^{-2}
Luminosity, $L_\odot$	3.85×10^{26} J s^{-1} $=$ 3.85×10^{26} W $=$ 3.85×10^{33} erg s^{-1}
(from solar constant, distance and Stefan-Boltzmann law)	

Principal chemical constituents *(from analysis of Fraunhofer lines)*	(By Number of Atoms)	(By Mass fraction)
Hydrogen	92.1 percent	X = 70.68 percent
Helium	7.8 percent	Y = 27.43 percent
All others	0.1 percent	Z = 1.89 percent

Age *(from age of oldest meteorites)*	4.6 billion years
Density (center)	1.513×10^5 kg m^{-3}
Pressure (center)	2.334×10^{16} Pa
(photosphere)	10 Pa = 0.0001 bar
Temperature (center)	1.56×10^7 K
(photosphere)	5780 K
(chromosphere)	$6 \times 10^3 - 2 \times 10^4$ K
(transition region)	$2 \times 10^4 - 2 \times 10^6$ K
(corona)	$2 \times 10^6 - 3 \times 10^6$ K
Rotation Period (equator)	26.8 days
(30° latitude)	28.2 days
(60° latitude)	30.8 days
Magnetic field (sunspots) *(from Zeeman effect)*	0.1 to 0.4 T $= 1 \times 10^3 - 4 \times 10^3$ G
(polar)	0.001 T = 10 G

[a]Mass density is given in kilograms per cubic meter, or kg m^{-3}; the density of water is 1000 kg m^{-3}. The unit of pressure is Pascal, abbreviated Pa, where 1.013×10^5 Pa = 1 bar and one bar is the pressure of the Earth's atmosphere at sea level.

duced in the core fusion reactions, collides with electrons and ions to be re-radiated in the form of less energetic radiation. Outside the radiative zone is a convective zone in which currents of gas flow upward to release energy at the photosphere before flowing downward to be reheated. The visible disk, or photosphere, from which the light we see comes, is some hundreds of kilometers thick, and white-light images of the photosphere reveal the granules and supergranulation that mark the top of convection cells. It takes about 170 thousand years for radiation to work its way out from the Sun's core to the bottom of the convective zone, and only about 10 days for the heated material to carry energy through the convective zone to the photosphere. Sunlight travels from the photosphere to the Earth in 499 seconds. The internal structure of the Sun can be determined by observing the five-minute oscillations in the photosphere, using the technique of helioseismology. The GONG network of Earth-based instruments and instruments aboard the *SOlar and Heliospheric Observatory*, abbreviated *SOHO*, have been used in helioseismology. The convective zone and photosphere undergo differential rotation, in which the equatorial middle regions rotate faster than the polar regions; at the equator the rotation period is 26.8 days and at 60 degrees latitude it is 30.8 days. The atmospheric layer immediately above the photosphere is the chromosphere. The tenuous, million-degree outermost layers, forming the solar corona, expand outward to form the solar wind. The *Ulysses* spacecraft has examined the solar wind all around the Sun, including above its polar regions for the first time, showing that the high-speed wind emanates from polar coronal holes. The corona is visible edge-on during a total eclipse of the Sun or with a coronagraph; it can be seen across the face of the Sun at X-ray wavelengths, as was done with the *Yohkoh* spacecraft. *SOHO* has been used to examine the solar atmosphere from the photosphere to the Earth, including powerful explosions on the Sun, known as coronal mass ejections and flares, which travel out into interplanetary space and are triggered and powered by the Sun's magnetic fields. These magnetic fields are generated by dynamo action inside the Sun, at the base of the convective zone, rising to pervade the solar atmosphere and produce sunspots and the coronal loops that connect them. The Zeeman effect is used to measure the direction and strength of the magnetic field in the photosphere, showing that the magnetism in sunspots is thousands of times more intense than the Earth's magnetic field. *See* absorption line, chromosphere, convective zone, corona, coronagraph, coronal hole, coronal loop, coronal mass ejection, flare, Fraunhofer line, GONG, granulation, helioseismology, Hertzsprung-Russell diagram, main sequence, photosphere, proton-proton chain, solar activity cycle, *SOlar and Heliospheric Observatory*, solar constant, solar eclipse, solar wind, spectroheliograph, sunspot, sunspot cycle, supergranulation, *Ulysses*, *Yohkoh*, and Zeeman effect.

Sunspot: A dark, temporary concentration of strong magnetic fields in the Sun's photosphere. A sunspot is cooler than its surroundings and therefore appears darker. A typical spot has a central umbra surrounded by a penumbra, although either feature can exist without the other. In the umbra, the effective temperature can be about 4000 kelvin compared with 5780 kelvin in the surrounding photosphere. Sunspots are associated with strong magnetic fields of 0.2 to 0.4 Tesla, and vary in size from about 1000 to 50 000 kilometers across. They occasionally grow to about 200 000 kilometers

TABLE 54   Sun – abundant elements

Element	Symbol	Atomic Number Z	Abundance[a] (logarithmic)	Date of Discovery on Earth
Hydrogen	H	1	12.00	1766
Helium	He	2	[10.93 ± 0.01]	1895[b]
Carbon	C	6	8.39 ± 0.05	(ancient)
Nitrogen	N	7	7.78 ± 0.06	1772
Oxygen	O	8	8.66 ± 0.05	1774
Neon	Ne	10	[7.84 ± 0.06]	1898
Sodium	Na	11	6.17 ± 0.04	1807
Magnesium	Mg	12	7.53 ± 0.09	1755
Aluminum	Al	13	6.37 ± 0.06	1827
Silicon	Si	14	7.51 ± 0.04	1823
Phosphorus	P	15	5.36 ± 0.04	1669
Sulfur	S	16	7.14 ± 0.05	(ancient)
Chlorine	Cl	17	5.50 ± 0.30	1774
Argon	Ar	18	[6.18 ± 0.08]	1894
Potassium	K	19	5.08 ±0.07	1807
Calcium	Ca	20	6.31 ± 0.04	1808
Chromium	Cr	24	5.64 ± 0.10	1797
Manganese	Mn	25	5.39 ± 0.03	1774
Iron	Fe	26	7.45 ± 0.05	(ancient)
Nickel	Ni	28	6.23 ± 0.04	1751

[a] Logarithm of the abundance in the solar photosphere, normalized to hydrogen $=12.00$, or an abundance of 1.00×10^{12}. Indirect solar estimates are marked with [ ]. The data are courtesy of Nicolas Grevesse, Université de Liège.

[b] Helium was discovered on the Sun in 1868, but it was not found on Earth until 1895.

in size, becoming visible to the unaided eye. Their duration varies from a few hours to a few weeks, or months for the very biggest. The number and location of sunspots depend on the 11-year solar activity cycle or sunspot cycle. They usually occur in pairs or groups of opposite magnetic polarity that move in unison across the face of the Sun as it rotates. The leading, or preceding, spot is called the P spot; the following one is termed the F spot. *See* Hale's law, penumbra, solar activity cycle, Spörer's law, sunspot cycle, umbra, and Zeeman effect.

Sunspot cycle: The recurring, eleven-year rise and fall in the number and position of sunspots discovered in 1843 by the German astronomer **Samuel Heinrich Schwabe (1789–1875)**. The conventional onset for the start of a sunspot cycle is the time when the smoothed number of sunspots, the 12-month moving average, has decreased to its minimum value. At the commencement of a new cycle sunspots erupt around latitudes of 35 to 45 degrees north and south. Over the course of the cycle, subsequent spots emerge closer to the equator, continuing to appear in belts on each side of the

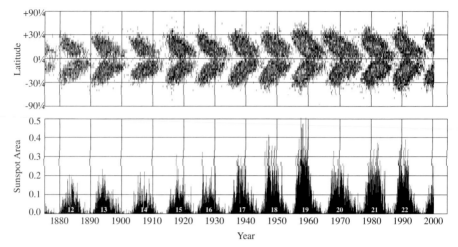

FIG. 50 Sunspot cycle The location of sunspots (*upper panel*) and their total area (*bottom panel*) have varied in an 11-year cycle for the past 100 years, but this activity cycle varies both in cycle length and maximum amplitude. In 1843 the German astronomer **Samuel Heinrich Schwabe (1789–1875)** first measured the periodicity, at about ten years. As shown in the upper panel, the sunspots form at about 30 degrees latitude at the beginning of the cycle and then migrate to near the Sun's equator at the end of the cycle. Such an illustration is sometimes called a "butterfly diagram" because of its resemblance to the wings of a butterfly. The total area of the sunspots (*bottom panel*), given here as a percent of the visible hemisphere, follows a similar 11-year cycle. (Courtesy of David Hathaway, NASA/MSFC.)

equator and finishing at around 7 degrees north and south. This pattern can be demonstrated graphically as a butterfly diagram. All forms of solar activity vary in step with the eleven-year sunspot cycle, including the number and frequency of coronal mass ejections and solar flares and the X-ray intensity of the Sun. *See* coronal mass ejection, flare, solar activity cycle, solar flares, Spörer's law, and X-rays.

Supercluster: A collection of clusters of galaxies, typically about one hundred million, or 10^8, light-years across and containing tens of thousands of galaxies. The Milky Way and the entire Local Group are part of the Local Supercluster, which also includes the Virgo cluster. *See* cluster of galaxies and Local Group.

Supergiant: One of the largest and brightest stars, designated by luminosity class I and up to thousands of times brighter than a giant star of the same spectral type. They are located at the top of the Hertzsprung-Russell diagram. *See* Antares, Betelgeuse, Deneb, Hertzsprung-Russell diagram, and Rigel.

Supergranulation: A system of convective cells, called supergranules, with diameters of about 30 000 kilometers and lifetimes of 1 to 2 days, which cover the visible disk of the Sun, the photosphere. The supergranulation was discovered in 1960 by the American physicist **Robert Leighton (1919–1997)** and his colleagues using an instrument developed by them, which produced an image, called a Dopplergram, of motions in the photosphere. The gas flows horizontally from the center to the edge of each supergranule, carrying the photosphere magnetic field to the edges of the supergranules and forming polygonal-shaped structures called the magnetic network that is also detected as bright calcium emission in the overlying chromosphere. The spicules are also concentrated at the boundaries of the supergranule cells. *See* chromospheric network and spicule.

Super Kamiokande: A massive underground neutrino detector in Japan filled with pure water, replacing the Kamiokande detector. *See* Kamiokande.

Superluminal motion: An apparent motion of a cosmic object that exceeds the velocity of light. Very Long Baseline Interferometry of quasars and radio galaxies have revealed superluminal motion, which is attributed to relativistic high-speed jets that are directed toward the observer and moving at nearly the speed of light, but not exceeding it. *See* Very Long Baseline Interferometry.

Supermassive black hole: A black hole possessing as much mass as a million to a billion stars like the Sun. Supermassive black holes reside in the centers of galaxies and are the engines that power active galactic nuclei, radio galaxies and quasars. A supermassive black hole is located in the elliptical galaxy M87, which is associated with the radio galaxy named Virgo A or 3C 274. The supermassive black hole in M87 feeds the radio galaxy with jets that are detected at visible-light, radio and X-ray wavelengths. *See* active galactic nucleus, black hole, radio galaxy, quasar, and Virgo cluster.

Supernova: Abbreviated SN, the explosion of a star that destroys the star. For a week or so, a supernova may outshine all the other stars in its galaxy, becoming a thousand million, or a billion, times brighter than the Sun and a thousand times brighter than a nova. Type Ia supernovae are explosions of white dwarf stars that receive material from a nearby companion star and exceed the Chandrasekhar limit. Type Ib, Ic and II supernovae are explosions of massive stars that have depleted their nuclear fuel. The inward collapse and immense heat and pressure at the core result in a stellar explosion that ejects most of the mass of the star into interstellar space, forming a supernova remnant. A neutron star can be formed at the center of such a supernova. The Crab Nebula, M1, is the expanding remnant of a supernova observed in 1054 by Chinese astronomers; the supernova remnant is centered on a neutron star, which is also a pulsar. Stars greater than roughly six solar masses may evolve so fast that they become unstable and explode to form a Type Ib, Ic or Type II supernova. Type Ia, Ib, and Ic supernovae have no hydrogen in their spectra, whereas type II supernovae do have hydrogen in their spectra. In 1572 the Danish astronomer **Tycho Brahe (1546–1601)** observed a supernova in the direction of the constellation Cassiopeia, and in 1604 the German astronomer **Johannes Kepler (1571–1630)** observed one in Ophiuchus. The first supernova to be visible to the unaided eye since 1604 occurred on 23 February 1997 in the Large Magellanic Cloud, and is designated Supernova 1987A, or SN 1987A. Supernovae can nevertheless be routinely observed in other galaxies using dedicated telescopes and electronic detection equipment. *See* Chandrasekhar limit, Crab Nebula, Kepler's star and supernova remnant, neutron star, nova, supernova 1987A, supernova remnant, Tycho's star and supernova remnant, type Ia supernova and type II supernova.

TABLE 55   Supernovae – historical[a]

Date (AD)	Remnant Name	Maximum Apparent Visual Magnitude (mag)	Time Visible To Unaided Eye (months)	Distance[b] (kpc)	Radius[c] (pc)
185	RCW 86	−8.0	20	3.0	17.5
393	CTB 37A or B	0.0	8	10.4 ± 3.5	12.0
1006	PKS 1459−41	−9.5	>24	1.0	4.4
1054	Crab, 3C 144	−5.0	22	2.0 ± 0.1	1.95
1181	3C 58	0.0	6	2.6 ± 0.5	2.65
1572	Tycho, 3C 10	−4.0	16	2.3 ± 0.3	2.75
1604	Kepler, 3C 358	−3.0	12	4.4	1.9
1658[d]	Cas A, 3C 461	—	—	2.8 ± 0.6	2.65

[a] Adapted from Lang, Kenneth R., *Astrophysical Data: Planets and Stars*, New York, Springer-Verlag, 1992, p. 704, 705, 716. Galactic and celestial coordinates are given on pages 725 to 744 of that book.

[b] The distance is given in kiloparsecs, abbreviated kpc, where 1 kpc = 3260 light-years. Good distance estimates have their uncertainties given, while less certain but reasonable distances do not have the errors given.

[c] The radius is in parsecs, abbreviated pc, where 1 pc = 3.26 light-years = 3.086×10^{15} kilometers.

[d] The radio source Cassiopeia A, also known as 3C 461, is thought to be a supernova remnant, whose explosion date is inferred from a backward extrapolation of its expansion velocity, of 7000 km s^{-1}, its radius, and the assumption that it has not decelerated. No supernova explosion has been recorded in historical documents, perhaps because it was hidden by interstellar dust.

Supernova 1987A: The type II supernova explosion of a blue supergiant star in the Large Magellanic Cloud on 23 February 1997, the first supernova to be visible to the unaided, or naked, eye since 1604. Dubbed SN 1997A, the supernova was discovered by **Ian Shelten (1957–)**. The 20-solar-mass star exploded with a peak apparent visual magnitude of 6.4. At its distance of 170 000 light-years, the maximum absolute visible luminosity was –12.8 and the peak intrinsic visible luminosity was 10 million times that of the Sun. Underground neutrino detectors recorded neutrinos released during the collapse of the stellar core, about three hours before the first visible sighting of the supernova. The neutrino luminosity briefly equaled the visible luminous output of all the stars in the Universe. *See* supernova and type II supernova.

Supernova remnant: Abbreviated SNR, the expanding shell of matter thrown off into space during the outburst of a supernova. These remnants are often strong radio sources, emitting synchrotron radiation; the ejected material also collides with the surrounding interstellar gas and heats it up to about a million kelvin, emitting intense X-rays. *See* Crab nebula, supernova and synchrotron radiation.

Supersonic: Moving at a speed greater than that of sound, or with a velocity that exceeds the velocity of sound. In air on Earth, the velocity of sound is about 340 meters per second, but the velocity of sound depends on both the temperature and composition of the gas. *See* velocity of sound.

TABLE 56   Supernova 1987A (SN 1987A)[a]

Progenitor Star

Name	Sk$-69°202$ = Sanduleak $-69°202$
Location	LMC = Large Magellanic Cloud
Right Ascension	R.A.(1950.0) = 05^h 35^m 49.992^s
Declination	Dec.(1950.0) = $-69°$ $17'$ $50.08''$
Distance	D = 50 kpc = 170 000 light-years
Distance Modulus	m $-$ M = 18.7 mag
Spectral Type	Sp = B3 Ia (blue supergiant)
Apparent Visual Magnitude	V = 12.4 mag
Color Index	B $-$ V = $+0.04$ mag
Visual Absorption	$A_v \approx 0.5$ mag
Bolometric Correction	B.C. = $+1.15$
Bolometric Absolute Magnitude	M_{bol} = -7.9
Absolute Luminosity	L = 4.6 x 10^{38} erg s$^{-1} \approx 10^5$ L$_\odot$
Effective Temperature	T_{eff} = 16 000 K
Radius	R = 3 $\times$ 10^{12} cm $\approx$ 50 R$_\odot$
Initial Main Sequence Mass	M = 20 M$_\odot$
Evolved Helium Core Mass	M_{He} = 6 M$_\odot$
Evolved Hydrogen Envelope Mass	$M_H \approx$ 10 M$_\odot$

Neutrino Burst

Kamiokande II Detector

Event Start Time	t_0 = 07^h 35^m 35^s UT ($\pm$1 min) on 23 Feb 1987
Number of Anti-Neutrinos Detected	N = 11
Energy Range of Anti-Neutrinos	ΔE = 7.5 to 36 MeV
Burst Duration	Δt = 12.5 s

Irvine – Michigan – Brookhaven - (IMB) Detector

Event Start Time	t_0 = 07^h 35^m 41.37^s UT ($\pm$10 ms) on 23 Feb 1987
Number of Anti-Neutrinos Detected	N = 8
Energy Range of Anti-Neutrinos	ΔE = 20 to 40 MeV
Burst Duration	Δt = 5.6 s

Neutrino Burst

Total Anti-Neutrino Energy	E $\approx$ 3 $\times$ 10^{52} erg
Total Neutrino Energy	E $\approx$ 2.5 $\times$ 10^{53} erg
Total Neutrino Luminosity	L $\approx$ 10^{55} erg s^{-1} $\approx$ 10^{22} L$_\odot$
Average Neutrino Temperature	T = 4 MeV $\approx$ 10^{10} K
Number of Neutrinos Produced	N = 10^{58} neutrinos
Neutrino Flux at Earth	F $\approx$ 5 x 10^{10} cm^{-2}

(*continued*)

TABLE 56   Supernova 1987A (SN 1987A) (continued)

Neutrino Burst

Inferred Neutrino Mass Limit	$m \leqslant 16\,eV$
Core Collapse Time	$\Delta t = 0.01\,s = 10\,ms$
Collapsed Remnant Mass	$M = 1.4\,M_{\odot}$
Collapsed Remnant Radius	$R = 10\,km$
Gravitational Potential Energy Released	$E = 2.5 \times 10^{53}\,erg$

Visible Explosion[b]

Optical Discovery	$V = 5.0$ mag on 24.122 Feb. 1987 $(t \approx 19\ \text{hours})$ (Ian Shelton)
Explosive Optical Brightening	$V = 6.36$ mag on 23.444 Feb. 1987 $(t \approx 3\ \text{hours})$
	$M_{bol} = -12.8$ mag
	$L = 4.2 \times 10^{40}\,erg\,s^{-1} \approx 10^{7}\,L_{\odot}$
Secondary Visual Maximum	$V = 2.97$ mag on 20 May 1987 $(t \approx 90\ \text{days})$
	$M_{bol} = -16.2$ mag
	$L \approx 10^{42}\,erg\,s^{-1} \approx 10^{8}\,L_{\odot}$

[a] Adapted from Lang, Kenneth R., *Astrophysical Data: Planets and Stars*, New York, Springer Verlag, 1992, pp. 706, 707. Here $L_{\odot}$, $R_{\odot}$, $M_{\odot}$ respectively denote the absolute luminosity, radius and mass of the Sun.
[b] Time t from neutrino burst at $t_0 = 23.316$ Feb 1987.

Swift: NASA launched the **Swift** spacecraft on 20 November 2004. Its three instruments work together to observe gamma-ray bursts and afterglows at gamma ray, X-ray and visible light wavelengths, providing new insights to the origin and evolution of these powerful, explosive bursts.

Swift-Tuttle Comet: After Halley's Comet, the second brightest periodic comet, discovered in 1862 by the American astronomers **Lewis Swift (1820–1913)** and **Horace Tuttle (1837–1923)**. As noticed in 1866 by the Italian astronomer **Giovanni Schiaparelli (1835–1910)**, the Perseid meteor shower occurs when the Earth's orbit intersects the orbit of Comet Swift-Tuttle, which has distributed meteoroids along its orbit. *See* meteor shower and meteor stream.

Symbiotic star: A binary star system in which a cool, red giant star sheds mass onto a more compact, hotter companion star. The giant stars are of late spectral type G, K or M, and the hot components may be main-sequence stars or an accreting white dwarf star.

Synchronous orbit: The orbit of a satellite whose orbital period is equal to the rotation period of the planet it is rotating. Artificial satellites in synchronous orbit around the Earth remain above the same location on the planet. *See* geosynchronous orbit.

Synchronous rotation: A satellite in synchronous rotation has a rotation period that is equal to its orbital period, and it always presents the same face toward the planet it

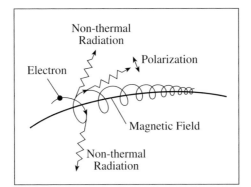

FIG. 51 **Synchrotron radiation** High-speed electrons moving at velocities near that of light emit a narrow beam of synchrotron radiation as they spiral around a magnetic field. This emission is sometimes called non-thermal radiation because the electron speeds are much greater than those of thermal motion at any plausible temperature. The name "synchrotron" refers to the man-made, ring-shaped synchrotron particle accelerator where this type of radiation was first observed; a synchronous mechanism keeps the particles in step with the acceleration as they circulate in the ring.

is revolving around. Our Moon is in synchronous rotation, always presenting its near side to the Earth. The four largest moons of Jupiter, the Galilean satellites, are also in synchronous rotation.

Synchrotron particle accelerator: *See* particle accelerator.

Synchrotron radiation: Electromagnetic radiation emitted by an electron traveling almost at the speed of light in the presence of a magnetic field. Such radiation is linearly polarized. The name arises because it was first observed in 1948 when electrons were accelerated in the General Electric synchrotron particle accelerator. The English physicist **George A. Schott (1868–1937)** discussed such a radiation process by electrons in a paper published in 1907 and in his 1912 book on *Electromagnetic Radiation*. The acceleration of the electrons causes them to emit radiation that is strongly polarized and increases in intensity at longer wavelengths. The wavelength region in which the emission occurs depends on the energy of the electron – 1 MeV electrons radiate mostly in the radio region. In 1950 the German astronomer **Karl O. Kiepenheuer (1910–1975)** correctly proposed that the radio emission of the Milky Way is due to the non-thermal synchrotron radiation of cosmic ray electrons spiraling about an interstellar magnetic field. The Russian astrophysicist **Iosif S. Shklovskii (1916–1985)** proposed that both the optical, or visible-light, and radio emission of the Crab Nebula supernova remnant are due to synchrotron radiation, and three years later his Russian colleague **Vitalii L. Ginzburg (1916–)** proposed that cosmic rays are accelerated in the stellar explosions known as supernovae. Synchrotron radiation was subsequently used to explain the intense radio emission of radio galaxies and quasars. In 1974 the American astronomer **Roger D. Blandford (1949–)** and English astronomer **Martin Rees (1942–)** proposed that the radio lobes of radio galaxies are fed by jets of high-energy electrons sent out along the rotation axis of a rotating, supermassive black hole. *See* black hole, Crab Nebula, particle accelerator, radio galaxy, and supernova remnant.

Synodic month: The period between two identical phases of the Moon, equal to 29.530 59 mean solar days.

Synodic period: The period of apparent rotation or orbital revolution as observed from the Earth.

Syrtis Major: A dark triangular feature, now known as Syrtis Major Planitia, or the Gulf of Sirte Plains, that extends from the south polar cap on Mars to across its equator. The Dutch astronomer **Christiaan Huygens (1629–1695)** first sketched Syrtis Major in 1659, and used his observations of the feature to conclude that the rotation period of Mars is about 24 hours.

Syzygy: A syzygy occurs when the Earth and Sun are aligned with a third celestial body, such as the Moon or a planet. A syzygy occurs at the conjunction or opposition of a planet and at full and new Moon. *See* conjunction and opposition.

T

TAI: Acronym for Temps Atomique International. *See* international atomic time.

Tarantula Nebula: An emission nebula, designated NGC 2070 and also known as 30 Doradus, located in the Large Magellanic Cloud at a distance of about 170 000 light-years in the constellation Doradus. The Tarantula Nebula measures half a degree and about 1000 light-years across, far bigger and brighter than any emission nebula in our Galaxy. The Tarantula Nebula has an estimated mass of about 500 000 times the mass of the Sun, and it gets its name from its spidery appearance. *See* emission nebula.

Tectonics: The crustal deformations of a planet caused by internal heat rising to the surface. The churning convection of internal molten rock drives tectonic plates sideways across the Earth, producing continental drift and earthquakes. Hot rising material within Venus produces volcanoes, rifts, and buckled, fractured features known as arachnids and coronae. *Tectonics* is the Greek word for "carpenter or building." *See* arachnid, continental drift, coronae, earthquake, and plate tectonics.

Telescope: An instrument for collecting and magnifying electromagnetic radiation from a cosmic object. A refracting telescope uses a lens to collect light, while a reflecting telescope uses a mirror. The Dutch spectacle maker **Hans Lippershey (c. 1570–1619)** applied for a patent on one of the first telescopes in 1608, but such spyglasses soon became quite common and no patent was awarded. The Italian astronomer **Galileo Galilei (1564–1642)** built his own telescopes, and was the first to use them, in 1609, for astronomical observations. Telescopes of different design operate at most electromagnetic wavelengths, including the X-ray, ultraviolet, visible, infrared and radio regions of the spectrum. *See* Cassegrain telescope, Gregorian telescope, *Hubble Space Telescope,* Keck telescopes, Newtonian telescope, radio telescope, reflecting telescope, refracting telescope, Schmidt telescope, *Solar and Heliospheric Observatory, Spitzer Space Telescope, Ulysses,* Very Large Array, and *Yohkoh.*

Temperature: A measure of the heat of an object and the average kinetic energy of the randomly moving particles in it.

Terminator: The boundary between the illuminated and dark hemispheres of the Moon, another satellite, or a planet when viewed from the Earth.

Terrestrial planets: The four rocky planets in the inner part of the solar system: Mercury, Venus, Earth and Mars.

Tesla: The Systeme International, or SI, unit of magnetic flux density, named after **Nikola Tesla (1856–1943)**, a Croatian-born American pioneer in the fields of alternating-current electricity, and electrical power generation and distribution. The Tesla is a measure of the strength of a magnetic field. The centimeter-grams-second, or c.g.s., unit

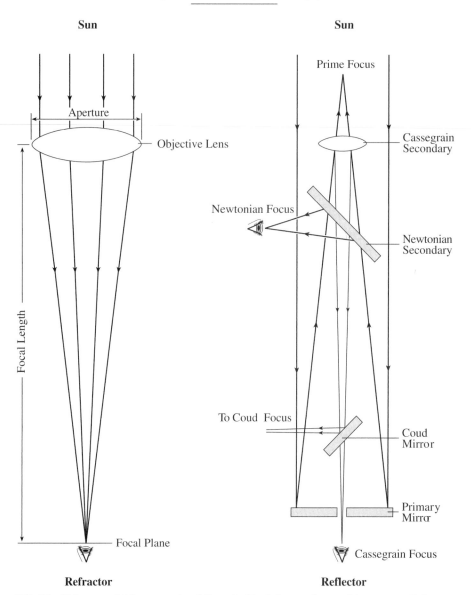

Sun   **Sun**

Refractor   **Reflector**

FIG. 52   Telescopes Light waves that fall on the Earth from a distant object are parallel to
one another, and are focused to a point by the lens or mirror of a telescope. Early telescopes
were refractors (*left*). The curved surfaces of the convex objective lens bend the incoming
parallel light rays by refraction, and bring them to a focus at the center of the focal plane,
where the light rays meet and an image is created. A second, smaller lens, called the eye-
piece, was used to magnify the image in the early refractors; later versions placed photographic or electronic detectors at the focal plane. In 1670 the English physicist **Isaac Newton
(1642–1727)** constructed the first reflecting telescope (*right*), which uses a large, concave, or
parabolic, primary mirror to collect and focus light. A small, flat secondary mirror, inclined at
an angle of 45 degrees to the telescope axis, reflects the light sideways, at a place now known
as the Newtonian focus. Other light-deflecting mirror arrangements can be used to obtain any
desired focal length, which varies with the curvature and position of small convex mirrors.

TABLE 57 Telescopes – biggest optical[a]

Mirror Diameter[b]	Name	Location/Comments
10.0-meter 10.0-meter (394-inch)	Keck I Keck II	Mauna Kea, Hawaii/two telescopes that operate independently and can be connected as an interferometer. Each telescope mirror consists of 36 hexagonal segments. The W. M. Keck Observatory is operated by the California Association for Research in Astronomy.
9.2-meter (362-inch)	Hobby-Eberly	Mt. Fowlkes, Texas/A joint project of five universities.
8.2-meter (323-inch)	Subaru	Mauna Kea, Hawaii/A project of the National Astronomical Observatory of Japan
8.2-meter (323-inch)	Antu, Kueyen, Melipal, Yepun	Cerro Paranal, Chile/European Southern Observatory, four independent telescopes that will be connected as an interferometer called the Very Large Telescope.
8.1-meter 8.1-meter (320-inch)	Gillett Gemini South	Mauna Kea, Hawaii/Gemini North Cerro; Pachón, Chile/ Gemini South. Twin telescopes managed by the Association of Universities for Research for Astronomy.
6.5-meter (256-inch)	MMT	Mt. Hopkins, Arizona/Formerly the Multiple Mirror Telescope, or MMT, now a single mirror, operated by the Fred L. Whipple Observatory and the University of Arizona.
6.5-meter (256-inch)	Walter Baade Landon Clay	La Serena, Chile/Magellan I La Serena, Chile/Magellan II Twin telescopes at the Las Campanas Observatory.
6.0-meter (236-inch)	Bolshoi Teleskop Azimutalnyi	Mount Pastukhov/Zelenchukska Observatory, Special Astrophysical Observatory, Russian Academy of Sciences.
5.0-meter (200-inch)	Hale	Palomar Mountain, California
4.2-meter (165-inch)	William Herschel	La Palma, Canary Islands/Osservatorio del Roque de los Muchachos, Spain.
4.1-meter (162-inch)	SOuthern Astrophysical Research, SOAR	Cerro Pachón, Chile/Cerro Tololo Inter-American Observatory, Brazil, U.S.
4.0-meter (158-inch)	Victor Blanco	Cerro Tololo, Chile/Cerro Tololo Inter-American Observatory.

(continued)

TABLE 57 Telescopes – biggest optical[a] (continued)

Mirror Diameter[b]	Name	Location/Comments
3.9-meter (150-inch)	Anglo-Australian	Siding Spring Mountain, New South Wales.
3.8-meter (150-inch)	Mayall	Kitt Peak, Arizona/National Optical Astronomy Observatory.
3.8-meter (150-inch)	United Kingdom Infra-Red Telescope UKIRT	Mauna Kea, Hawaii/Dedicated to infrared.
3.6-meter (142-inch)	3.6m Telescope	La Silla, Chile/European Southern Observatory.
3.6-meter (142-inch)	Canada-France-Hawaii	Mauna Kea, Hawaii.
3.6-meter (142-inch)	Telescopio Nazionale Galileo	La Palma, Canary Islands/Italian

[a]Largest telescopes operating at optical, or visible-light, wavelengths, and also sometimes at infrared wavelengths. Bigger telescopes are under construction.

[b]Diameter of the effective aperture of the primary mirror, giving its light-collecting area.

of magnetic field strength, often used in astrophysics, is the Gauss, where 1 Tesla = 10 000 Gauss = 10^4 Gauss. *See* gamma and Gauss.

Theory: A hypothesis used to explain some facts.

Thermal bremsstrahlung: Emission of radiation by energetic electrons encountering the field of a positive ion in a gas that is in thermal equilibrium. *See* bremsstrahlung and thermal equilibrium.

Thermal energy: Energy associated with the motions of molecules, atoms, or ions.

Thermal equilibrium: A physical system in which all parts have exchanged heat and are characterized by the same temperature at all points. In such equilibrium, a single temperature characterizes the velocity distribution. *See* Gaussian distribution and Maxwellian distribution.

Thermal gas: A collection of particles that collide with each other and exchange energy frequently, giving a distribution of particle energies that can be characterized by a single temperature. *See* non-thermal particle.

Thermal particle: A particle that is part of a thermal gas. *See* non-thermal particle.

Thermal radiation: Electromagnetic radiation emitted by a gas in thermal equilibrium. Thermal radiation arises by virtue of an object's heat, or temperature. Energetic electrons that are not necessarily in thermodynamic equilibrium emit non-thermal radiation. *See* blackbody radiation and non-thermal radiation.

Thermodynamics: The study of the behavior of heat and other related forms of energy in changing systems.

Thermonuclear fusion: The fusion of atomic nuclei at high temperatures to form more massive nuclei with the simultaneous release of energy. Thermonuclear fusion is the power source at the core of the Sun. *See* fusion.

Tides: The Moon's gravitational attraction draws the Earth's oceans into the shape of an egg, causing two high tides as the planet's rotation carries the continents past the two tidal bulges each day. The Sun's gravity also contributes to the tides, but the Moon's contribution is 2.2 times that of the Sun. As the tides flood and ebb, they produce friction and dissipate energy at the expense of the Earth's rotation, causing the length of the day to increase by about 0.002 seconds per century. This interaction also results in an outward motion of the Moon of about 0.04 meters per year. The height of the tides and the phase of the Moon depend on the relative orientations of the Earth, Moon and Sun. When the tide-raising forces of the Sun and Moon are in the same direction, they reinforce each other, making the highest high tides and the lowest low tides. These spring tides occur at new or full Moon. The range of tides is least when the Moon is at first or third quarter, producing the neap tides when the tide-raising forces of the Sun and Moon are at right angles to each other. Tides can also be produced in the solid body of an object, such as Jupiter's tidal in-and-out flexing of its satellite Io, heating the satellite's interior and producing its volcanoes. *See* Io, and Moon.

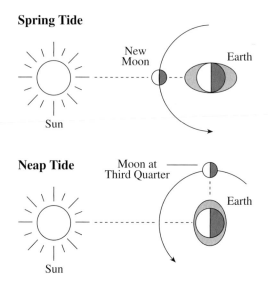

FIG. 53 Tides The height of the tides and the phase of the Moon depend on the relative positions of the Earth, Moon and Sun. When the tide-raising forces of the Sun and Moon are in the same direction, they reinforce each other, making the highest high tides and the lowest low tides. These spring tides (*top*) occur at new or full Moon. The range of tides is least when the Moon is at first or third quarter, and the tide raising forces of the Sun and Moon are at right angles to each other. The tidal forces are then in opposition, producing the lowest high tides and the highest low tides, or the neap tides (*bottom*). The height of the tides has been greatly exaggerated in comparison to the size of the Earth.

Time: A dimension that distinguishes between past, present, and future. *See* day, Greenwich Mean Time, International Atomic Time, month, Universal Time, and year.

Time dilation: The slowing down of the time separating two events on a moving object, as detected by an external observer not participating in the motion. Time dilation would make a moving clock appear to go slower, and it can noticeably lengthen the apparent lifetime of elementary particles. The effect becomes important at speeds approaching the velocity of light. *See Special Theory of Relativity.*

Titan: The largest satellite of Saturn, discovered in 1655 by the Dutch astronomer **Christiaan Huygens (1629–1695)**. Titan has a radius of 2575 kilometers, just over twice the radius of the Earth's Moon and it is the second largest satellite in the Solar System. The mass density of Titan is 1880 kilograms per cubic meter, nearly twice that of water, so it is most likely composed of nearly equal amounts of water ice and rock. In 1980 instruments aboard the **Voyager 1** spacecraft showed that Titan has a substantial atmosphere whose surface pressure is about 1.5 times the air pressure at sea level on Earth. Nitrogen molecules are the main constituents of Titan's atmosphere, as they are in the Earth's air, and methane is the second most abundant ingredient of Titan's atmosphere, at 1 to 6 percent. The surface temperature of Titan is 94 kelvin, at which methane can be solid, liquid or gas, so methane could play the same role on Titan as water does on Earth, producing clouds, rains, rivers and seas. A veil of orange smog hides the surface of Titan from view, but the *Huygens Probe* descended to its surface on 14 January 2005, obtaining evidence for flowing liquid, perhaps due to rivers or seas of methane. *See Huygens Probe.*

Titius-Bode law: The approximate, relative distances of the planets from the Sun is given by the sequence 0, 3, 6, 12, 24, ... , adding 4 and dividing by 10. The German mathematician **Johann Daniel Titius (1729–1796)** first noticed the law in 1770, and the German astronomer **Johann Elert Bode (1749–1826)** subsequently popularized it. The formula predicted the distance of Uranus and the first asteroids, which were discovered after the law was stated, and it also predicted the approximate distance of Neptune.

TRACE: Acronym for ***Transition Region And Coronal Explorer***. *See **Transition Region And Coronal Explorer***.

Transit: The slight dimming of starlight caused by a planet passing between it and the Earth. The planets Mercury and Venus occasionally transit the Sun. The word transit is also used for the passage of a celestial object across the observer's meridian or the passage of a satellite across a planet, when viewed from Earth.

Transition Region And Coronal Explorer: NASA launched its ***Transition Region And Coronal Explorer***, abbreviated ***TRACE***, spacecraft in April 1998. It provided images of the solar atmosphere at temperatures from 10 thousand to 10 million kelvin with high temporal resolution and second of arc angular resolution using observations at ultraviolet and extreme ultraviolet wavelengths. ***TRACE*** showed that the million-degree corona is comprised of thin magnetic loops that are naturally dynamic and continuously evolving.

Trapezium: A multiple star system that lies at the center of the Orion Nebula and illuminates it. The name comes from the pattern formed by four of its brightest stars, which have apparent visual magnitudes of 5.1, 6.7, 6.7 and 8.0 and are visible in a small telescope. The Trapezium stars are very hot, about 50 000 kelvin, and emit most of their energy as ultraviolet radiation that makes the Orion Nebula glow. A larger telescope reveals the presence of two other, fainter stars. The technical designation for the Trapezium is Theta^1 Orionis, or θ^1 Orionis. *See Orion Nebula.*

Triangulation: Measurement of the distance of a planet or nearby star by sighting its apparent position against background objects from two or more separate locations. *See parallax.*

Triangulum Galaxy: A nearby spiral galaxy in the constellation Triangulum, designated M33 and sometimes called a Pinwheel Galaxy. The Triangulum Galaxy lies 2.6 million light-years away, has a diameter of about 40 000 light years, containing about 15 billion stars, about half the size and one-tenth the total mass of the Milky Way Galaxy. The Triangulum Galaxy is the third largest member of the Local Group, after the Andromeda galaxy and our Milky Way Galaxy. *See* Local Group and Messier Objects.

Trifid Nebula: An emission nebula, designated M20 or NGC 6514, located about 5200 light-years away in the direction of the constellation Sagittarius. The Trifid Nebula measures 27 by 29 minutes of arc across, and it gets its name from the dark lanes that trisect it. *See* emission nebula.

Trigonometric parallax: Another name for annual parallax. It is a method of determining the distance to a nearby star by triangulation from a baseline of twice the astronomical unit, using observations separated by six months. *See* annual parallax, astronomical unit, and parallax.

Trillion: A million million, and a thousand billion or 10^{12}.

Triple-alpha process: Nuclear reactions that permit the formation of carbon nuclei from triple collisions of helium nuclei, or alpha particles, in the cores of giant stars. The process was first suggested by the Estonian astronomer **Ernst Öpik (1893–1985)** in 1951, and independently presented in greater detail the following year, in 1952, by the American astrophysicist **Edwin E. Salpeter (1924–)**. In 1954 the English astrophysicist **Fred Hoyle (1915–2001)** showed that carbon can only be produced in substantial amounts within giant stars if a resonant reaction is involved, and the predicted excited state of carbon was observed in the terrestrial laboratory a few years later by the American astrophysicist **William A. "Willy" Fowler (1911–1995)** and his colleagues.

Triton: The largest satellite of Neptune, discovered in 1846 by the English astronomer **William Lassell (1799–1880)** soon after the discovery of the planet in the same year. Triton is the only large satellite in the Solar System to circle a planet in the retrograde direction, opposite to the planet's direction of rotation, and its orbital plane is highly inclined by 157 degrees to the planet's equatorial plane. The retrograde and inclined orbit suggest that Triton did not originate with Neptune, but was captured into its unusual orbit from the nearby Kuiper belt by Neptune's gravitational forces, which are now slowly pulling Triton in toward a collision with the planet in the very distant future. With a radius of 1352 kilometers, Triton is about three-quarters the size of the Earth's Moon, and its mass density is 2060 kilograms per cubic meter. Measurements from the *Voyager 2* spacecraft in 1989 indicated that Triton has a surface temperature of 38 kelvin, the coldest measured surface of any natural body in the Solar System, and that Triton has a very tenuous, nitrogen-rich atmosphere, bright polar caps of nitrogen and methane ice, frozen lakes flooded by past volcanoes of ice, and towering geysers that may now be erupting on its surface. *See* Kuiper belt.

Trojan asteroid: One of the asteroids that move along Jupiter's orbit, preceding and following the giant planet near one of the two points where the gravitational force of Jupiter and the Sun are equal, known as the forth and fifth Lagrangian points. The gravitational perturbations of the inner planets produce slight swinging motions about these two points. The first known Trojan asteroid was 588 Achilles, discovered in 1906 by the German astronomer **Maximilian Wolf (1863–1932)**. Hundreds of them are now known. As with Achilles, they are all named after heroes of the Trojan War, and they are therefore collectively known as the Trojan asteroids. *See* asteroid and Lagrangian point.

Tropical year: Also known as the solar year, the tropical year is the time for the Earth to revolve once around the Sun relative to the Vernal Equinox, and equal to 365.242 mean solar days. Due to precession, the tropical year is about 20 minutes shorter than the sidereal year. *See* precession, sidereal time, Vernal Equinox, and year.

Troposphere: Lowest level of the Earth's atmosphere, where most of the weather takes place, from the ground to about 15 kilometers above the surface. The temperature of the troposphere decreases from 290 kelvin at the ground to 240 kelvin at its top. The troposphere is any region of a planetary atmosphere in which convection normally takes place. *See* convection.

T Tauri: The prototype for a distinct class of stars that are embedded in nebulous clouds of gas and dust, and that exhibit irregular and unpredictable changes of light, first described in detail in 1945 by the American astronomer **Alfred Joy (1882–1973)**. The T Tauri stars appear to be in an early stage of stellar life, still in gravitational contraction or just settling down on the main sequence of the Hertzsprung-Russell diagram. T Tauri has a variable apparent visual magnitude of 9.6, lies at a distance of 460 light-years, and has an absolute visual magnitude of about 2.0. *See* Hertzsprung-Russell diagram.

Tully-Fisher relation: A correlation between the intrinsic luminosity of an isolated spiral galaxy and its rotational velocity as measured from the width of the 21-centimeter hydrogen line. This relation, discovered in 1977 by the American astronomers **R. Brent Tully (1943–)** and **J. Richard Fisher (1935–)**, can be used to determine the distances of galaxies, from their apparent luminosity and the absolute, or intrinsic, luminosity inferred from the Tully-Fisher relation. *See* inverse-square law and twenty-one centimeter line.

Tunguska explosion: An atmospheric explosion that occurred on 30 June 1908 between 5 and 10 kilometers above the Podkamennaya Tunguska River in central Siberia. The energy produced was equivalent to the aerial explosion of the nuclear bomb that leveled Hiroshima, Japan. The Tunguska explosion has been attributed to collision with an asteroid about 50 meters across, which would have sufficient internal strength to penetrate this deeply into the atmosphere and explode without creating a crater in the ground below.

Tunnel effect: A quantum mechanical effect that permits an energetic nuclear particle to escape the nucleus of a radioactive atom. It also enables two colliding protons in the Sun to overcome the electrical repulsion between them, resulting in their nuclear fusion. The tunnel effect involves a quantum leap over, or under, an otherwise insurmountable barrier.

Turbulence: The chaotic mass motions associated with convection.

Twenty-one centimeter line: In 1949 the Dutch astronomer **Hendrik C. "Henk" van de Hulst (1918–2000)** predicted that the electron in each cold, interstellar hydrogen atom could undergo a transition between the two possibilities of its spin, giving rise to emission or absorption lines at a wavelength of twenty-one centimeters. In 1951 Harvard graduate student **Harold I. "Doc" Ewen (1922–)** and his advisor American physicist **Edward M. Purcell (1912–1997)** detected the twenty-one centimeter line from interstellar neutral hydrogen using a novel "switched-frequency" detection mode suggested by Purcell. They delayed publication of their discovery until it was confirmed by the Dutch astronomers **Karl Müller (1927–)** and **Jan Oort (1900–1992)** and in a cable from Australian radio astronomers. Observations of the twenty-one centimeter line have traced out the spiral arms of our Milky Way Galaxy and used to infer substantial amounts of dark matter in the extensive halos of our Galaxy and other spiral galaxies. *See* dark matter.

Tycho's star and supernova remnant: A "new star" observed in 1572 by the Danish astronomer **Tycho Brahe (1546–1601)** in the constellation Cassiopeia at a place where no star had been seen before. It was brighter than Venus, near apparent visual magnitude of – 4, and visible in full daylight, most likely a supernova of type Ia. Although the exploded star faded from visibility in 1574, astronomers still observe the radio and X-ray emission of the expanding debris, called Tycho's supernova remnant. *See* supernova, supernova remnant, and type I supernova.

Type Ia supernova: The explosion of a white dwarf star that receives material from a nearby companion star, exceeds the Chandrasekhar limit, and blows up, completely annihilating itself. Type Ia supernovae have no hydrogen in their spectra. In 1998 two groups of astronomers reported that observations of type Ia supernovae indicate that the expansion of the Universe is now speeding up, with some sort of mysterious dark energy counteracting the pull of gravity. *See* Chandrasekhar limit, dark energy, supernova and Tycho's star and supernova remnant.

Type II supernova: The explosion of an isolated, massive star with more than eight solar masses, which runs out of fuel, hurling most of the star into interstellar space as a supernova remnant. The collapsing core of the type II supernova can become a neutron star or a black hole. Type II supernovae have hydrogen in their spectra. *See* Crab Nebula, neutron star, supernova, supernova 1987A, and supernova remnant.

U

Uhuru Mission: The ***Uhuru*** satellite was launched from Kenya on 12 December 1970, on the seventh anniversary of Kenyan independence, when it was named ***Uhuru,*** the Swahili word for "freedom". ***Uhuru*** provided a comprehensive, all-sky survey of cosmic X-ray sources, including black holes and neutron stars in binary stellar systems, supernova remnants, Seyfert galaxies, and clusters of galaxies. It discovered the diffuse X-ray emission from clusters of galaxies, which accounts for some of the so-called dark matter in the Universe. The mission ended in March 1973. *See* Centaurus X-3, Cygnus X-1, Hercules X-1 and Scorpius X-1.

Ultraviolet radiation: Abbreviated UV, electromagnetic radiation with a slightly higher frequency and somewhat shorter wavelength than visible blue light. The German physicist **Johann Wilhelm Ritter (1776–1810)** discovered ultraviolet light in 1801, when examining the spectrum of sunlight at wavelengths just a bit shorter than the wavelength of violet light. Ultraviolet radiation has wavelengths between about 10^{-8} and 3.5×10^{-7} meters, or from 10 to 350 nanometers, with the extreme ultraviolet lying in the short wavelength part of this range. The Earth's atmosphere absorbs most cosmic ultraviolet radiation, so thorough studies of the ultraviolet output of cosmic objects must be conducted from space. Such observations have not led to as many discoveries of totally new cosmic objects as the first detections of the radio and X-ray Universe. Observations of the extreme ultraviolet radiation from the Sun were pioneered by NASA's ***Orbiting Solar Observatories,*** launched from 1962 to 1971, and continued to the contemporary instruments aboard the ***SOlar and Heliospheric Observatory***, or ***SOHO*** for short. The ***International Ultraviolet Explorer***, launched in 1978 and abbreviated ***IUE***, made years of ultraviolet observations of stars and other cosmic objects, followed by the ***Extreme UltraViolet Explorer***, abbreviated ***EUVE*** and launched in 1992. Hot, blue stars produce substantial amounts of ultraviolet radiation, which ionizes the surrounding interstellar hydrogen, producing an emission nebula or H II region. *See* emission nebula, ***Extreme UltraViolet Explorer***, extreme ultraviolet radiation, H II region, ***International Ultraviolet Explorer***, ***Orbiting Solar Observatory***, ***SOlar and Heliospheric Observatory***, and Strömgren radius.

Ulysses Mission: The ***Ulysses*** spacecraft was a joint undertaking of ESA and NASA, launched by the ***Space Shuttle Discovery*** on 6 October 1990 to study the interplanetary medium and the solar wind at different solar latitudes. It provided the first opportunity for measurements to be made over the poles of the Sun, using the gravity assist technique to take it out of the plane of the Solar System. After an encounter with Jupiter in February 1992, the spacecraft moved back towards the Sun to pass over the solar south pole in September 1994 and the north pole in July 1995, and then again over the south pole in November 2000 and the north pole in December 2001. Comparisons of ***Ulysses*** observations with those from other spacecraft, such as ***Yohkoh*** and ***SOHO***, showed that much, if not all, of the high-speed solar wind comes from open magnetic fields in polar coronal holes, at least during the minimum of the Sun's 11-year cycle of magnetic activity.

Umbra: The dark inner core of a sunspot with a penumbra, or a sunspot lacking a penumbra, visible in white light. Also the inner part of the shadow cast during an eclipse, where a total eclipse is visible. *See* lunar eclipse, penumbra, solar eclipse, and sunspot.

Uncertainty principle: Proposed in 1927 by the German physicist **Werner Heisenberg (1901–1976)**, the uncertainty principle states that the position and momentum, or velocity, of a quantum-mechanical particle cannot be simultaneously known with complete certainty or precision. *See* quantum mechanics.

United Kingdom InfraRed Telescope: Abbreviated UKIRT, a 3.8-meter (150-inch) infrared telescope in operation since 1979 and located near the summit of Mauna Kea, Hawaii.

United States Naval Observatory: Abbreviated USNO, the United States Naval Observatory was established in 1830 to assist timing and navigation at sea. Today the mission of the USNO includes astrometry and timing data essential for accurate navigation and the support of communications on Earth and in space. It includes determining the positions and motions of the Earth, Sun, Moon, planets, stars and other celestial objects, providing astronomical data; determining precise time; measuring the Earth's rotation; and maintaining the Master Clock for the United States. USNO scientists have also played important pioneering and continuing roles in extreme ultraviolet and X-ray observations of the Sun. USNO scientists discovered and confirmed the existence of Pluto's satellite Charon. *See* Pluto.

Universal time: The Earth's rotation with respect to the Sun is the basis of Universal Time, abbreviated UT, and also called Solar Time. Universal Time is used for astronomical purposes throughout the world, and is popularly known as Greenwich Mean Time, or GMT for short. The time UT0 is the observed rotation of the Earth relative to the Sun, and UT1 is UT0 corrected for the Earth's polar motion. Time signals distribute a Coordinated Universal Time abbreviated UTC for the French wording. The UTC is a regular system of time based on observations of atomic clocks and routinely kept in step with UT1 to the nearest second. *See* Greenwich Mean Time and Royal Greenwich Observatory.

Universe: All observable and potentially observable phenomena.

Uranus: The seventh major planet from the Sun, and the second-smallest of the four giant planets, discovered on 13 March 1781 by the German-born English astronomer **William Herschel (1738–1822)** using a homemade reflecting telescope. It is just visible to the naked eye under good seeing conditions, with a visual apparent magnitude of 5.5. In contrast to all the other planets in the Solar System, Uranus is tipped sideways so its spin axis lies nearly within the planet's orbital plane. Unlike the other three giant planets, Uranus has no strong, detectable internal source of heat, and its winds and storms are relatively mild. Most of the interior of Uranus probably consists of a vast internal ocean of water, methane, and ammonia. The planet's magnetic axis is tilted

TABLE 58 Uranus[a]

Mass	868.32×10^{23} kilograms $= 14.535$ M_E
Radius	25 559 kilometers $= 4.007$ R_E
Mean Mass Density	1318 kilograms per cubic meter
Rotation Period	17.24 hours
Orbital Period	84.01 Earth years
Mean Distance from Sun	19.19 AU
Atmosphere	83 percent hydrogen 15 percent helium
Energy Balance	less than 1.4
Effective Temperature	59.3 kelvin
Temperature at 1-bar level	76 kelvin
Central Temperature	5000 kelvin
Magnetic Dipole Moment.	50 D_E
Equatorial Magnetic Field Strength	0.23×10^{-4} tesla

[a]The Earth's mass, M_E, is 59.743×10^{23} kilograms, the Earth's equatorial radius, R_E, is 6378 kilometers, the astronomical unit, denoted AU, is the mean distance between the Earth and the Sun with a value of 1.496 $\times 10^{11}$ meters. The energy balance is the ratio of total radiated energy to the total energy absorbed from sunlight, the effective temperature is the temperature of a blackbody that would radiate the same amount of energy per unit area, and D_E is the magnetic dipole moment of the Earth.

by 58.6 degrees to the rotation axis, perhaps due to the origin of its magnetism in an internal shell of ionized water. The American astronomer **James L. Elliot (1943–)** unexpectedly discovered the thin, widely spaced rings of Uranus in 1977, just before and after observing a star pass behind the planet. These narrow, dark rings were confirmed when the *Voyager 2* spacecraft arrived at Uranus in 1986, while also discovering more of them.

UT: Abbreviation for Universal Time. *See* Greenwich Mean Time and Universal Time.

UV: Abbreviation for ultraviolet radiation. *See* ultraviolet radiation.

UV Ceti: A binary star in the constellation Cetus, consisting of two 13th magnitude red dwarf stars that exhibit large flares every few hours. UV Ceti is at a distance of 8.7 light-years, and is the prototype for a class of variable stars called UV Ceti stars, also known as flare stars. *See* flare stars.

Vacuum: The supposedly empty space between cosmic objects or fundamental particles. According to quantum physics, the vacuum might be occupied by virtual particles. *See* virtual particles.

Valles Marineris: An enormous system of interconnected canyons stretching 4500 kilometers across Mars, just south of its equator.

Van Allen belts: Two concentric, ring-shaped regions of high-energy charged particles that girdle the Earth's equator within the Earth's magnetosphere, named after the American scientist **James A. Van Allen (1914–)**, who discovered them with his colleagues in 1958–59, using the first two successful United States artificial satellites *Explorer 1* and *3*. The inner belt lies between 1.2 and 4.5 Earth radii, measured from the Earth's center, and the outer belt is located between 4.5 and 6.0 Earth radii. The inner Van Allen belt contains protons with energies greater than 10 MeV and electrons exceeding 0.5 MeV. The outer belt also contains protons and electrons, most of which have energies under 1.5 MeV. The outer belt contains mainly electrons from the solar wind, and the inner belt mainly protons from the solar wind. Another source of radiation-belt particles is neutrons produced when cosmic rays and energetic particles bombard Earth's atmosphere; some of these neutrons decay into protons and electrons that are trapped by the Earth's magnetic field. Within the inner belt is a radiation belt consisting of particles produced by interactions between the solar wind and heavier cosmic ray particles. Because the Earth's magnetic field is offset from the planet's center by about 500 kilometers, the inner belt dips down towards the Earth's surface in the region of the South Atlantic Ocean, off the coast of Brazil. *See* radiation belt and South Atlantic Anomaly.

Variable star: A star whose brightness varies with time. The variation may be intrinsic to the star, such as a pulsating star with regular changes in brightness, or it might be due to periodic eclipses from a companion star. In both of these cases, there is a regular variation in the light we see, with a well-defined period. Eruptive, or cataclysmic, variable stars, such as novae, flare stars and supernovae briefly brighten without a well-defined periodicity. *See* Algol, Cepheid variable, Delta Cephei, eclipsing binary, flare star, Mira, novae, RR Lyrae, and supernovae.

Vega: The second brightest star seen from the Northern Hemisphere and the fifth brightest star in the night sky. Vega is the brightest star in the constellation Lyra, and hence also named Alpha Lyrae. It is a white, A0 dwarf, or main-sequence, star with an apparent visual magnitude of +0.03, which appears so bright because it is relatively nearby, at a distance of only 25.3 light-years. Its absolute visual magnitude is +0.65 and its intrinsic luminosity is about 54 times that of the Sun. In 1983, the American astronomers **Hartmut "George" Aumann (1940–)** and **Frederick Gillett (1937–2001)** reported that NASA's *InfraRed Astronomical Satellite*, abbreviated *IRAS*, had discovered infrared radiation from a circumstellar disk around Vega, extending out to 85 AU and perhaps the site of planetary systems in formation.

Vela pulsar: A radio pulsar, designated PSR 0833–45, with a period of 89 milliseconds, discovered in 1968 in the Vela supernova remnant. The Vela pulsar has been observed in detail at X-ray wavelengths with the **Chandra X-ray Observatory** and also in visible light. *See* **Chandra X-ray Observatory** and Vela supernova remnant.

Vela supernova remnant: The expanding debris of the supernova explosion of a star in the constellation Vela, which occurred about 12 thousand years ago. The Vela supernova remnant now subtends an angle of about 255 minutes of arc; it is located at a distance of about 1600 light years. A radio pulsar with a period of 89 milliseconds was discovered in the Vela supernova remnant in 1968; it is known as the Vela pulsar and also designated as PSR 0833–45.

Velocity: A quantity that measures the rate of movement and the direction of movement of an object.

Velocity dispersion: The spread of velocities for a given group of stars or galaxies, indicating how they move relative to one another.

Velocity of light: The fastest speed that anything can move. The velocity of light, denoted by c, equals 299 792.458 kilometers per second.

Velocity of sound: Denoted by c_s, the velocity of sound is proportional to the square root of the gas temperature, T, and inversely proportional to the square root of the mean molecular weight, μ, or $c_s \propto (T/\mu)^{1/2}$.

Venera Missions: A series of spacecraft launched by the Soviet Union to the planet Venus between 1961 and 1983. *Venera 4* was the first probe to be parachuted into another planet's atmosphere, confirming in 1967 that carbon dioxide is the main ingredient of the atmosphere of Venus. *Venera 7* was the first spacecraft to land on the surface of another planet, measuring the high surface temperature and pressure in 1970. Subsequent landers, *Veneras 9, 10, 11, 12, 13* and *14* obtained photographs of the surface of Venus and determined the composition of its rocks. *See* Venus.

Venus: The most brilliant of the planets, and the brightest object in the night sky, after the Moon. Venus is visible at the edges of night, known as the so-called morning star or evening star, seen low in the sky near either dawn or dusk. Venus is the nearest planet to Earth, with a mass and size that are comparable to those of the Earth. The surface of Venus is always hidden by a thick, impenetrable atmosphere of carbon dioxide with clouds of sulfuric acid. In 1964 the American scientist **Irwin I. Shapiro (1929–)** reported that radar measurements from the **Arecibo Observatory** indicate that Venus spins in a backward, retrograde direction, opposite to that of its orbital motion, with a rotation period of 243 Earth days that is longer than the planet's orbital period of 224.7 Earth days. In 1970 the Russian probe *Venera 7* achieved a soft landing on Venus, measuring a surface temperature of 735 kelvin and a surface pressure of 92 times the atmospheric pressure at sea level on Earth. As foreseen in 1940 by the German-born American astronomer **Rupert Wildt (1905–1976)**, the surface temperature is raised to high levels by the greenhouse effect of the planet's thick carbon dioxide atmosphere. Radar maps of its surface, from the *Magellan* spacecraft that orbited Venus for four years in the early 1990s, reveal rugged highlands, smoothed-out plains, extensive vol-

TABLE 59   Venus[a]

Mass	48.685×10^{23} kilograms = 0.815 M_E
Mean Radius	6051.84 kilometers = 0.949 R_E
Mean Mass Density	5204 kilograms per cubic meter
Rotation Period	243.025 Earth days, retrograde
Orbital Period	224.7 Earth days
Mean Distance from Sun	$1.081\ 57 \times 10^{11}$ meters = 0.723 AU
Age	4.6×10^9 years
Atmosphere	96 percent carbon dioxide, 3.5 percent nitrogen
Surface Pressure	92 bar
Surface Temperature	735 kelvin
Magnetic Field Strength	less than 3×10^{-9} tesla or 10^{-5} B_E

[a] The symbols M_E, R_E, and B_E denote respectively the mass, radius and magnetic field strength of the Earth.

canoes, and sparse, pristine impact craters. *See* arachnid, Arecibo Observatory, coronae, greenhouse effect, *Magellan Mission,* radar astronomy, and *Venera Missions*.

Vernal Equinox: The point where the Sun crosses the celestial equator going northward in spring, around March 21. The Vernal Equinox is also known as the spring equinox and the First Point of Aries. It is the zero point of the celestial coordinate called right ascension. The location of the Vernal Equinox is slowly moving, or precessing, westward at the rate of 50.3 seconds of arc per year or about one-seventh of a second of arc daily. *See* equinox, precession and right ascension.

Very Large Array: Abbreviated **VLA,** a radio interferometer near Socorro, New Mexico consisting of twenty-seven dishes of 25 meters diameter, moveable along the arms of a giant Y with dish separations of up to 34 kilometers. The VLA operates at wavelengths between 0.018 and 0.90 meters. *See* aperture synthesis, interferometer, National Radio Astronomy Observatory, and radio telescope.

Very Large Telescope: Abbreviated **VLT,** a system of four unit telescopes, each 8.2-meters (323-inches) in diameter at Cerro Paranal, Chile, operated by the European Southern Observatory. The telescopes can be used at visible and infrared wavelengths, either individually or together in a **Very Large Telescope Interferometer**, abbreviated **VLTI**, with a resolution equivalent to a 16-meter (630-inch mirror). The four unit telescopes are named *Antu, Kueyen, Melipal,* and *Yepun,* the names for the Sun, the Moon, the Southern Cross, and Sirius in the local Mapuche language. Four auxiliary telescopes, each of 1.8-meters (72-inches) in diameter will complement the VLTI. *See* European Southern Observatory and interferometer.

Very Long Baseline Array: Abbreviated VLBA, a system of ten radio-telescope antennas, each 25-meter (82 feet) in diameter, controlled remotely from Socorro, New Mexico by the National Radio Astronomy Observatory. Dedicated in 1993, it is the world's largest full-time astronomical instrument. The telescopes are located in sites stretching across the United States from Mauna Kea, Hawaii to St. Croix in the United

States Virgin Islands, spanning more than 8000 kilometers (5000 miles) and providing the sharpest vision, up to 0.001 seconds of arc, of any telescope on Earth or in space. *See* aperture synthesis, interferometer, National Radio Astronomy Observatory, and Very Long Baseline Interferometry.

Very Long Baseline Interferometry: Abbreviated VLBI, the combination of radio signals recorded at widely separated radio telescopes observing the same cosmic object at the same time, to achieve an angular resolution of a radio telescope with a diameter equal to the distance between the component telescopes, and providing the sharpest vision of any single telescope on Earth or in space. An angular resolution as fine as 0.001 seconds of arc has been obtained at radio wavelengths using VLBI observations with radio telescopes separated by as much as 8000 kilometers (5000 miles). VLBI baselines, or telescope separations, are extended beyond the diameter of the Earth by combining signals from radio astronomy satellites and ground-based radio telescopes. *See* Australia Telescope, interferometer, MERLIN, National Radio Astronomy Observatory, Parkes Observatory, and Very Long Baseline Array.

Vesta: The fourth asteroid to be discovered, by the German astronomer **Heinrich Wilhelm Olbers (1758–1840)** in 1807, with a radius of 265 kilometers and a mass density of 3440 kilograms per cubic meter. Vesta is the brightest and third largest of the asteroids, whose spectrum shows the absorption signature of volcanic basalt, indicating that its surface is covered with lava-flows. *See* asteroid.

Viking Mission: NASA's ***Viking*** mission to Mars was composed of two spacecraft, ***Viking 1*** and ***Viking 2***, each consisting of an orbiter and a lander. The orbiters imaged the entire surface of Mars at a resolution of 150 to 300 meters, showing volcanoes, lava plains, immense canyons, cratered areas, wind-formed features and evidence for catastrophic floods of water across the Martian surface in the past. The landers found no detectable organic molecules in the Martian surface, which means that this part of the surface now contains no detectable cells, living, dormant or dead. ***Viking 1*** was launched on 20 August 1975 and arrived at Mars on 19 June 1976. On 20 July 1976 the ***Viking 1 Lander*** touched down at Chryse Planitia. ***Viking 2*** was launched on 9 September 1975, and entered Mars orbit on 7 August 1976. The ***Viking 2 Lander*** touched down at Utopia Planitia on 3 September 1976. The ***Viking 1*** and ***2 Orbiters*** were powered down in 1980 and 1978, respectively, and their landers ended communications in 1982 and 1980, respectively, after providing years of information about the local weather on Mars.

Virgo cluster: A nearby cluster of galaxies in the constellation Virgo, containing about 3000 galaxies and a total of 16 bright Messier objects. The Virgo cluster of galaxies includes the giant elliptical galaxy M87, also designated NGC 4486, located about 45 million light-years away and associated with the intense radio galaxy named Virgo A or 3C 274. M87 contains a supermassive black hole that feeds the radio galaxy, with high-speed jets that are detected at visible-light, radio and X-ray wavelengths. *See* black hole, cluster of galaxies, radio galaxy, and supermassive black hole.

Virial theorem: For a bound gravitational system, the long-term average of the kinetic energy is one-half of the potential energy.

Virtual particles: Short-lived particles that arise from a vacuum.

Visible light: The form of electromagnetic radiation that can be seen by human eyes, occupying a narrow range of wavelengths in the electromagnetic spectrum. Visible light is also known as optical radiation since optics is used to detect it. Visible light extends roughly from violet wavelengths at 385 nanometers, or 3.85×10^{-7} meters, to red wavelengths at 700 nanometers, or 7.0×10^{-7} meters, and lies between the ultraviolet and infrared parts of the electromagnetic spectrum. *See* electromagnetic radiation, invisible radiation, optical astronomy, optical radiation, and optics.

Visible radiation: *See* visible light.

Visual magnitude: The absolute or apparent magnitude of a celestial object in the color region to which the human eye is most sensitive, at wavelengths near 560 nanometers or 5.6×10^{-7} meters. The absolute visual magnitude of the Sun is $+4.82$ magnitudes and its apparent visual magnitude is -26.72 magnitudes. *See* absolute magnitude, apparent magnitude, and magnitude.

VLA: Acronym for the **Very Large Array**, United States. *See* Very Large Array.

VLBA: Acronym for **Very Long Baseline Array**: *See* Very Long Baseline Array.

VLBI: Acronym for **Very Long Baseline Interferometry**. *See* Very Long Baseline Interferometry.

VLT: Acronym for Very Large Telescope. *See* European Southern Observatory and Very Large Telescope.

Volcanism: The release of molten material, called magma, from the interior to the surface of a planet or satellite. The magma cools to become lava. Volcanism flooded the maria on the Moon, and the vast plains of Venus. Volcanoes are ubiquitous throughout the Solar System, from Earth, Mars and Venus to Jupiter's satellite Io. Neptune's satellite Triton may contain volcanic geysers of ice. *See* Io, maria, Mars, Moon, and Venus.

***Voyager 1* and 2:** Two almost identical planetary probes launched by the United States in 1977. *Voyager 1* passed Jupiter in 1979, discovering the volcanoes of Io and Jupiter's ring, and met Saturn in 1980, discovering the thick nitrogen atmosphere around Titan. *Voyager 2* also passed near Jupiter and Saturn, and went on to pass Uranus in 1986 and Neptune in 1989, confirming the thin, widely spaced rings of Uranus and Neptune, and discovering storms on Neptune and geysers on Neptune's satellite Triton. *See* Io, Jupiter, Neptune, rings, Saturn, and Uranus.

Vulcan: A hypothetical planet supposedly revolving around the Sun at somewhat less than half of Mercury's mean orbital distance. Vulcan was proposed by the French mathematician **Urbain Jean Joseph Leverrier (1811–1877)**, arguing that its gravitational attraction would pull Mercury ahead in its orbit and account for its unexplained motion. The planet Vulcan was never found, and in 1915 the German physicist **Albert Einstein (1879–1955)** explained the anomalous motion of Mercury as a consequence of the Sun's curvature of nearby space. *See General Theory of Relativity* and Mercury.

Water: A substance, denoted by H_2O, composed of atoms of hydrogen, H, and oxygen, O, two of the most abundant elements in the Universe. Water can exist on Earth as a liquid, gas or ice, and in its liquid form it is vital for life. Water molecules have been observed in the atmosphere of Mars and large amounts of frozen water are now found in the surface of Mars. Water molecules are also found in interstellar space.

Wavelength: The distance between successive crests or troughs of an electromagnetic or other wave. Wavelengths are inversely proportional to frequency. The longer the wavelength, the lower the frequency. The product of the wavelength and the frequency of electromagnetic radiation is equal to the velocity of light. *See* velocity of light.

Waves: Propagation of energy by means of coherent vibration.

Weak nuclear force: Fundamental force of nature that governs the process of radioactivity. *See* neutrino and radioactivity.

Westerbork Synthesis Radio Telescope: Abbreviated WSRT, a 3-kilometer east-west array of fourteen 25-meter (83 feet) radio telescopes combined to achieve the resolution of a single large radio telescope using the technique of aperture synthesis. It is also part of the European Very Long Baseline Interferometry network of radio telescopes. The WSRT is located near Groningen in the Netherlands, and is operated by the Netherlands Foundation for Research in Astronomy. *See* aperture synthesis and Very Long Baseline Interferometry.

Whirlpool galaxy: The spiral galaxy, designated as M51 and NGC 5194, located about 25 million light-years away in the constellation Canes Venatici. Its spiral shape was discovered in 1845 by the Irish astronomer **William Parsons (1800–1867)**, the third Earl of Rosse using his 1.8-meter (72-inch) telescope known as the Leviathan of Parsonstown.

White dwarf: A very small, dense star, about the size of the Earth, but with a mass about that of the Sun. The first two white dwarf stars, 40 Eridani B and Sirius B, were discovered in 1914 and 1915, respectively, when the American astronomer **Walter S. Adams (1876–1956)** showed that these faint stars are hot, white stars of spectral type A0. The high disk temperature and low intrinsic luminosity meant that these stars had to be very small, only about the size of the Earth. In 1926 the English physicist **Ralph H. Fowler (1889–1944)** showed that a white dwarf star is supported against gravity by degenerate electron pressure, which does not depend on the temperature and is related to the exclusion principle proposed in 1925 by the Austrian physicist **Wolfgang Pauli (1900–1958)**. When first formed, a white dwarf is a faint, hot star with an effective temperature similar to or hotter than that of the Sun, but because it cannot sustain thermonuclear reactions it shines by leftover heat and gradually cools into an invisible black dwarf. A white dwarf is the endpoint of the evolution for stars with a mass comparable to that of the Sun. Such a star evolves from the main sequence into a red giant, then sheds its atmosphere and exposes its hot core, which becomes a white dwarf. A white dwarf star cannot have a mass greater than 1.4 solar masses, the so-called Chandrasekhar limit; more massive stars collapse under their own weight to form a neutron star or black hole at the endpoints of stellar evolution. A type Ia supernova occurs when material from a nearby companion adds mass to a white dwarf, forcing it above the Chandrasekhar limit and causing the entire star to explode. *See* black hole, Chandrasekhar limit, degenerate matter, exclusion principle, 40 Eridani, Hertzsprung-Russell diagram, neutron star, planetary nebula, Sirius, and type Ia supernova.

White light: The visible portion of sunlight that includes all of its colors. Sunlight integrated over the visible portion of the spectrum, from 400 to 800 nanometers, so that all colors are blended to appear white to the eye.

White light flare: An exceptionally intense and rare solar flare that becomes visible in white light, or in all the colors of the Sun combined.

Widmanstätten pattern: When an iron meteorite is cut and polished and then etched with acid, a Widmanstätten pattern of crystalline structure emerges. The sizes and shapes of these crystals indicate that they grew very slowly, and that the material must have been hot, almost to the melting point, for tens of millions of years, suggesting that the meteorite was once inside a parent body a few hundred kilometers in radius and the size of a large asteroid. The pattern is named for **Alois Josep von Widmanstätten (1754–1849)**, a Viennese scientist who discovered it in 1808.

Wilkinson Microwave Anisotropy Probe: NASA launched its *Wilkinson Microwave Anisotropy Probe*, abbreviated *WMAP*, on 30 June 2001, publishing definitive full sky maps of the faint anisotropy, or temperature variations, of the cosmic microwave background radiation in early 2003. When combined with the results of other diverse cosmic measurements, the *WMAP* data suggest that the Universe is 13.7 billion years old, that the first stars ignited about 200 million years after the Big Bang, that the radiation in the *WMAP* images is from 379 000 years after the Big Bang, that the Hubble constant has a value of 71 kilometers per second per Megaparsec, and that the Universe is composed of 4 percent atoms, 23 percent cold dark matter and 73 percent dark energy. The probe is named in honor of American astronomer **David T. Wilkinson (1935–2002)**, a pioneer in cosmic microwave background research and the *MAP* instrument scientist.

William Herschel Telescope: A 4.2-meter (165-inch) telescope at La Palma in the Canary Islands. It was used in 1997 to discover the first optical counterpart of a gamma-ray burst. The telescope is named for the famous English astronomer **William Herschel (1738–1822)**, and it is owned and operated by the Royal Observatory, Edinburgh.

WIMP: Acronym for Weakly Interacting Massive Particle, a hypothetical subatomic particle that could explain dark matter and account for most of the mass of the Universe. *See* dark matter.

WMAP: Acronym for *Wilkinson Microwave Anisotropy Probe*. *See Wilkinson Microwave Anisotropy Probe*.

Wolf-Rayet stars: A type of stars whose visible-light spectra contain broad and intense emission lines, indicating very hot disk temperatures of 25,000 to 50,000 degrees kelvin and intrinsic luminosities between 100 thousand and 1 million times that of the Sun. The French astronomers **Charles Wolf (1827–1918)** and **Georges Rayet (1839–1906)** discovered the first examples of this type of star in 1867, noting their emission lines of hydrogen and helium, so they are now known as Wolf-Rayet stars. Emission lines of nitrogen, carbon and oxygen have also been discovered. Many of the central stars of planetary nebulae are Wolf-Rayet stars.

World line: The path traced out in four-dimensional space-time by a given object or particle. *See* Minkowski metric, and *Special Theory of Relativity.*

X: The mass fraction of hydrogen. Observations of sunlight and meteorites indicate that X = 0.706 ± 0.025 for the solar material outside its core. Owing to nuclear reactions, there is now less hydrogen in the Sun's core.

XMM: Acronym for *X-ray Multi-Mirror. See X-ray Multi-Mirror – Newton.*

X-ray: *See* X-ray radiation.

X-ray binary: A double star system that often contains a neutron star or black hole and a nearby optical, or visible-light, companion, and is an intense source of X-rays. The two stars are in a close orbit, so mass from the visible component spills over its Roche lobe onto the other one, heating up and emitting X-rays as it is accreted by the invisible companion. The high-mass X-ray binary systems contain a massive, bright visible-light star as the mass donating component, and their invisible components include black holes, such as Cygnus X-1, and X-ray pulsars, like Centaurus X-3 and Hercules X-1. Low mass X-ray binaries are binary systems with an accreting component that is thought to be a neutron star, and a mass, donating optical secondary that is a low mass star, equal to or less than the mass of the Sun; they exhibit non-periodic, bursting behavior at X-ray wavelengths. *See* accretion, black hole, Centaurus X-3, Cygnus X-1, Hercules X-1, pulsar, and Roche lobe.

X-ray Multi-Mirror – Newton observatory: ESA launched its *X-ray Multi-Mirror – Newton*, abbreviated *XMM–Newton*, spacecraft on 10 December 1999, with a designed 10-year lifetime. It provides X-ray images and spectroscopy of selected X-ray objects, such as accreting black holes, active galactic nuclei, clusters of galaxies, and supernovae. It is named after the English mathematician and astronomer **Isaac Newton (1643–1727)** who invented spectroscopy and is best remembered for his laws of motion and gravity.

X-ray radiation: An energetic form of electromagnetic radiation that has short wavelengths, between those of gamma rays and ultraviolet radiation. X-rays were discovered in 1895 by the German physicist **Wilhelm Konrad Röntgen (1845–1923)**, while cosmic X-ray sources were discovered in 1962 by the Italian-born American astronomer **Riccardo Giacconi (1931–)** and his colleagues with instruments aboard a five-minute rocket flight. X-ray radiation covers the wavelength range from about 10^{-8} to 10^{-11} meters and a photon energy range between 0.1 and 100 keV. Soft X-rays have lower energy, between 1 and 10 keV, and hard X-rays have higher energies ranging from 10 to 100 keV. Because X-rays are absorbed by the Earth's atmosphere, X-ray astronomy is performed in space using satellites such as the *Chandra X-ray Observatory, Uhuru* and *Yohkoh. See Chandra X-ray Observatory*, electromagnetic spectrum, hard X-rays, soft X-rays, *Uhuru Mission*, and *Yohkoh Mission*.

Y

Y: The mass fraction of helium. Observations of sunlight and meteorites indicate that $Y = 0.274 \pm 0.026$ for solar material outside the core. Owing to nuclear reactions, there is now more helium in the Sun's core.

Year: The length of time that a planet takes to revolve around its star. The Earth's year is 365.25 Earth days, where one Earth day is 24 hours long, the time to complete one rotation of our planet. In contrast, Mercury's year is only 88 Earth days, and Jupiter's year is 11.86 Earth years. By definition, one Julian century has 36 525 days, so the Julian year on Earth has 365.25 days, which is equal to 8766 hours and to 525 960 minutes. The sidereal year is the time required for the Earth to complete one revolution around the Sun relative to the fixed stars, and it is equal to 365.25636 mean solar days. The tropical year, defined as the mean interval between Vernal Equinoxes, is equal to 365.242 189 7 mean solar days. The Gregorian calendar year is 365.242 5 days long. *See* precession, sidereal time, tropical year, and Vernal Equinox.

Yerkes Observatory: In the 1890s the American astronomer **George Ellery Hale (1868–1938)** persuaded a wealthy Chicago financier **Charles T. Yerkes (1837–1905)** to provide $349,000 to build a 1-meter (40-inch) diameter refractor telescope for the University of Chicago, which would be the largest in the world. It was finished in 1897 and is located at the University of Chicago's Yerkes Observatory in Williams Bay, Wisconsin, remaining the world's largest refractor even today.

Yohkoh Mission: The Japanese Institute of Space and Astronautical Science, abbreviated ISAS, launched its *Solar-A* spacecraft on 30 August 1991, renaming it *Yohkoh*, which means "sunbeam" in English. *Yohkoh* collected high-energy radiation from the Sun at soft X-ray and hard X-ray wavelengths, with high angular and spectral resolution, for more than a decade, providing new insights to the mechanisms of solar flare energy release and the heating of the million-degree solar corona. It was a collaborative effort of Japan, the United States and the United Kingdom. *See* hard X-rays, HXT, soft X-rays and SXT.

Z: The mass fraction of elements heavier than hydrogen and helium. Observations of sunlight and meteorites indicate that Z = 0.01886 ± 0.0085 for solar material. A capital Z also denotes the atomic number. *See* atomic number.

Zeeman components: The linearly polarized, π, and circularly polarized, σ, components which comprise a line split in the presence of a strong magnetic field. *See* Zeeman effect and Zeeman splitting.

Zeeman effect: A splitting of a spectral line into components by a strong magnetic field, named after the Dutch physicist **Pieter Zeeman (1865–1943)**, who discovered the effect in 1897. If the components cannot be resolved, there is an apparent broadening or widening of the spectral line. The amount of splitting measures the strength of the magnetic field, and the direction of the magnetic field can be inferred from the polarization of the components. The Zeeman effect is thereby used to determine the direction, distribution and strength of the longitudinal magnetic fields in the photosphere. It has also been used to measure the magnetic fields in other stars and in the interstellar medium. In the simplest case, denoted as normal Zeeman splitting, a line splits into three components. One component, the π component, is not displaced in wavelength or frequency, and is linearly polarized. Two components, the σ components, are shifted by equal amounts to lower and higher wavelengths or frequencies, the magnitude of the shift being proportional to the magnetic field strength. In the general case, the σ components are both circularly and linearly polarized. *See* magnetograph, Zeeman components, and Zeeman splitting.

Zeeman splitting: The splitting of atomic spectral lines in a magnetic field. *See* Zeeman components and Zeeman effect.

Zelenchukskaya Observatory: Located near the Zelenchukskaya River in the Caucasus Mountains of southern Russia, the Zelenchukskaya Observatory includes the 6-meter (236-inch) visible-light Bolshoi Teleskop Azimutalnyi on Mount Pastukhov and the relatively nearby **RATAN-600 radio telescope**, consisting of a ring of reflecting panels 600 meters (2000 feet) in diameter. RATAN is an acronym for Radio Astronomical Telescope Academy Nauk. The Special Astrophysical Observatory of the Russian Academy of Sciences operates both telescopes.

Zodiac: An imaginary band in the night sky about 8 degrees to each side of the ecliptic, which encompasses the paths of all the major planets and is divided into 12 constellations, each 30 degrees wide: Aquarius, the water bearer, Aries, the ram, Cancer, the crab, Capricorn, the goat, Gemini, the twins, Leo, the lion, Libra, the scales, Pisces, the fish, Sagittarius, the archer, Scorpius, the scorpion, Taurus, the bull, and Virgo, the virgin. A thirteenth constellation, Ophiuchus, crosses the ecliptic between Scorpio and Sagittarius. The ancient Greeks named the signs of the zodiac for these constellations, but because of precession they no longer coincide with the constellations of the same name. *See* precession.

Zodiacal cloud: Cloud of interplanetary dust in the Solar System, lying close to the ecliptic plane. The dust in the zodiacal cloud comes from both comets and asteroids.

Zodiacal light: A faint conical glow in the night sky caused by sunlight scattering off interplanetary dust near the plane of the ecliptic. A luminous pyramid of light that appears brightest and widest in the direction of the Sun, stretching along the ecliptic or zodiac from the western horizon after evening twilight or from the eastern horizon before morning twilight. It is visible at all seasons in the tropics in the absence of moonlight. The zodiacal light is an extension of the dust component of the solar corona, and results from scattering of sunlight by dust particles in the plane of the ecliptic. The zodiacal dust cloud probably originates from both matter ejected by the Sun and from the decay of comets and asteroids.

Zone of avoidance: The region coinciding with the plane of our Milky Way Galaxy in which very few galaxies are found, due to absorption of their light by interstellar dust.

REFERENCES

This book has been prepared by reference to the following books.

Dreyer, John: *A History of Astronomy from Thales to Kepler,* first published in 1906, reproduced New York, Dover Publications 1953.

Duhem, Pierre: *To Save the Phenonmena: An Essay on the Idea of Physical Theory from Plato to Galileo,* first published in French in 1908, English translation Chicago, University of Chicago Press, 1969.

Duhem, Pierre: *Medieval Cosmology,* first published in French in 1916, abridged English translation Chicago, University of Chicago Press, 1985.

Hoskin, Michael (ed.): *The Cambridge Illustrated History of Astronomy,* New York, Cambridge University Press 1997.

Kaler, James B.: *The Hundred Greatest Stars,* New York, Copernicus Books, Springer 2002.

Koyré, Alexandre: *From the Closed World to the Infinite Universe,* Baltimore, Johns Hopkins University Press, 1957.

Lang, Kenneth R.: *Astrophysical Formulae I, II,* New York, Springer-Verlag 1999.

Lang, Kenneth R.: *The Sun From Space,* New York, Springer-Verlag 2000.

Lang, Kenneth R.: *The Cambridge Encyclopedia of the Sun,* New York, Cambridge University Press, 2001.

Lang, Kenneth R.: *The Cambridge Guide to the Solar System,* New York, Cambridge University Press, 2003.

Lang, Kenneth R.: *Parting the Cosmic Veil,* New York, Springer, 2006

Lang, Kenneth R.: *Sun, Earth, and Sky,* Second Edition, New York, Springer, 2006

Lang, Kenneth R. and Gingerich, Owen (eds.): *A Source Book in Astronomy and Astrophysics 1900–1975,* Cambridge, Mass., Harvard University Press 1979.

Murdin, Paul (ed.): *Encyclopedia of Astronomy and Astrophysics.* London, Institute of Physics Publishing 2001.

Park, David: *The How and the Why: An Essay on the Origins and Development of Physical Theory,* Princeton, New Jersey, Princeton University Press, 1990.

Porter, Roy and Ogilvie, Marilyn (eds.): *The Biographical Dictionary of Scientists I, II,* New York, Oxford University Press 2000.

Ridpath, Ian and Woodruff, John: *Philip's Astronomy Dictionary,* London, George Philip Ltd 1999.

Toulmin, Stephen and Goodfield, June: *The Discovery of Time,* Chicago, University of Chicago Press, 1965.

Many of the seminal articles in astronomy and astrophysics can be found in Lang, Kenneth R. and Gingerich, Owen (eds.), *A Source Book in Astronomy and Astrophysics, 1900–1975,* Cambridge, Mass., Harvard University Press 1979, and Helmut A. Abt (ed.), *The Astrophysical Journal, American Astronomical Society Centennial Issue,* Chicago, University of Chicago Press 1999. They can also be found on the Web at the NASA Astrophysics Data System at the URL http://adswww.harvard.edu/index. html.

AUTHOR INDEX

SUBJECT INDEX